THIRD EDITION

Dance Anatomy

Jacqui Greene Haas

Library of Congress Cataloging-in-Publication Data

Names: Haas, Jacqui Greene, 1958- author.
Title: Dance anatomy / Jacqui Greene Haas.
Description: Third edition. | Champaign, IL : Human Kinetics, 2025. |
  Includes bibliographical references.
Identifiers: LCCN 2023036521 (print) | LCCN 2023036522 (ebook) | ISBN
  9781718219915 (paperback) | ISBN 9781718219939 (epub) | ISBN
  9781718219946 (pdf)
Subjects: LCSH: Dance--Physiological aspects. | BISAC: PERFORMING ARTS /
  Dance / General | SCIENCE / Life Sciences / Human Anatomy & Physiology
Classification: LCC RC1220.D35 H33 2025 (print) | LCC RC1220.D35 (ebook)
  | DDC 617.1/0275--dc23/eng/20230925
LC record available at https://lccn.loc.gov/2023036521
LC ebook record available at https://lccn.loc.gov/2023036522

ISBN: 978-1-7182-1991-5 (print)

Senior Acquisitions Editor: Michelle Earle; **Senior Developmental Editor:** Cynthia McEntire; **Managing Editor:** Shawn Donnelly; **Copyeditor:** Jenny MacKay; **Permissions Manager:** Laurel Mitchell; **Senior Graphic Designers:** Nancy Rasmus and Sean Roosevelt; **Cover Designer:** Keri Evans; **Cover Design Specialist:** Susan Rothermel Allen; **Cover illustration:** © Human Kinetics/Heidi Richter; **Photographs (for illustration references):** Jason Allen and Peter Mueller; **Photo Production Specialist:** Amy M. Rose; **Photo Production Manager:** Jason Allen; **Senior Art Manager:** Kelly Hendren; **Illustrations:** © Human Kinetics/Fran Milner, Molly Borman, and Heidi Richter; **Printer:** Versa Press

Human Kinetics
1607 N. Market Street
Champaign, IL 61820
USA

United States and International
Website: **US.HumanKinetics.com**
Email: info@hkusa.com
Phone: 1-800-747-4457

Canada
Website: **Canada.HumanKinetics.com**
Email: info@hkcanada.com   E8992 (paperback) / E8993 (loose-leaf)

THIRD EDITION

Dance Anatomy

CONTENTS

PREFACE

The field of performing arts medicine has grown significantly over the past 35 years and is made up of committed health care professionals with a passion for helping dancers, musicians, actors, and circus artists, to name just a few. As a dancer, you can find more information regarding injury risk education, injury management, rehabilitation, strength training, and body conditioning than ever before. The field of performing arts medicine is devoted to improving your health care. Professional dance companies, college dance departments, and local dance studios are developing health care programs that focus on exceptional patient care to get you back to dancing after an injury. As competition in the field of dance continues to grow, so does unique and creative choreography that challenges your body. The more you know about how to care for your body and mind, the better chance you have of tackling innovative and challenging choreography. To compete in this high-performance market, you must be mentally and physically fit. The need to impress audiences has never been greater, and extreme choreography sells tickets and wins competitions. The third edition of *Dance Anatomy* will give you more tools for enhancing your strength while providing injury risk education and pertinent information on which muscles are working for various challenging dance movement patterns.

The ways in which dancers move and the ways in which they express themselves through that movement keep audiences connected to the beauty of dance. Live dance enhances creativity and has been used as a form of creative self-expression for generations. It allows you to use your creative and critical thinking skills to execute challenging movement patterns. Strong and talented performers can emotionally connect with their audience. As a dancer, you can communicate an idea through movement to enhance your audience's experience. You can change an audience's mood by your self-expression. Dance involves an extraordinary display of physical skill that can convey both raw energy and charming delight. The hallmarks of this art form include chiseled poses, innovative choreography, and striking images. Moreover, all forms of dance rely on impeccable balance, precise muscular control, grace, rhythm, and speed.

There are more than 50,000 dance studios in the United States alone. The rigor of classes, rehearsals, performances, and competition schedules can be overwhelming. You are working harder than ever and doing your best to understand and soak in every cue that your teachers give you. Your instructors, meanwhile, face the demands of teaching technique, artistry, musicality, and tricky choreography, as well as marketing their businesses and managing their teaching schedules.

In this hectic environment, the details of technique are sometimes over-looked. Dance technique has been passed down over the years with very little anatomical analysis. This tradition may have worked for earlier generations, but for you to meet the demands of today's challenging choreography, you must receive the most proficient training and be stronger than ever, which requires you to understand basic anatomy and the benefits of strength training. *Dance Anatomy* will help you recognize your muscles and understand how they function. A better understanding of your bones, joints, and muscles is also critical for injury risk education.

Your technique class should emphasize the development of muscular strength to enable you to control and protect your joints as well as prepare you for rehearsals and performances. Sometimes, technique classes may not provide enough muscle loading to protect your joints, increase strength, or enhance performance. Furthermore, the repetitive nature of dance and the training intensity put you at risk for injury. Supplemental training can give you tools needed to protect joints, increase strength, and enhance performance. This edition of *Dance Anatomy* offers a large variety of exercises that can be used as supplemental training, which, when added to your schedule, may help reduce injury rates while building strength. The more than 250 medical illustrations provide visual diagrams of anatomy to enhance your learning.

Each chapter addresses a key principle of movement to help you improve performance. Chapter 1 provides the book's foundation. It highlights three beautiful positions of dance that show the entire body and its musculature. This chapter also emphasizes the importance of developing a basic under-standing of how your body works through descriptions of anatomy, movement planes, and muscular actions. Details of your muscles in action are illustrated beautifully in every image.

While performing arts medicine has advanced during the past 35 years, mental health awareness for dancers still has a long way to go. Chapter 2 introduces you to the neurological connections between your brain and your body. Chapter 2 also addresses chronic stress and anxiety to provide you with some tools for overcoming stage fright or performance anxiety. If you suffer from heightened anxiety prior to competitions, performances, or auditions, it would be advantageous to better understand how your brain reacts and the physiological changes that occur. Learning more about the changes that occur in your body when you are under stress can help you develop coping strategies.

Chapter 3 educates you on injury risk, including the intrinsic and extrinsic factors associated with injury. A segment on the healing journey as well as tips for designing an injury care plan for dance studios have been added in this edition.

Chapter 4 covers spinal alignment and placement—where it all begins. This chapter addresses the spinal curves and all movements of the spine. It also presents specific exercises devoted to placement of the spine, along with detailed medical illustrations highlighting specific deep musculature. The exercises included in this chapter are not meant to be challenging; rather, they will help you develop muscular awareness and understand your muscles' role in supporting and stabilizing the spine for better alignment.

Chapter 5 focuses on the physiological benefits of efficient breathing to enhance your technique, and chapter 6 emphasizes your core and the importance of strengthening your abdominals. New and advanced core exercises have been included in this edition. Breathing efficiently with your diaphragm increases intra-abdominal pressure so your abdominals can shorten and lengthen, which helps to improve your core strength. Better breathing patterns improve muscle function while you exercise and increase oxygen levels in your blood. Both chapters include medical illustrations depicting the relationship between quality breathing, abdominal muscle contraction, and healthy spine movement. Basic dance classes may not address all layers of abdominal muscles or their importance in enhancing technique; therefore, you must include supplemental core conditioning in your training.

Chapter 7 details the musculature of the shoulders and arms and presents exercises to help you improve your port de bras and lifting or partnering skills. As break dancing, hip-hop, and street styles continue to gain popularity, upper-body strength is important to execute various freeze-type movements. This edition includes strengthening exercises to safely execute break dance freeze movements.

Chapter 8 focuses on hip dissociation, or the ability to maintain stability in the spine and pelvis while moving your leg at your hip joint. If you are unable to efficiently move your legs separately from your spine, then your body will begin to compensate and use other joints, particularly your lower back. Overuse of the lower segments of your spine puts you at risk for lower-back injuries. Chapter 8 will provide a better understanding of pelvic and lumbar stability. This edition also includes more information regarding turn-in and turn-out exercises. Chapter 9 presents exercises that focus on elegance and power in the legs. Each illustration shows muscle originations and insertions to help you fine-tune muscle contractions to improve your développé.

Most dance injuries occur in the ankles and feet. Chapter 10 emphasizes conditioning for the lower legs to help reduce the risk of foot and ankle injuries. The foot includes 26 bones and 33 joints, thus creating multiple movement possibilities. These small joints are responsible for weight transfer, push-off, and landing. Without sufficient strength in the muscles that support the small joints, alignment and technique will be compromised. This chapter includes detailed exercises for strength, alignment, balance, and flexibility of the lower legs, ankles, and feet.

In conclusion, chapter 11 presents exercises that involve multiple areas of the body. In addition to strengthening your body, these exercises promote your body's ability to work as a cohesive unit to accomplish your positions and movements. A section on the benefits of periodization has been added to this edition. Periodization refers to an organized, planned, gradual increase in training while alternating exercise and rest periods. How can periodization help? The goal is to increase strength, flexibility, and balance leading up to the dance performance. To benefit from the exercises offered in this book, you must develop an effective conditioning program that takes into consideration your changing cycles of classes, rehearsals, and times of layoff. The goals are to limit the volume of ineffective training and improve the quality of effective training.

To progress as a dancer, you need to be organized and precise in the overall appearance of your movement. Your body must exhibit definitive direction in the space it uses. If your movement is clean, it will be more rhythmic, dynamic, and musical. Whether you are competing in front of a panel of judges, performing on stage, or taking a technique class, the observers (whether judges, audience members, or instructors) want to see strength, power, and clean lines. All exercise descriptions in this edition provide instruction in proper breathing, education in recruiting the core muscles for improved placement, and important safety tips. The lists of muscles used in the exercises are accompanied by detailed illustrations highlighting the muscles in the dance positions. You can see the relationship between a given exercise and dance position; these relationships apply to all forms and styles of dance.

The exercises presented in *Dance Anatomy* will help you put more practical thought into your dance work without compromising the beauty of the art form. You can use this text as a tool for understanding corrections and the mechanics of your own body movements. You will continue the process of refining your physique and improving your technique for the moment when the director picks you for the leading role!

ACKNOWLEDGMENTS

It is dance that brings joy, and for that I am grateful. Dance has brought me lifelong friendships, employment, performing opportunities, worldwide travel, education, emotional growth, valuable lessons, and the incredible experience of building a loving family, and for that I am forever grateful.

Thank you to Planet Dance Cincinnati and Exhale Dance Tribe for the use of their wonderful studios for our very fun photo shoot.

Thank you to the lovely dancers Julius Jones and Sarah Rolfsen for their beauty as models.

Thank you to Human Kinetics for the opportunity to share my work.

CHAPTER 1
The Dancer in Motion

Motion has been defined as any physical movement or change in position. When you watch a dancer in motion, you see much more than physical changes in position. You see vibrant visual art made up of brief images created by balance, strength, and grace. *Balance* is a key word and is emphasized throughout this book. It can be defined as equal distribution of weight, a state of equilibrium, or harmonious arrangement of proportion. You must understand muscular balance to become the best dancer you can be. Of course, the aesthetics of dance as an art form must never be sacrificed for scientific analysis, but learning basic movement principles to maintain healthy muscular balance allows you to move your body both effectively and safely. This chapter demonstrates movement principles through illustrations of three fundamental dance positions—jazz layout, attitude derrière, and split jump.

Bones, Joints, and Skeletal Muscles

To understand movement, you must develop a basic awareness of bones, joints, and muscles. These parts of the body form the building blocks that enable you to create human motion. Your body is an amazing, evolving gift of energy and information! Learning how to organize the building blocks will give you muscular balance and fresh energy; it will also enhance your skills as a dancer.

Bones

Your body includes 206 bones, which provide support and protection and serve as levers for your muscles. There are five different types or shapes of bones: flat, long, short, sesamoid, and irregular shaped. The flat bones make up the bones in your skull, ribs, and pelvis. Your skull bones protect your brain, and your ribs and pelvis protect your internal organs. The long bones, which are in your arms and legs, play a role in movement, serving as levers for your muscles. Short bones, located in your feet and wrists, provide some movement but also provide stability. You also have sesamoid bones, which are free-floating bones within tendons and provide shock absorption. Your irregularly shaped bones include vertebral bones, which are designed to protect your spine.

Your bones are made up of calcium, which gives them strength, and collagen, which gives them flexibility. Calcium feeds your bones and helps create healthy muscle contractions; low calcium intake can lead to bone weakness and put you at risk for stress fractures. The U.S. Department of Health and Human Services tells us that boys and girls between the ages of 9 and 18 need 1,300 milligrams of calcium per day, and at least 1,000 milligrams per day is needed for adults between the ages of 19 and 50. Men over the age of 50 need 1,000 milligrams daily, and women over 50 need 1200 milligrams a day (Office of Dietary Supplements 2022). Where can you get your daily calcium, aside from supplements? Good sources include milk, cheese, yogurt, leafy greens (e.g., kale and spinach), fortified cereals, and calcium-fortified orange juice. These sources can help you maintain your calcium balance and keep your bones strong.

Movement involves the use of leverage. A lever is a rigid bar that moves a fixed point when effort, or force, is applied to it. The effort is used to move a resistance, or load. In your body, the fixed points are your joints, the levers are your bones, and the effort is provided by muscle contractions. Consider, for example, the jazz layout position, which is illustrated in figure 1.1. Focus on the gesture leg: The fixed point is the hip joint, the lever is the femur (thigh bone), and effort is provided by contraction of the hip flexors. These relationships are enabled by the fact that your muscles are attached to your bones by tendons, and your bones are attached to each other by strong ligaments.

Tendons are fibrous cords of dense connective tissue. They are flexible but strong, and they transmit forces when a muscle contracts. Certain tendons are surrounded by a sheath, which helps keep the tendon in place and allows it to

Figure 1.1 Jazz layout position.

glide easily. Tendon sheaths can become inflamed from overuse or overtraining, and this condition is referred to as *tendonitis* or *tenosynovitis*. You can reduce your risk for various tendon injuries by maintaining a healthy balance of strength training, stretching, and proper nutrition. More information about injuries and injury risks are presented later in the text.

Ligaments are also strong cords of fibrous connective tissue, but they connect bone to bone and hold the joints together. Made up of strong collagen fibers, they keep your joints stable while you are dancing. Ligaments can be sprained, and a severe stretch of a ligament can cause a tear. Many sprains and tears happen when a performer is descending from a jump, possibly twisting the ankle or knee. These injuries would require rest and healing time, along with physical therapy to regain strength.

Joints

Joints exist where two bones meet, and they work smoothly thanks to cartilage, which is soft, smooth tissue located on the ends of bones. Years of dance and joint overuse can cause the cartilage to break down, resulting in chronic inflammation. As cartilage breaks down in a joint, the body compensates and overuses other joints, thus creating imbalance in how you work. You can help reduce injury risks (and enhance your balance) by maintaining strong muscles to support your joints and getting adequate sleep to aid muscle regeneration.

You need to be familiar with several types of joints; the main types discussed in this text are ball-and-socket, gliding, and hinge. All movements occurring at the joints have specific names, and most come in pairs, which typically describe movements made in the same plane but in opposite directions. For example, flexion at the knee involves bending of the knee, whereas extension at the knee involves straightening of the knee (see table 1.1).

Table 1.1   Joint Movements

Action	Movement	Example
Flexion	Bending or folding of a joint	Front of hip bending or flexing with grand battement devant
Extension	Straightening of a joint	Elbow straightening from push-up position
Abduction	Moving away from center	Arms in à la seconde moving from alongside the body to second position
Adduction	Moving toward center	In assemblé, the legs coming together
External rotation	Rotating outward	In turnout, the thighs externally rotating from the hip to achieve grande plié in second position
Internal rotation	Rotating inward	Shoulder joint internally rotating to place the hand on the hip
Plantar flexion	Pointing the foot	In relevé, rising en pointe
Dorsiflexion	Flexing the foot	Rocking back on the heels and lifting the forefoot
Pronation	Rolling the foot inward	Dropping or collapsing the arches, flat feet
Supination	Rolling the foot outward	Higher arches rotating to the outer edge of the foot

Ball-and-socket joints include the hip and shoulder joints. In these joints, one end of the bone is rounded, which fits into a cup-shaped indentation of another bone. In terms of the hip joint, this information is important for improving turnout and développé; this concept is explored further in chapters 8 and 9. The hip joint has a deeper cup-shaped indentation than does the shallow shoulder joint. Look closely at figure 1.2, in which the standing (supporting) leg's hip joint shows how the femoral head fits into the acetabulum. Visualize how movement occurs at this joint; note that it involves rotational action as well as flexion and extension.

Your hips and shoulders work very hard in creating beautiful lines for any choreography you are asked to perform. Your hips carry your body weight in pliés, jumping exercises, and leg work. You can reduce the risk of chronic hip injury by maintaining pelvic stability and a healthy balance of the muscles that support your hips, which chapter 8 further discusses. Your shoulders also need to be strong and stable because the shoulder socket is so shallow. Shoulder dislocations do happen in dance, and you can reduce your risk of this type of injury by strengthening your shoulder joint, which is discussed further in chapter 7.

Gliding joints are made up of bones in which both ends are relatively flat; they allow for very little movement. For example, the point where each rib meets the spinal vertebrae is a gliding joint, as shown in figure 1.3. The fact that these joints allow very little movement helps us understand the lack of good mobility throughout the midspine (thoracic) region, which is addressed further in chapter 4.

Figure 1.2 Attitude derrière position.

Figure 1.3 Split jump.

In a hinge joint, a bone with a slightly concave end meets a bone with a convex end; one example is the knee. When the knee flexes and extends, it allows movement primarily in one plane; however, as addressed later, it also allows slight rotational movement. Look back at figure 1.1 and notice that the supporting leg shows knee flexion, while the gesture leg shows knee extension.

Skeletal Muscles

Skeletal movement is initiated by skeletal muscles, which are composed of connective-tissue partitions containing muscle cells, fibers, and numerous nerves. When the nerves are stimulated by your brain, a chemical reaction causes the involved muscle to contract. Each muscle has both an origination point on a bone, a belly or center of the muscle, and an insertion point on a bone. What does that really mean? The origination of the muscle–tendon unit attachment is relatively stationary, while the insertion is more mobile. On contraction, the muscle fibers shorten, creating tension, which allows the insertion point of the muscle to be pulled toward the point of origin. Let's use relevé as an example. When you rise into a half pointe position, your calf muscles contract, pulling on the Achilles tendon to elevate your heel. The calf muscles originate behind the knee (femoral condyles) and insert into the heel (calcaneus) via the Achilles tendon. On contraction, the origination is relatively stable, while the insertion brings you into relevé.

A muscle's reaction to a stimulus depends on the muscle's characteristics. Each muscle contains two basic types of fibers: slow-twitch (or type I) and fast-twitch (or type II). Slow-twitch fibers contract slowly and exhibit high resistance to fatigue; they are used primarily for placement and posture and for aerobic activities. Fast-twitch fibers contract more quickly and exhibit low resistance to fatigue; however, they can produce more power than slow-twitch fibers. Thus fast-twitch fibers are used for petit allegro, or short anaerobic movements. Regardless of your dance intensity level, slow-twitch fibers are recruited before fast-twitch fibers.

Your muscles have the capability to contract, or create tension, in either static or dynamic fashion. Static contraction, described as lack of movement, creates tension on the muscle without visible movement at the joint. Dynamic contraction, in contrast, is described as any type of tension on a muscle in which the length of the muscle changes; this type of contraction certainly creates movement at the joint. Dynamic contractions can be either concentric or eccentric. Concentric contraction involves shortening of a muscle to create movement, whereas eccentric contraction involves a forced lengthening of the muscle.

For example, during pointe tendu, as your foot points, your calf muscles shorten via concentric contraction. As your foot returns to the starting position, your calf muscles begin to lengthen; in other words, during this return phase, the muscles work eccentrically. The significance of this distinction comes into play especially when landing from jumps, wherein eccentric contraction of the relevant muscles helps you decelerate your body against gravity. Thus, even as you work hard to build strength and power to jump higher, you also need to work on controlling your landings to reduce your risk of injury. In another example, as you execute a relevé in first position and hold, the hold phase involves isometric contraction of the leg muscles. Your muscles contract concentrically to elevate you, contract isometrically to hold the position, and contract eccentrically to decelerate you.

As your muscles contract to produce movement, various muscles work together to achieve your movement goal. Dance movements can be carefully controlled because the muscles work so well together. From this perspective, skeletal muscles can be divided into four distinct categories: agonists, antagonists, synergists, and stabilizers.

- **Agonists.** The muscles that contract to produce movement are the movers, or agonists. The ones that are most effective in creating a movement are the primary movers. For example, the action of pointing your foot is created by the gastrocnemius and soleus muscles acting as the primary movers but are also assisted by other muscles acting as secondary movers.

- **Antagonists.** The muscles that oppose the primary movers are called antagonists. They may relax and lengthen somewhat while the prime movers are working; however, they can also contract with the primary movers and provide a cocontraction. As you might imagine, agonists and antagonists are located opposite each other. Look back at figure 1.2 and focus on the gesture leg in attitude derrière. The agonists here are the hamstrings and glutes that activate to move the leg to the back into hip extension. The antagonists are the hip flexors, or the muscles along the front of the hip and thigh, which lengthen while the primary movers contract. Now, imagine a grande plié in second position. As you are coming up, the quadriceps (agonists) work to straighten the knee, and the hamstrings (antagonists) can contract as well, thus providing a cocontraction to better support the knee joint.

- **Synergists.** Synergist muscles can be confusing, so let's break things down. Muscles that are synergists have two functions—to promote and neutralize movement. The key you as a dancer must understand is that the synergist muscles help you define your movement by counteracting any unwanted directional force. For example, returning to figure 1.2, focus on the right arm. When you forcefully lift your arm by flexing at the shoulder, why doesn't the humerus (upper arm) bone separate from the scapula (shoulder blade)? The answer lies in a small muscle hidden under the pectoralis major—the coracobrachialis—which displays synergistic qualities by contracting to help control the movement of the humerus in relation to the scapula. Although the primary mover muscles typically get all the credit, the synergists help the agonists establish smooth and coordinated movement.

- **Stabilizers.** Muscles that can fixate a joint are called stabilizers. They serve as anchors by holding a joint firm, allowing desired movement to occur.

Because stabilizing muscles play such a key role, they are reviewed repeatedly throughout this book. In figure 1.2, for example, the spine is stabilized and braced by contraction of the abdominals; without that contraction, the momentum and strength of the gesture leg moving backward would cause the spine to collapse. You are working so hard on the leg that creates most of the movement that you might forget the importance of the muscles that enable that movement by creating stability and holding you firm.

Body Composition

Let's take a moment to discuss body composition and how it affects your work as a dancer. Body composition, which relates directly to your level of fitness, is described as the body's ratio of fat to lean muscle. The International Association for Dance Medicine & Science suggests that a healthy proportion of body fat should range from 17 percent to 25 percent for women and just under 15 percent for men (Irvine, Redding, and Rafferty 2011). Traditionally, dancers have had less body mass than other athletes because the dance field has encouraged thinness. However, we all need a small amount of body fat to enable healthy muscle function and to help us resist fatigue during long rehearsals. Moreover, restricting calories to minimize fat increases one's risk of injury, amenorrhea, and poor bone health. Ask your health care provider how best to test your body composition. Several sophisticated methods exist, but the most economical and widely used test to determine your body fat percentage involves the use of skinfold calipers or body fat calipers. There are certain areas of your body that are measured by folding or pinching the skin away from the muscle; the caliper device then measures the thickness of the fold. The numbers are then converted into a percentage. Another method, hydrostatic testing, measures body composition under water. Hydrostatic testing is based on calculating body weight under water compared to land weight to determine the most accurate body composition. Again, speak with your health care provider if you are interested in determining your body composition.

Movement Planes

Motion involves changes in position and is created by force. For you, that force is created by the coordinated efforts of your body and mind. Focusing on the efforts of your body will help you to become familiar with some anatomical positions used in this text.

When a muscle contracts, it produces movement at a joint, which is the connector between bones—easy enough, right? Dance uses such actions to move you in all different directions, patterns, and shapes. Understanding how your body moves in space enables you to learn challenging choreography and execute movements with beautiful lines.

You can better understand movement by dividing the body into three imaginary planes—frontal (vertical), transverse (horizontal), and sagittal—which correspond to the three dimensions in space (see figure 1.4). The frontal plane, which divides the body into front and back sides, is represented in the figure by the legs moving directly to the sides. The transverse plane, which divides the body into upper and lower halves, is represented by the rotation through the trunk. And the sagittal plane, which divides the body into right and left sides, is represented by the arms, one in front and one in back. Dance is very multiplanar; choreography occurs in all planes of motion. Maintaining strength and flexibility in all planes reduces risk of injury, assists in changing directions quickly and safely, and promotes healthy, functional movement patterns.

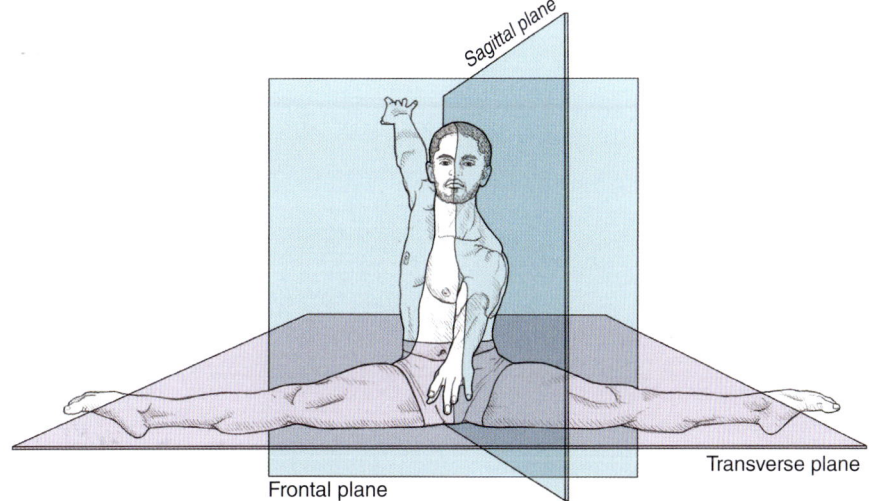

Figure 1.4   Three planes of movement.

Because you can change your orientation in space and your arms and legs can change position, it is important to organize positional directions of movement and to refer to your body in terms of a standard anatomical position. The standard anatomical position gives you a formal starting position to best describe your anatomy. As shown in figure 1.5, that position involves facing front, with your feet comfortably parallel, your arms by your sides, and the palms of your hands rotated to face front. This position enables all directional body movements to begin from a common starting point and all anatomical terminology to refer to the same starting point (table 1.2).

Visualize your standard anatomical position with the various imaginary planes within yourself. You are divided into upper and lower halves by a transverse plane, into equal right and left portions by a sagittal plane, and into front and back portions by a frontal plane. For instance, when you move your arms from en bas through first to high fifth position, you are moving

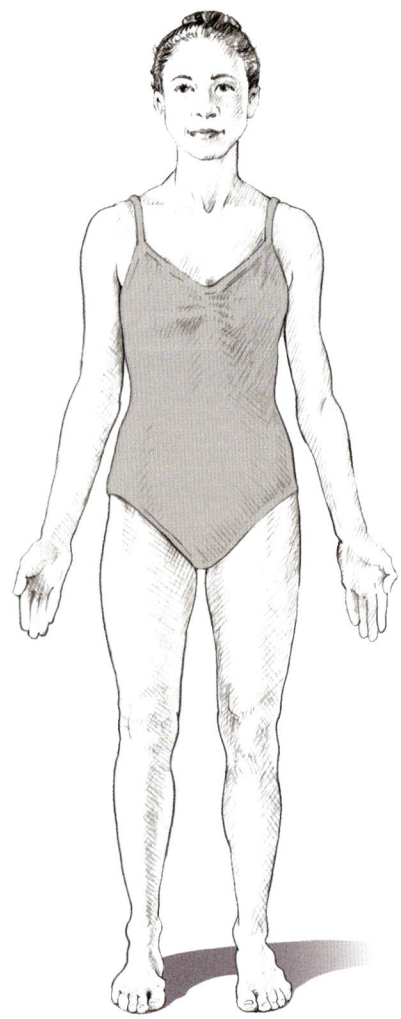

Figure 1.5 Standard anatomical position.

within your sagittal plane. This movement has a purpose; specifically, it works efficiently within an imaginary plane to high fifth, with no deviation and no incorporation of other movement.

In other examples, when you cambré to the side, you are moving in the frontal plane—moving directly to the side without any inefficient movement, as if you are side-bending along an imaginary pane of glass. Similarly, when you execute a plié with the legs turned out, your legs are moving directly to the side along the frontal plane. In contrast, in various hip-hop movements, the legs rotate in and out; in this movement, at the hip, each leg moves along the transverse plane. Similarly, when twisting from the waist, your trunk moves along the transverse plane.

Table 1.2   Anatomical Position and Directional Terminology

POSITIONAL TERMINOLOGY	
Term	Definition
Anatomical position	Standing with feet and palms facing front
Supine	Lying on the back
Prone	Lying facedown
DIRECTIONAL TERMINOLOGY	
Term	Definition
Superior	Above or toward head
Inferior	Below or toward feet
Anterior	Front side or in front of
Posterior	Back side or in back of
Medial	Closer to the median plane or toward midline
Lateral	Farther from the median plane or toward side
Proximal	Closer to root of limb, trunk, or center of body
Distal	Farther from root of limb, trunk, or center of body
Superficial	Closer to or on surface of body
Deep	Farther from surface of body
Palmar	Anterior aspect of hand in anatomical position
Dorsal (for hands or feet)	Posterior aspect of hand in anatomical position; top aspect of foot when standing in anatomical position
Plantar	Bottom aspect of foot when standing in anatomical position

Reprinted by permission from K. Clippinger, *Dance Anatomy and Kinesiology* (Champaign, IL: Human Kinetics, 2007), 18.

Look back at the split jump depicted in figure 1.4. In which plane are the legs moving? The frontal plane. If one leg was slightly forward, thus breaking the plane, then the movement would not produce the clean line that you strive for. Thus, you would need to repeat the jump until you got it correct. The resulting repetition and overrehearsing—necessitated by failure to understand where the legs should be—could lead to an overuse injury.

Mind and Body Connections

Your mind plays a key role in using dance anatomy to improve your technique. A healthy and well-balanced dancer in motion connects the mind and body. Mindfulness can be defined as being more aware, focusing on the present, and being alert to one's surroundings. Take a few moments to consider some coping skills that might help you be a more mindful dancer. Visualization, deep breathing, and positive self-talk are just a few tools you can use to improve your mind–body awareness and will be discussed in this section. Using the tools during the dance-focused exercises in this book can also help you pay attention to your body, focus on specific muscles, and truly connect to the movement. This discussion will be continued in the psychological awareness segment of chapter 2.

The connection between stress and injury has been examined by researchers. Various psychological factors can play a role in injuries. The demands of a dance career can create tension, anxiety, perfectionism, pressure to succeed, and disordered eating, which are associated with injury. High expectations from others, competition, and job insecurity can also create negative dance stress, which has been associated with injury. Stress occurs when the demands of a dance career outweigh your coping abilities (Mainwaring and Finney 2017). Like any other sport, dance requires intense training and conditioning to maintain the highest level of physical performance. Thus, you seek perfection and may push yourself beyond your limits. If you allow competition anxiety or fear of failure to overwhelm your mind, then you may lose the ability to cope and put yourself at risk of getting injured. The following are healthy coping skills to help improve your mind–body connections.

Visualization

Part of being a dancer in motion is understanding primary muscle actions, but you can also use visualization as a tool to help you dance more efficiently. How many times do you practice the act of développé? How many times do you feel gripping in the thigh, accompanied by anxiety, because you are unable to raise your leg higher? Imagine what it would be like to know which muscles need to contract, lengthen, and stabilize without gripping. Imagine elevating your leg higher without anxiety. This is possible when you use your mind along with your physical ability.

The practice of creating a picture in your mind without performing the pictured physical activity has been described by various terms, including *imagery*, *mental simulation*, and *visualization*. Of the many kinds of imagery, the focus here is on basic visualization skills for improving performance. To release unwanted tension, you can use simple positive images and focus on maintaining a calm center. Visualize exactly what you want your body to do and keep your thoughts positive. Eric Franklin is a master at visualization techniques. His term *seed imagery* from his book *Conditioning for Dance* (2019) refers to planting an intuitive thought and letting that image grow to enhance performance. Imagery can help you release tension, lengthen muscles, and better understand and connect to the requirements of the movement. Imagery is a practice that should be used regularly to produce positive change.

When you repeatedly train your actions (as you do in class and rehearsal), you induce physiological changes and increase accuracy. Take a little time every day to find a quiet spot, close your eyes, and just listen to yourself breathe. Now, imagine the dancer you want to be, and see yourself moving with ease. Focus on how clean your lines are. Visualize how much control you have in every combination you perform. You can see it in your mind, you can hear the music playing, and you can feel your body executing the sequences with detail. Let everything else go and focus on your technique. You are training the relationship between your mind and your muscles. They must work together to create balance and help you reach your goals. Mindful visualization is a practice that will get easier the more you participate in it.

Deep Breathing

Your state of mind influences the outcome of your work. If you prepare for a pirouette with tension in your upper body, nervousness about having to execute a double turn, or anxiety about losing your balance, how on earth can you turn? Instead, visualize beautiful multiple turns around a firm but calm center—and breathe! Deep breathing can help you relax and reduce stress. It also encourages mindfulness and allows your body to take in more oxygen. If you take a moment to practice deep breathing, you can slow your heart rate and create a sense of calm to prepare for your pirouettes. Dance your way into the pirouette, release the fear, use breathing to help you, and enjoy turning!

Positive Self-Talk

Some dancers learn to maintain a healthy, positive, internal conversation that creates motivation and encouragement. This positive inner dialogue can reduce tension and facilitate ease of movement. Remember, you are building a healthy, balanced connection between your mind and your body. Accept yourself, enjoy the process of learning, and love dancing—it's that easy! However, some individuals are full of criticism, doubt, and negative self-talk. Some inner dialogue automatically happens even when other, more deliberate self-talk is most beneficial. Using positive and deliberate inner dialogue can help you reduce anxiety and ultimately improve your performance. If you love to dance and want to improve, you must stop the negative self-talk and dissatisfaction with yourself. Stay away from telling yourself that you cannot do something or that a given movement is too hard. Instead, be firm and tell yourself that it's possible. If you still find yourself talking negatively, do your best to acknowledge the negativity and couple it with a positive thought. For example, if you have told yourself you can't do a double jazz turn, also tell yourself that today is the day you *will* do a double jazz turn. What if it all works out? You are now challenging your negativity and responding with helpful, positive thoughts.

Dance-Focused Exercise

There is a distinct relationship between the exercises and illustrations presented in this book. Throughout the exercises, visualize ease and balance in your neck, as well as stability throughout your center, and allow those qualities to carry over into your technique. For example, when performing the exercises for your legs, visualize ample joint mobility, not tension, in your hips. Remember to keep the images positive and brief.

Each exercise includes how and when to breathe for the most efficient movement. Use each breath to help connect with your core muscles. You can focus on using your breath to prepare and execute the movement, which builds a stronger mind–body connection.

After practicing visualization skills during the exercises, send those brief images through your mind before classes, rehearsals, and performances. Notice how your skills improve—how you work more efficiently with less gripping in your muscles. Continue using positive visualization skills; these exercises of the mind require practice. Don't let negative thoughts creep back in and ruin your technique. To help you build this positive habit, chapters 4 through 10 each include a section called Dance-Focused Exercise that guides you in applying these skills to each chapter's exercises.

Cardiorespiratory Fitness

Although this book focuses on dance-specific exercises, do not overlook the benefits of cardiorespiratory fitness, which increases the efficiency of your heart and lungs. More and more medical research on dance shows that dancers' cardiorespiratory capacities are like those of other athletes in nonendurance sports. In 2015, for example, Rodrigues-Krause, Krause, and Reischalk-Oliveira published a fantastic article titled "Cardiorespiratory Considerations in Dance: From Classes to Performances." This article emphasized the importance of aerobic conditioning for reducing fatigue-related injuries. Even though rehearsals and performances last for only brief periods—and thus constitute *anaerobic* activity—it is necessary to engage in aerobic training to improve your cardiorespiratory health, which improves blood circulation and oxygen supply to the cells.

More specifically, aerobic training increases heart size, which allows a larger volume of blood to be pumped through the body. Cardiorespiratory fitness allows for better transportation of oxygen and thereby increases your endurance. In turn, high cardiorespiratory endurance reduces physical and mental fatigue, either of which can lead to injury. As a result, the more aerobically fit you are, the longer you can rehearse before fatigue sets in. Your daily dance class, however, does not provide enough aerobic benefit. The best way to increase your aerobic capacity is to raise your heart rate to 70 to 90 percent of your maximum for at least 20 minutes. For example, you can improve your cardiorespiratory endurance by training on an elliptical machine, treadmill, or stationary bike—or by swimming—three or four times per week for at least 20 minutes per session.

Dance teachers can redesign portions of their classes to provide more aerobic benefit by allowing exercises to repeat or run longer with less rest time. They can also include longer jumping combinations in the center. Teachers owe it to their students to help them improve their cardiorespiratory fitness, which will help them reduce injury rates and improve their overall health.

Conditioning Principles

To define or enhance your conditioning plan, you should be familiar with the fundamentals of certain principles. First, you are developing strength not only in your muscles but also in your tendons and ligaments. Research increasingly shows that to become a stronger dancer, you must work to strengthen yourself outside of daily dance classes. For example, Koutedakis and Jamurtas (2004) published an article titled "The Dancer as Performing Athlete," which emphasizes that poor physical fitness relates to injuries and that fitness can be enhanced through additional monitored aerobic training. As always, artistry is important; you are, after all, an artist. However, to be a healthy and well-balanced dancer, you may need to consider additional training.

- **Functional training.** This type of training uses more than one joint at a time. The goal is to train multiple muscles for the purpose of improving dance technique. You can emphasize upper- and lower-body movement while challenging the core. Functional training incorporates multijoint, multiplanar, and multimuscle exercises. This book includes many functional training exercises to progress you from exercises on the floor to the barre and center.

- **Principle of overload.** If you want to increase strength, you must work the targeted muscle group past your normal load. The exercises are executed at maximal contraction through the entire range of motion. Typically, this type of training uses fewer repetitions and more resistance, and it works your muscles to fatigue. Don't worry that building strength will cause you to lose flexibility; research shows that you can improve your strength without losing the flexibility that you need for dance. An article published in the *Journal of Dance Medicine & Science* titled "The Significance of Muscular Strength in Dance" (Koutedakis, Stavropoulos-Kalinogiou, and Metsios 2005) reviews the importance of muscular strength in dance and that strength levels positively impact flexibility.

- **Principle of reversibility.** When conditioning is stopped, strength is quickly lost. Therefore, to maintain your fitness, continue your dance-specific conditioning (per the exercises presented in this book) at least four times per week even when you are not dancing (for instance, during a layoff or holiday break).

- **Principle of specificity.** You must condition dance-specific muscles to improve your technique. Your conditioning or training should be relevant to the dance technique you are looking to improve. For example, if you are interested in improving your jumping combinations, then you should consider a plyometric or jump-training program along with a cardiovascular training program. Your training should imitate the dance skills in which you are looking to excel.

- **Alignment.** All repetitions must be executed without sacrificing alignment, core control, or proper breathing. Your goal is to work efficiently. If you feel your alignment beginning to falter, take a moment to stop, reorganize, and then start again. As you perform each exercise, emphasize the main muscle movement but notice how it affects your entire body.

- **Warm-up and cool-down.** Each conditioning session should begin with a basic warm-up to increase blood flow, accelerate your breathing, and slightly raise your body temperature. If you warm up in this way, your exercises will be more effective. Take 10 minutes to include exercises from chapter 4 to get centered, then do some low-level jogging in place. After your conditioning session, perform a sufficient cool-down to allow your body to return to its resting state. This work can last for about 10 minutes and include the breathing exercises presented in chapter 5. You can also include some gentle stretching to reduce muscle soreness.

Your warm-up, cool-down, and exercise program should take about 50 minutes. Each exercise presented in this book is geared to a specific goal. All the exercises are designed to provide muscular balance and to require control through the full range of the movement. Avoid initiating the movement with momentum and then allowing gravity or loss of awareness to take over as you finish the movement. Instead, begin each exercise with slow, precise control and maintain that control throughout the movement. Each chapter allows you to work with specific muscle groups to enhance the intensity and deepen your awareness while also including functional training exercises. Keep your mind focused on safe skeletal alignment, which is emphasized throughout the book.

Because expert opinion varies, it is next to impossible to recommend a one-size-fits-all conditioning program with specific durations, repetitions, sets, and intensities. For general purposes, repeat each exercise presented here 10 to 12 times for three sets unless otherwise stated. Understand, however, that it may take some practice for you to determine your personal needs. If you are trying to build strength, you must execute maximal muscle contraction through the entire range of motion and overload the muscle in a progressive manner. Some of the exercises presented in this book use resistance bands or small weights for progressive resistance, but the goal is to maintain excellent alignment. You can add resistance gradually when your alignment is secure and a given exercise is no longer challenging. Emphasize balance and quality of movement.

CHAPTER 2
Brain Health

Your amazing brain oversees every dance move you make. Therefore, the healthier your brain and its connections to your muscles, the better your performance will be. Strong neurological connections promote better speed, agility, and balance. In fact, dance creates new brain cells through a process called *neurogenesis*; in other words, dance movement can, in effect, rewire your brain. Each time you learn new choreography, your brain develops new neural pathways for your body, and these pathways help you think and acquire knowledge.

How does all this happen? When your brain processes information better, your body can respond better. Greater accuracy in the nervous system allows for better turns and higher jumps and enables you to pick up choreography quickly and efficiently. One of your most important senses is your balance, or proprioception—basically, how you know where you are in space. You can learn to train your balance, your nervous system, and your muscle contractions to give you greater technical precision.

This chapter does not provide an exhaustive explanation of neurological science. Rather, it provides an overview of how your nervous system can help you improve technique and performance; in the process, it introduces the fundamentals of how messages are sent to and from your brain and muscles. The chapter ends with information on how to better understand stress and anxiety to help improve your performance.

Nervous System

The nervous system controls all movement, whether voluntary or involuntary, including dance movement. It also handles the cardiorespiratory and digestive systems, but for now, we will focus on the motor system and how movement occurs. Movement happens when the nervous system sends electrical signals to and from relevant parts of the body. Again, the more efficiently your nervous system works, the better your performance will be.

The nervous system is divided into two parts: central and peripheral (figure 2.1). The central nervous system is made up of the brain and spinal cord, whereas the peripheral nervous system includes the nerves that come from the brain and spinal cord. The peripheral nervous system itself is subdivided into two parts: the somatic nervous system, which primarily takes care of voluntary muscular movement, and the autonomic nervous system, which primarily takes care of digestion and the beating of your heart. The autonomic system breaks down into three more divisions: sympathetic, parasympathetic, and enteric. The autonomic system will be discussed in more detail later in this chapter.

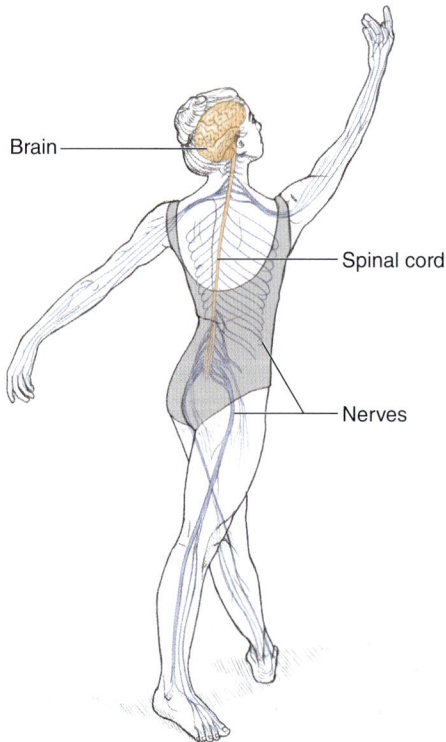

Figure 2.1 The nervous system is made up of the central nervous system and the peripheral nervous system.

Nerves are bundles of fibers that send impulses or signals to and from the brain and various body parts for sensory or motor purposes. Neurons are cells that create impulses for the nerves, each of which contains a cell body, a nucleus, dendrites, and an axon (figure 2.2). The surfaces of dendrites typically receive messages and send them to the cell nucleus. The longer extensions of the neuron are known as axons, which send messages away from the cell body and are protected by a myelin sheath. The sheath provides some insulation for the axon. At the end of each axon lie multiple axon terminals, which can send messages to other neurons. The axon terminal, also called the synaptic terminal, releases neurotransmitters that cross the synapse, the tiny space separating the dendrite from the axon of another neuron.

Thus, messages enter neurons through the dendrites, then travel to the cell body and on to the axon. Small gaps along the axon, known as the nodes of Ranvier, make it possible for the impulse to travel to more dendrites connected to the muscle. Once this action or impulse occurs, the muscle contracts. This whole process happens at about 400 miles (645 km) per hour!

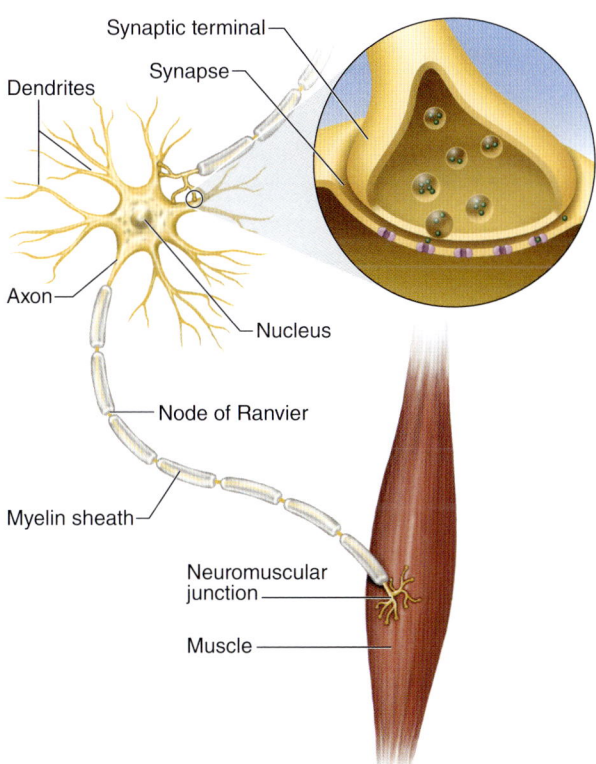

Figure 2.2   Motor neuron action.

There are three types of neurons: motor neurons, sensory neurons, and interneurons.

1. Motor neurons are found in *efferent* nerves, which send messages *from* your brain to your muscles and certain glands. The end of the axon of the motor neuron forms a neuromuscular junction with your muscle fibers to allow them to contract.

2. Sensory neurons stimulate skin, muscles, and joints. They can sense touch or pressure by sending pain signals to your brain. They can also sense light, smell, and taste, along with hot and cold temperature changes. These neurons are found in *afferent* nerves, which send signals *to* your brain.

3. Interneurons provide a connection point between motor and sensory neurons. They let the efferent and afferent nerves work together. Interneurons are only in the brain and spinal cord and are not part of the peripheral system. The information coming in from sensory nerves is transmitted through the interneurons to the motor neurons.

If you rehearse in new shoes and develop a blister, the afferent nerves around that blister send a sensory message to your brain that relays pain. In turn, efferent nerves send messages from your brain back to the muscles in your feet to alter or adjust your footwork and thereby prevent or minimize the pain associated with the blister. Afferent or sensory nerves take in information, whereas efferent or motor nerves send information out to your muscles.

Brain Function

Regular aerobic exercise can improve brain health, as well as memory and thinking skills. When dance movement is combined with music, the music stimulates the brain's reward or satisfaction center, while the dancing stimulates the brain's motor circuits, which help organize movement. Studies are increasingly showing that dance can reduce the risk of dementia-related symptoms because of both the social interaction involved and the mental effort required. Dance is a social activity that provides motor coordination and memory enhancement and can be used as an intervention for people with dementia due to its psychological benefits (Tao et al. 2021).

Dance is also being used more and more as therapy for patients with the motor disorder Parkinson's disease, particularly for improving gait and quality of life (Haputhanthirige et al. 2023). For example, the Brooklyn Parkinson Group offers dance classes taught by choreographer, dancer, and director Mark Morris. The classes stimulate participants' brains to help them improve their balance, control their movement, and build confidence. The Mark Morris Dance Group offers training and enrichment programs internationally to bring the joy of dance to individuals with Parkinson's.

As a dancer, you already know the benefits of dancing and how it makes you feel. You also know that repetition enhances learning, improves memory, and aids in improving technique. The more you rehearse, the more your body feels comfortable with a movement as you continue to perfect the artistry. The brain's memory centers are improved through dynamic movement—that is, movement involving multiple joints—which is exactly what dance is. In contrast, lack of healthy, high-quality movement causes neural connections to atrophy.

Let's continue this discussion of how dance, movement, and exercise benefit the brain by addressing some basics of how the brain functions. Do you have any idea how your brain sends messages to your legs and feet to help you perform a basic tendu or a challenging pirouette? How do you learn new combinations or creative choreography? Your brain oversees how you move; it is the choreographer for your dance steps. Even a simple movement, such as a pointe tendu, gives your brain a very complex job to do. Your brain must decide which muscles need to fire to execute the tendu, as well as how much force your movement needs. This complex job is referred to as *motor function*, which can be defined briefly as how movement occurs with the help of the nerves.

Basic Anatomy

The brain and spinal cord make up the central nervous system. The brain is surrounded and protected by the bones of the skull, or cranium, and contains more than 100 billion nerves. Several sections of the brain are responsible for movement, sight, emotion, and other functions.

The main parts of the brain (figure 2.3) include the cerebrum, cerebellum, and brain stem.

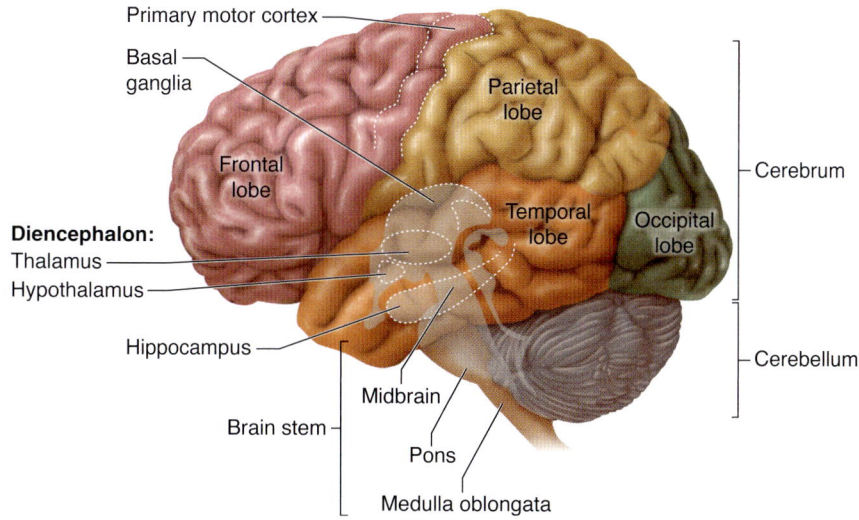

Figure 2.3   The human brain.

The largest portion of the brain is the cerebrum, which has an outer surface called the cortex. The cerebrum is divided into four lobes: frontal, parietal, occipital, and temporal. The frontal lobe is responsible for controlling motor skills, problem solving, impulse control, and judgement. The parietal lobe is responsible for processing pain, the occipital lobe for interpreting visual information, and the temporal lobe for processing memory. Another part of the cerebrum, the prefrontal cortex, is located directly in front of your brain; it is responsible for thinking, making choices, and expressing your personality.

The area of the cerebrum that initiates movement is the primary motor cortex. This strip, located along the center of the cerebrum, has connections to every part of your body. All voluntary movement is controlled through your primary motor cortex.

Deep within the brain, between the cerebrum and the brain stem, there is a small section known as the diencephalon. Sometimes referred to as the inter-brain, it consists of the thalamus and hypothalamus, which together process hunger, thirst, emotional responses, memory, pain, and physical pressure.

Another part of the cerebrum is the limbic system, which oversees emotional responses; this system includes the thalamus and hypothalamus. The limbic system plays a role in how you handle stress. The thalamus and brain stem work with the basal ganglia, which lie deep in the base of the forebrain, to help control voluntary movement so that you can dance with ease. More specifically, messages are sent from the prefrontal cortex to the basal ganglia, which pass the messages to the correct motor system to initiate smooth movement.

Another part of the limbic system is the hippocampus, which is located deep in the temporal lobe and plays a huge role in emotion, motivation, and memory. When you learn to execute a basic tendu, for example, your hippocampus stores that information and helps you remember how to execute it again.

The second major part of the brain—the cerebellum—is not as large as the cerebrum, but it contains 50 percent of the brain's neurons. It is in the back of the brain, between the brain stem and the cerebrum and beneath the occipital lobe. It receives messages and controls balance, posture, and movement coordination, as is needed when learning a new dance step. Thus, this portion of the brain is crucial for motor learning and fine-tuning movements repeatedly until the desired movement pattern is reached.

The third major area of the brain is the brain stem, which includes three parts—the midbrain, the pons, and the medulla, which connects to the spinal cord. The midbrain helps control body movement. The pons can carry messages into the brain from the nerves as well as send messages from the brain to the nerves. The medulla is responsible for your breathing and the beating of your heart. The brain stem functions as a business center by relaying messages from the cerebrum and cerebellum to the spinal cord.

Spinal Cord

The spinal cord's job is to connect the brain with the nerves. It consists of a long, cylinder-shaped cord located deep inside the spine and running from the base of the brain to the first and second lumbar vertebrae. It is protected by the vertebrae surrounding it.

The spinal cord serves as a conduction pathway from which 31 pairs of nerves originate. It includes both ascending nerve tracts and descending nerve tracts.

Each spinal nerve consists of both afferent and efferent nerves. Spinal nerves can be either somatic or autonomic. Somatic nerves send messages to and from the muscles, tendons, and joints. Autonomic nerves relay messages to and from your heart and other glands.

Balance System

As a dancer, you need excellent balance skills. Without them, how could you balance on the tip of a pointe shoe? Balance depends, in considerable part, on proprioception. Derived from the Latin word *proprius*, the term *proprioception* basically means "one's own"; for our purposes here, it refers to your sense of where you are in space. Proprioception also helps you get on stage in the dark without losing your balance before the lights and curtain go up. It also helps you learn new choreography without losing your balance. Proprioception helps you turn and coordinate movement while different body parts perform different actions at the same time.

Proprioception and balance depend on three bodily systems:

1. Vestibular
2. Motor
3. Visual

Your vestibular system is in the inner ear (figure 2.4). Your ear performs two jobs: hearing and proprioception. Your outer ear sends sound waves to the tympanic membrane, or eardrum, where your auditory nerve (cochlea nerve) picks up those sound signals from the tympanic membrane and carries them to the brain. Your vestibular system is made up of semicircular canals deep inside your ears that are filled with fluid and lined with tiny hairs called cilia. They help you with rotation movements such as pirouettes and fouettés. Balance for linear movements, on the other hand, is enabled by an organ called the otolith, which contains small calcium carbonate crystals that stimulate the cilia.

Your inner ear also includes the push–pull system, in which the semicircular canals work with each other to help you balance during horizontal head movement. The left canal is stimulated when you turn to the left, and the right canal is stimulated when you turn to the right. Your fluid-filled ear canals, as well as cilia, coordinate signals to help you keep your head upright. When all inner ear systems are functioning properly, you can practice turning with ease.

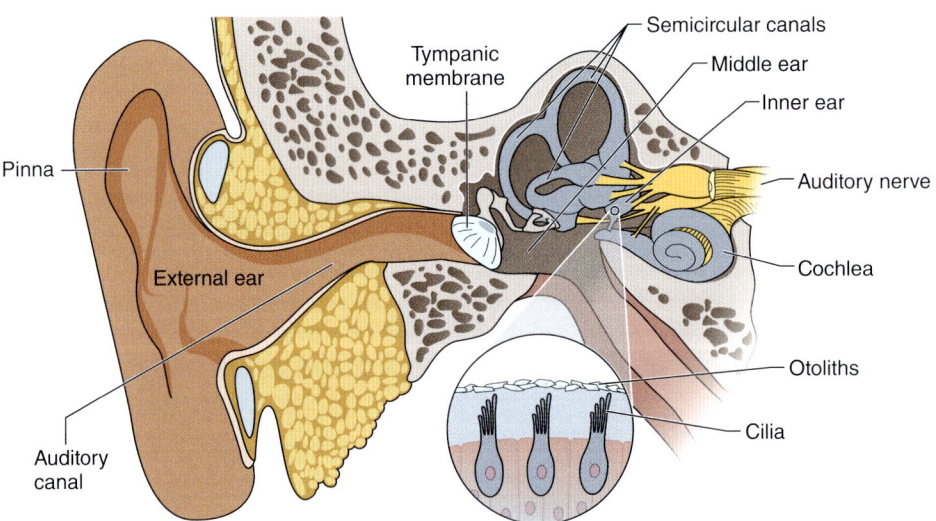

Figure 2.4   Ear anatomy.

The motor system is the second system that is critical to proprioception and balance. It includes sensory receptors that stimulate key body parts to help you maintain balance. The sensory receptors are in your joints, muscles, ligaments, and tendons and are constantly working to provide you with spatial orientation. Strong proprioceptive information from the sensory receptors is carried through the spinal cord to your brain to help you maintain your balance. Weak proprioception skills lead to poor balance, uncoordinated movement patterns, and inaccuracy. This deficit can also lead to malalignment and, eventually, muscle weakness. Thus, as you can imagine, it sets you up for injury.

The third system important for proprioception and balance is the visual system, which coordinates with motor function to help you move effectively. For example, when you want to try a new piece of choreography, where do you begin? Believe it or not, the instant you see the choreography, you begin to feel it in your nerves and muscles even before you take a step. This strong connection comes from the accuracy of your visual system. When teachers and choreographers demonstrate for their students, the students will see the movement to begin the learning process. Your eyes see the choreography and immediately send signals to the visual cortex in your brain for interpretation. This system works quickly and efficiently through specific neurons that organize body movement, balance, and head movement.

When all systems work well together, your turns will be outstanding. For example, your spotting is efficient because your eyes and horizontal head movement are coordinated to maintain control. Spotting, which is essential for maintaining balance and preventing dizziness, depends on learning to rotate your body at a certain speed while your head rotates at a faster speed. Your eyes then stop on a focal point before the rotation process begins again. The moment in which your head stops for your eyes to focus is crucial because it allows your body to establish stability before the turn begins anew.

It is imperative for teachers to teach students to keep a level head, literally, while learning to turn. Any tilting of the head—whether to the side, up, or down—confuses the vestibular and visual systems, which then overcompensate to aid balance. Students must learn to see a focal point and stay on that point while their bodies begin to rotate. Once they can no longer maintain a visual connection with the focal point, they must quickly turn the head around, before the body rotates, so that the eyes and the visual system can find the same focal point again to reestablish balance and stability. This spotting efficiency can be learned—the body can adapt to it—and it is needed for exceptional multiple turns, such as fouettés and chaînés.

All three balance systems send and receive messages to and from the brain about how to maintain balance. You can improve your balance by challenging these systems. For example, if you perform balancing exercises on one foot and then advance to an unsteady surface, you challenge your motor-system proprioceptors to send messages to your brain to strengthen your inner sense of balance. If you practice balancing with your eyes closed, thus taking your visual system out of the equation, your motor and vestibular systems work harder. You can also challenge these two systems by balancing with your eyes open while moving them from side to side. Another way to advance your exercises is by trying to relevé at the barre with your eyes closed. Notice which way you sway and how long it takes to lose your posture before you grab the barre.

In summary, balance skills can be learned, practiced, and advanced to enhance neuromuscular coordination. Balance-specific training is discussed further in chapter 11.

Movement Execution

Movement execution is aided by three nerve pathways, or tracts:

1. Corticospinal tract
2. Cerebellar system
3. Extrapyramidal tract

The corticospinal tract is a motor tract that is responsible for voluntary skilled and precise muscle movement. It begins in the center of the cerebral cortex and travels through the brain stem to the spinal cord. It is responsible for directing very detailed movements. The cerebellar system helps coordinate the muscles so that they work well together to execute movements. The extrapyramidal tract sends messages to allow for good postural awareness and balance skills.

With this background in mind, let's consider what happens when you execute a simple tendu. Organized movement begins in the motor cortex and then travels down about 20 million nerve fibers in your spinal cord. The motor cortex is divided into various sections, each of which is responsible for a different part of the body. Your primary motor cortex, located in the frontal lobe, generates neural impulses that cross your body's center to activate muscles on the opposite side of the body. In other words, the right side of the brain handles the left side of the body and vice versa. The following list suggests just how busy your nervous system is when executing a tendu:

1. As you watch your teacher demonstrate a tendu, activation begins in your nerves and muscles.
2. Your visual system quickly sends signals to your visual cortex in the occipital lobe of the cerebrum.
3. Planning for the activation of the tendu begins in the frontal lobe of the cerebrum within the primary motor cortex.
4. Messages from the primary motor cortex are sent to the basal ganglia, where motor control and motor learning information is sent back to the motor cortex via the thalamus.
5. Your motor cortex sends messages through the corticospinal tract with help from the efferent nerves.
6. The messages are relayed by the thalamus.
7. The hippocampus is stimulated to memorize the tendu.
8. The cerebellum receives messages to stimulate your balance system.
9. The brain stem receives the stimulation and passes the signal to your spinal cord.
10. The spinal cord sends the signal to your hip, leg, ankle, and foot, thus stimulating the muscles needed to execute the tendu.
11. Your balance systems continue to relay proprioceptive input.

The whole process results in a simple, organized coordination of your nervous system to execute a basic tendu.

Psychological Awareness

Now, let's revisit our earlier discussion of how dance and exercise affect the brain. During exercise, your brain releases several chemicals that stimulate your reward center and change your mood and outlook. These chemicals are released in larger quantities during aerobic dance training than during anaerobic dance; this chemical production is further stimulated by rehearsing longer pieces that generate faster heart rates. Key chemicals in this process include various neurotransmitters, which send nerve impulses across synapses. Four types of these chemicals—endorphins, serotonin, brain-derived neurotrophic factor, and dopamine—are introduced briefly in the following paragraphs. The stress hormones cortisol and adrenaline will also be discussed. Table 2.1 summarizes these chemicals.

choreography. Dancers who compete with a team may feel anxious about not being able to meet the demands and possibly letting their teammates down. Remember, when heightened performance anxiety triggers your body's fight-or-flight mechanism, your brain perceives a threat and releases adrenaline and cortisol that can cause the following physical changes:

1. Adrenaline relaxes the lungs to allow for more oxygen intake, speeding up the heart rate; this can make you feel lightheaded or dizzy.
2. Adrenaline increases blood flow to deliver more blood sugar to your muscles, which can cause tension and involuntary muscle shaking.
3. Cortisol releases glucose (sugar) throughout your body for fast energy.
4. Cortisol causes the body to increase production of stomach acid, which can cause nausea or an upset feeling in the stomach.
5. When the heart rate and breathing rate increase, your body heats up and you begin to sweat.

The release of hormones produces normal changes that occur during anxiety. Stress can have a positive or a negative impact on your performance. Chronic high levels of stress before or during competitions can make it challenging to perform well due to tension, stomach pain, and difficulty breathing. However, low levels of short-term anxiety can help you focus, make you more alert, and motivate you to reach your competitive goals. The key is to understand this anxiety or stress response, develop coping skills, and use these skills to improve your performance.

If you suffer from any type of stage fright, learning to manage stress is important to improve your performance. Chapter 1 covered visualization, deep breathing, and positive self-talk. These coping skills can help reduce the negative effects of the stress response. Meditation can also help to lower levels of adrenaline and cortisol by creating a deep sense of calmness and tranquility. Meditation can lower your heart rate and reduce negative emotions. The overall goal of any type of meditation is to incorporate relaxed breathing, mindfulness, and an open mindset to redirect anxious thoughts into positive ones. Practicing gratitude is another way to manage stress and can help promote optimism, empathy, and selflessness. Expressing gratitude helps your brain produce the feel-good hormones dopamine and serotonin, which immediately improve your mood.

If you are a dancer and you feel uncomfortable or suffer from heightened anxiety before competitions, performances, or auditions, it would be beneficial to understand the effects of stress and develop coping skills to help manage the symptoms. If you are a dance instructor, it would be advantageous to offer psychological awareness workshops to help your students better understand the effects of anxiety and learn stress management skills. You could have students name several things they are grateful for at the end of each class, teach meditation exercises as part of the rehearsal process, or remind your students of the benefits of positive self-talk to help build their confidence.

Dancer's Preperformance Tool Kit for Reducing Anxiety

- Be prepared: make a list of all the items you need to reduce the anxiety of forgetting something.
- Practice deep breathing to lower your heart rate.
- List reasons to be grateful, which will improve your mood.
- Visualize yourself enjoying your performance to reduce tension.
- Repeat positive affirmations to give yourself more confidence.
- Find time for mindful meditation to create a sense of calmness.

Brain Fuel

Your brain works for you 24 hours a day, sends and receives multitudes of signals, handles every emotion—and weighs only 3 pounds (1.4 kg)! It also serves as the control center for your every move. How can you take good care of it? Key steps include handling stress effectively, sleeping well, eating healthfully, and drinking plenty of water to stay hydrated.

Stress has been associated with synapse and neural dysfunction due to interference with neurotransmitters. It also causes the release of stress hormones that make your heart pound faster, your blood pressure rise, and your breathing change. Chronic stress is quite disruptive and can lead to cognitive and memory deficits; signs of chronic stress include nervousness, anxiety, and depression. To minimize stress, avoid overscheduling yourself and stay away from negative self-talk and negative studio drama. Also be mindful of internal causes of stress, such as excessive worrying and unrealistic expectations. Make time to relax by, for example, listening to music or spending time outside. Take note of the Dancer's Preperformance Tool Kit for Reducing Anxiety and find the right coping skills just for you.

A lack of sleep interrupts coordination, concentration, and memory; it also slows down reaction time and increases confusion. It is difficult to perform when you can't concentrate. When you are sleep deprived, sections of your cortex overwork to compensate for lack of rest. Even so, a tired brain cannot help you perform as well as when you are rested. When you sleep, your brain remains active and works to reenergize neural pathways. Nap if you need to so you feel rested during the day. People need different amounts of sleep to feel energized; if you are exhausted and your brain feels foggy, you probably need more sleep.

A healthy diet is crucial for warding off free radicals, which are highly reactive atoms that can damage body tissues on the cellular level. Free radicals are unstable and look to bond with other molecules to grow, causing further tissue damage. You can minimize the effect of free radicals by eating foods rich in vitamins, minerals, and antioxidants and by drinking plenty of healthy fluid. While individual needs vary, you should drink water throughout the day, during rehearsals, and during meals. If the frequency and duration of rehearsals and performances increase, it will have an impact on hydration. If you are performing outside under hot and humid conditions, you will need more fluids. The most efficient way to monitor hydration is by checking urine color in the morning; pale shades of yellow indicate proper hydration. For more information on staying hydrated, read the International Association for Dance Medicine & Science's Nutrition Resource Paper (Challes and Stevens 2019). It is also important to consume enough antioxidants. Although antioxidants can be found in vitamin supplements, the body absorbs them better from food. Good food sources of antioxidants include cranberries, plums, blackberries, blueberries, beans, artichokes, walnuts, and pecans.

Healthy eating fuels your brain and your body. Here are some more healthy food choices:

Whole-grain cereal, bread, rice, and pasta provide the brain with a steady source of energy in the form of glucose to help you focus.

Fish (e.g., wild salmon) provides oily omega-3 fatty acids that can help produce serotonin and may help ward off Alzheimer's disease.

Dark red and purple fruits, especially blueberries, have been shown to help with short-term memory loss and protect the brain from the negative effects of stress.

Green, leafy vegetables and asparagus are great sources of vitamin E and can help ward off cognitive decline. Other good sources of vitamin E include almonds, flaxseeds, walnuts, and peanuts.

Avocados have been associated with better blood flow, which can aid brain health; they may also help reduce high blood pressure.

Beans can stabilize blood glucose levels. Since the brain needs glucose but cannot store it, beans are a great choice for providing the brain with energy.

Last, but not least, dark chocolate can enhance your concentration and help you focus. It also stimulates the production of endorphins, which improve mood.

Medical studies show that dancing can stimulate cognitive function and muscle memory (Bergland 2013). Good cognitive function enables us to acquire knowledge, engage in reasoning, and pay attention. Dance movement involves the motor cortex, the basal ganglia, and the cerebellum; their participation improves your memory. Moreover, when you rehearse or perform, you must make quick decisions to organize your movement, stay with the music, maintain your balance, and remember the choreography; this process improves your neural connections. Even when you mark through your dance choreography, you can improve your complex skills and the memorization of those skills. These connections can be described as your wiring—that is, the connections between the neurons that send information along the paths of your nervous system.

An article published in *Neuroscience and Biobehavioral Reviews* (Teixeira-Machado, Mario Arida, and Mari 2019) reviewed eight studies and found that dance altered cognitive function by improving memory, attention, and balance, which in turn increases neuroplasticity. When the brain integrates movement and sound with creativity and performance, stimulation occurs in different cognitive and motor areas of the brain. The connectivity between the cerebral hemispheres is strengthened. Several of the studies looked at ballroom, salsa, jitterbug, and rumba dance styles, while another study looked at line dancing, jazz dance, and square dancing. Most results showed positive functional changes in the brain. While there will always be room for more research, the results are promising.

Although most studies include older populations, there is evidence that dance benefits children as well. Dance for children provides physical and mental stimulation, and it also increases social interaction. Participating in dance has reduced daytime fatigue and improved sleep patterns in children (Sandberg et al. 2021). When children enjoy dance, their confidence and self-esteem improve. Dance has been associated with reducing emotional distress and depressive symptoms in young female students (Tao et al. 2022). Dance increases body awareness and improves whole-body movement skills. Learning new choreography improves memory capacity. Teachers see the difference dance can make in improving mood and creating joy in their students. Dance improves brain health. Dance can also increase your self-esteem and confidence simply by letting you express yourself through movement and music.

So, whether you are enjoying a class in hip-hop dance, classical ballet, or Zumba, it's all dance—and it's all fantastic for your brain health!

CHAPTER 3
Injury Awareness

To reduce the risk of injuries, you need to understand how they occur. Lifetime injury rates are high for dancers, so let's look at factors, general awareness, and basic information on how to reduce the risks. Although injury risk education is addressed throughout the book, this chapter focuses specifically on evidence-based information to help you understand how and why injuries occur. You can make proactive use of this understanding by incorporating injury risk education into your work as a dancer, instructor, or choreographer.

What is the definition of an injury? Primarily, injury is defined as a physical impairment, whether acute or chronic, resulting in not being able to participate in dance for 24 hours or more.

Dance-related injuries have been the subject of more than 2,500 articles. Nearly two-thirds of these injuries are caused by overuse, and acute injuries account for one-third of those reported (Ramkumar et al. 2016). Lifetime injury rates among ballet dancers are estimated to be as high as 90 percent.

Dance-related injuries are influenced by many variables, and when they come together, you are at risk of hurting yourself. Injuries can be either intrinsic or extrinsic. Intrinsic injuries relate to something over which you have control, such as poor technique. An extrinsic variable, on the other hand, lies beyond your control; for example, dancing on a slippery surface could cause a fall. Moreover, as you might expect, if you dance on a slippery surface while using poor technique, you are really setting yourself up for an injury!

Intrinsic Factors

The intrinsic factors that can contribute to dance injuries include poor technique, which is associated with muscular imbalance, weakness, and poor alignment. Fatigue and poor cardiovascular fitness are also associated with injuries, as are inadequate nutrition, lack of quality sleep, and psychological factors.

Poor Technique

Many injuries can happen to weaker dancers who exhibit poor technique. When you try to perform challenging movements without sufficient strength, proper alignment, and good flexibility, you can set yourself up for an injury. As a dancer, you are responsible for listening to and learning from your instructor about proper alignment and using the correct muscles to achieve a given movement. The instructor's responsibility, in turn, is to help students advance their technique by educating them about wellness, body structure, and good alignment.

Dancing is associated with various muscular imbalances, which can be identified through screenings that facilitate awareness and education. For example, studies show that dancers tend to be weak in the deep abdominals, which increases their risk for low back pain. Lack of core strength can lead to an anterior tilt of the pelvis, which puts stress on the lower spinal joints (Kline et al. 2013). Maintaining strength in the gluteus maximus, gluteus medius, hip adductors, and hip external rotators is important for maintaining pelvic stability while you are executing dance movement on one leg. With this very basic knowledge, you can practice warm-ups that emphasize strengthening of the deep abdominals (transversus abdominis), gluteus medius, hip adductors, and turnout muscles.

Injuries can also relate to lower levels of muscular strength. For instance, one of the most common injuries for dancers is an ankle sprain, and research has identified a correlation between chronic ankle sprain and weakness in the hip abductor muscles (Friel et al. 2006). Therefore, it is crucial for you to understand that what happens at the ankle is connected to what happens at the hip. We also know that in patients with lower-back pain, the spine stabilizers are weak; the most basic spine stabilizers are the deep abdominals (transversus abdominis) and the multifidi. Spine dysfunction, or insufficient movement, is correlated with muscle imbalances and muscle weakness.

Improper alignment, particularly rolling in or overpronating at the foot and ankle, is associated with injuries. Specifically, when you overpronate to force turnout and use the floor for friction, you are setting yourself up for midfoot injuries, Achilles tendonitis, or plantar fasciitis. Forcing turnout incorrectly can also cause you to screw the knee, which applies intense torsional stress to the knee joint. This action can create tracking problems in the patella (kneecap), as well as injuries of the medial collateral ligament.

Fatigue

Dance injuries that tend to occur in the afternoon, evening, or toward the end of the performance season might suggest fatigue as a cause of injury. When you are tired, you have a harder time controlling your balance and may lose the ability to land jumps with appropriate alignment.

Sleep variables are also associated with injuries. Having trouble getting to sleep, waking up during the middle of the night, and experiencing sleepiness during the day have a negative impact on your ability to perform. Quality sleep is important for restorative and recovery processes. Lack of quality sleep also affects coordination, memory, and your ability to learn new choreography. Irregularities in your sleep–wake cycle allow for fatigue and increase your cravings for sugar, carbohydrates, and caffeine to stay awake during the day. The excess sugar and carbohydrates will not adequately fuel your body for classes, rehearsals, and performances. Experts recommend maintaining regular sleeping patterns and sleeping for approximately 7 to 9 hours a night to reduce your risk of injury but as many as 8 to 10 hours during high training periods (Nicholls 2022).

Cardiovascular Weakness

While some dance performances may provide high-intensity levels of exercise, technique classes focus more on artistic aesthetics rather than cardiovascular conditioning. Class and rehearsals do *not* improve cardiovascular fitness, because the exercises you perform are so intermittent that your heart rate does not reach maximum levels. Incorporating cross training and fitness classes that support aerobic conditioning is associated with reducing fatigue and therefore has a positive impact on reducing injury rates.

Nutrition Factors

Poor nutrition, eating disorders, and disordered eating are also associated with injuries. Poor nutrition may result from restricting calories, a lack of healthy choices in your diet, or the overconsumption of sugar, saturated fats, and sodium. Eating disorders such as bulimia or anorexia should be diagnosed and treated by a qualified health care provider. Disordered eating may cause a dancer to avoid major food groups, eat only certain foods, or engage in emotional eating. Disordered eating could lead to an eating disorder. Dancers, teachers, studio owners, and choreographers are encouraged to receive nutritional education from qualified health care providers to have a better understanding of how healthy choices can reduce injury risks.

Psychological Factors

Stress and poor psychological coping skills are associated with injuries. If you are struggling to meet the demands of your rehearsal schedule, expectations of others, or competitions, then you are probably putting yourself at risk for injury. The psychological distress of anxiety is also a precursor to injury. Take a moment to review the Psychological Awareness section in chapter 2 to gain a better understanding of stress management. If you are having chronic difficulty coping with the demands of your rehearsal, performance, and competition schedule, please seek advice from a health care provider.

Sustaining a significant injury will take you out of the life you love. In fact, regardless of how significant the injury is, it can wear on you, affecting you mentally, physically, and emotionally. Some dancers become angry and sad; some struggle with depression after suffering an injury. It is difficult to stand on the sidelines and watch someone else perform your role. Do not be afraid to ask for help.

To minimize your risk of injury, be proactive about taking care of yourself (figure 3.1). Stay on top of training your abdominals to support a healthy spine. Make high-quality, healthy food choices, and organize your sleeping habits to give yourself 7 to 9 hours of sleep a night. Make time for your cardiovascular health as well. Maintain strong legs and feet, focusing on alignment for technical efficiency so that you can continue to be the best dancer possible during a long, healthy career.

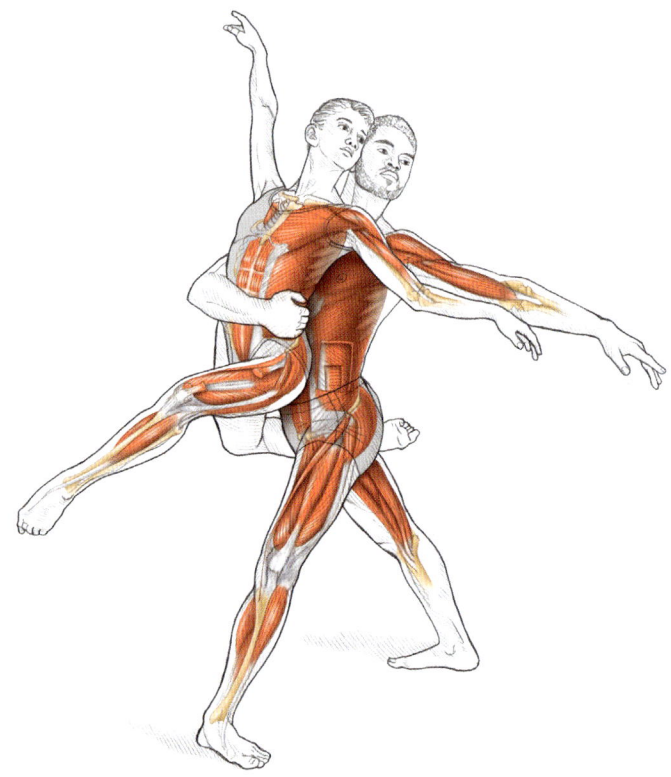

Figure 3.1 Be proactive about taking care of your mind and body to attain balance for a long, healthy career.

Extrinsic Factors

Extrinsic factors are the general hazards of the profession that can contribute to injury. They include hard floors, drastic changes in flooring, various shoe types, costumes, inappropriate training methods, and extreme changes in the number of hours danced per week.

Floors

Typically, you would like to dance on a sprung floor, which helps distribute the forces involved in landing. Sprung floors are made of wood placed over a sub-floor frame of pine board. The frame is covered with plywood and hardwood panels, and the top layer is made of vinyl or "marley." A sprung floor reduces the risk of knee and ankle injury by absorbing some of the force involved in landing from jumps. In contrast, injuries have been reported on hardwood floors laid over concrete, on carpet laid over concrete, and on floors that are either too slippery or too sticky.

Shoes and Costumes

Current research is sparse on how shoes affect injury rates among dancers. We do know, however, that ballet slippers, Irish dance soft shoes, and some jazz shoes provide little support for the midfoot or arch. Of course, even less support is available when dancing barefoot, which provides no shock absorption when landing from jumps. In contrast, lower injury rates are reported in tap dancers, perhaps because tap shoes provide more foot support. Pointe shoes, on the other hand, lack shock absorption, and the ribbons do not provide ankle stability. When the pointe shoe begins to break down and you continue to dance, you are at risk for compensatory malalignment and injury. Proper fit, foot and ankle strengthening, taping techniques, and padding may help reduce injury incidence.

Ornate, embellished, and complex costumes may create a fall risk. Costumes with long trains or baggy pants can also create a trip or fall risk. Dancers have been known to slip after tripping on beads that have come loose from decorated costumes. Costumes, masks, or headpieces that might obstruct your view can create a hazard. Heavy wigs and headpieces can put added stress on your neck, shoulders, and upper back. If elaborate costumes and headpieces are part of your performance, then remember to schedule plenty of dress rehearsals so you have time to adapt to your costume. Occupational accidents can also happen due to variables associated with dancing with props. Integrating props (property) into the choreography can enhance the performance but can also create safety concerns. Rehearsing with hats, canes, tambourines, chairs, swords, ribbons, umbrellas, or hula hoops, just to name a few, can certainly cause accidents if another dancer is struck by a prop that flew out of your hand! If a prop is accidentally dropped, it also becomes quite a distraction. Dancing with props takes skill and a lot of practice. Conditioning and ample rehearsals with your prop of choice are the best way to avoid accidents; you need to be able to use your prop safely. Are you able to execute your movement well? Are you able to see well? Is your prop too heavy? These are appropriate questions to ask yourself when integrating props into the choreography so you can have a safe and healthy performance.

Training

Many dance injuries are caused by overuse, overtraining, or overrehearsing. For example, if you ramp up rehearsals when you are tired—or go from four or five classes per week to four or five classes per day during intensive summer programs—you invariably stress your body, which may result in any number of injuries. The typical full day of rehearsals for competition dancers, university dancers, or professionals may not provide enough rest to reduce injury risks.

Screening and Assessment

Screenings can help you learn more about your body. They can also help instructors learn how best to promote health and wellness at the studio. A screening, or dance medicine physical, assesses your anatomy to identify any characteristics that could lead to injury by evaluating your alignment, strength, and range of motion. Such a screening can be performed by a health care provider who has experience in working with dancers. For example, an athletic trainer or physical therapist can administer a screening while educating you about your anatomy, strengths, and weaknesses.

A screening prior to going en pointe is another educational tool to promote health and wellness for young dancers. Age may not be a true determining factor for pointe readiness, because every dancer matures at different times. A pre-pointe screening can help dance instructors and parents have a better understanding of postural alignment, spine stability, range of motion at the foot and ankle, and functional turnout. Objective testing will assess balance skills, calf strength, abdominal strength, and alignment.

A personalized assessment is valuable for determining any anatomical errors that could potentially lead to an injury. The pre-pointe screening should not be used as a pass–fail exam but as an educational session to promote alignment, posture, strength, and endurance. Each student should also receive exercises or stretches specific to their needs, depending on the results of the screening. This is a win-win for all involved! Dancing en pointe is a totally different experience than dancing in ballet slippers. The more all parties are educated about pointe work, the less risk of injury.

Screenings may produce findings such as the following:

- Weakness in the deep spine stabilizers
- Tightness in the hip flexors
- Weakness in the hip abductors
- Weakness in the hamstrings
- Tightness in the calf muscles

- Excessive foot and ankle pronation
- Limitations in turnout

Let's take quick look at each of these possible findings.

In the screening process, you will learn how weaknesses in your core muscles might be attributing to back pain. The spine and core chapters of this book (chapters 4 and 6) will give you a better understanding of how the deep stabilizers work to support your spine and pelvis and provide exercises that can help reduce back injury risks. For example, a great introduction to engaging your deep stabilizing muscles is provided by the leg glide exercise presented in chapter 4. In addition, exercises presented in chapter 6 illustrate how to engage your core for spinal support; examples include abdominal bracing and the side plank with passé.

A screening will provide objective testing to look for tightness in hip flexors and weakness in hip abductors and adductors. Tightness in the hip flexors can pull your pelvis forward or create an anterior tilt that puts unnecessary stress on the lower segments of your spine. Anterior tilt of the pelvis is addressed in the spine chapter (chapter 4), which presents the locating-neutral exercise to help you with correct placement of the pelvis. Another relevant exercise is the hip flexor stretch presented in chapter 8.

Chapter 8 also discusses the hip abductors and their importance in hip stability. To help you isolate and strengthen your hip abductors, the last chapter of the book includes an exercise known as the lateral leg lift.

The significance of hamstring strengthening is discussed in chapter 9. The hamstring muscles can be isolated and strengthened through the kneeling hamstring curl and the supported hamstring lift.

In the calves, flexibility of the gastrocnemius is crucial to maintaining muscular balance; it can also help reduce the risk of ankle sprains, shin splints (medial tibia stress), and Achilles tendonitis. For a good calf-stretching exercise, see the elevé with ball over the edge, presented in chapter 10. The doming exercise presented in chapter 10 can help you strengthen the muscles within your arches for better midfoot support.

A good screening can help you understand how to use your feet efficiently by evaluating whether you exhibit too much foot and ankle pronation (rolling in). As discussed in chapter 10, you need a small amount of pronation to move efficiently; however, overpronating has been associated with shin splints, plantar fasciitis, and Achilles tendonitis. Overpronation can develop if you lack high-quality turnout from the hip and use your feet to compensate. Proper alignment is enabled by turning out from the hips; see the discussion titled Rotation of the Femur in chapter 8.

It is beneficial to learn all that you can about your physical strengths and weaknesses, and screenings can help you do so. Screenings also help educate health care practitioners about you and your dance technique. By promoting effective communication between dancers and health care providers, screenings increase education for both parties.

Warming Up

If you skip your warm-up because you're running late for a performance or competition, you might be putting yourself at risk. It is important to warm up your body before any performance or competition. To get the best performance outcomes, you must prepare your body. If the muscles that control the movement at a given joint are not warm enough to support the joint, then you are at risk for injury.

To avoid this pitfall, engage regularly in a routine of exercises for strengthening followed by stretching. It takes at least 30 minutes to complete the warm-up process, and the benefits are worth it.

- **Increased muscle temperature.** Warm muscles contract and relax more efficiently, enhancing both speed and strength.
- **Increased body temperature.** Warming up slightly elevates your overall body temperature, which improves your muscle elasticity, thus increasing your flexibility and reducing your risk of muscle strain.
- **Improved range of motion.** Increasing the range of motion around a joint enhances mobility, which in turn allows you to safely execute movement patterns without straining your muscles.
- **Mental prep.** The warm-up is a good time to prepare mentally for class, competition, or performance, thus increasing your focus and building your concentration.

If you have at least 30 minutes to warm up before a competition or performance, you can raise your heart rate by going through some of the plyometric exercises presented in chapter 11 for about 5 minutes. You could also perform 30 elevés with the ball (chapter 10), followed by 15 repetitions of the leg glide exercise (chapter 4) with each leg to warm up your deep spine stabilizers and hip flexors. Next, warm up your abdominal muscles by adding the trunk curl marching and functional obliques in second position from chapter 6 (10 to 15 repetitions each). You can then warm your upper body, as well as your core, by performing the plank to star exercise (chapter 7) 6 to 8 times on each side. Continue with 10 repetitions of the bridge exercise (chapter 4) for your hamstrings and gluteus maximus muscles. If you would like to work the gluteus medius, move on to the lateral leg lift exercise from chapter 11 and perform it at least 15 times on each side. Then move to dynamic stretches: thigh to chest (chapter 11) followed by the hip flexor stretch (chapter 8) for 30 seconds each. You will be so prepared for your performance!

Growth-Related Issues

Young dancers experience challenging physical changes during adolescence, typically between the ages of 11 and 14 for girls and between the ages of 13 and 16 for boys. Growth spurts can cause changes in balance and flexibility that affect technique. In such cases, it can be hard at first to understand why technique seems to be getting worse despite more and more practice. It is normal, however, to see a decline in technical performance during rapid growth. Adolescence is a tough time, during which a young person may feel awkward and weak. These challenges may be the reason that 55 percent of dancers quit during adolescence. When hormones are added to the mix, this transition period can really challenge a person's self-esteem.

Changes related to rapid growth may include the following:

- Bones growing faster than soft tissue
- Leg and arm bones growing faster than the trunk
- Weight changing
- Muscles and ligaments tightening
- Balance and coordination being compromised
- Thoracic spine growing faster than lumbar spine

The body is particularly vulnerable in the growth plates, which are areas of cartilage located at the ends of long bones. Because growth plates are soft, they are weak and vulnerable, especially during periods of rapid growth; the pull exerted on growth plates increases as muscles tighten. Injury in these areas can occur either acutely or through overuse, and growth plate injury can affect how bones grow. Depending on the severity of injury, growth plates may close prematurely, thus cutting off blood flow and causing the injured side to be shorter than the unaffected side. Higher-level stress fractures can occur in the lower spine, tibia, femur, and fifth metatarsal. Without proper evaluation and sufficient healing time, such injuries may result in bone deformity.

Thus, if you are a young dancer, it is vital to avoid these injuries now to be free of complications as you age. Once you stop growing, your growth plates will harden and turn to bone. Try to remember that you will grow out of this stage! It may take a year or two, but don't let that discourage you. Instead, be patient; this is a natural process of growing up.

Here are some actions you can take to reduce your risk of injury as you grow:

- After a warm-up, perform daily static stretching, which involves taking your muscles to a stretched position and holding it. Focus on feeling a good stretch and holding it for about 30 seconds; repeat at least three times per leg twice a day. Include stretches for your calves, hamstrings, quadriceps, and hip flexors, which can experience tightness during adolescence.
- After a warm-up, incorporate dynamic stretching, or actively moving through your stretches while slowly increasing the range of motion. For example, leg swings—standing on one leg and gently swinging the other leg front and back—can help stretch your quadriceps, hamstrings, and hip flexors.

- Limit jumping movements to reduce the impact on your joints. Instead, use that time in class to practice a combination of static and dynamic stretches.
- Perform abdominal work for spine stabilization. Chapter 6 presents several choices for abdominal exercise, including the trunk curl marching, bear to plank, and side plank to passé.
- Incorporate balance training to maintain your balance skills as you grow. Fundamental training exercises for proprioception and balance are included in chapter 11.
- Communicate with your teachers. Reach out to your instructors and explain to them your frustrations with growth-related discomfort. Let them know that you are trying to maintain your strength and balance but are struggling with pain and tightness. Talk to them about the need to do fewer jumping combinations and add more stretches to your classes and rehearsals.

Female Athlete Triad

Although dancers may benefit from hard work, long hours in the studio, and a lean body, these measures can put females at risk for what is known as the female athlete triad—that is, disordered eating, amenorrhea, and bone loss. If you are pushing yourself to lose weight in hopes that it will help you be a better dancer, think again. Losing weight does not improve your performance, and you could end up starving your muscles and thus reducing your strength. More generally, developing the female athlete triad puts you at risk for stress fractures and, in the long term, bone weakness and heart problems. Keep track of your periods, focus on healthy eating habits, see a nutritionist if you feel you can't maintain a healthy weight, and take good care of yourself.

Injury Physiology

What really happens when you get hurt? Unfortunately, at some point you are going to suffer some type of injury, and the more you understand about what happens to your body during the healing process, the better your healing journey will be. With that reality in mind, let's look at the fundamentals of injury physiology, specifically, the example of a basic ankle sprain.

Say that you have been rehearsing for several hours when you come down from a jump and twist your ankle by landing on the lateral (outside) edge of your foot (figure 3.2). This type of injury involves what is commonly referred to as an inversion mechanism. In this case, you hear a pop, feel immediate pain in your ankle, and find it very difficult to put weight on the injured foot. You have probably suffered an ankle sprain involving the lateral ligaments (which connect bone to bone) of your ankle.

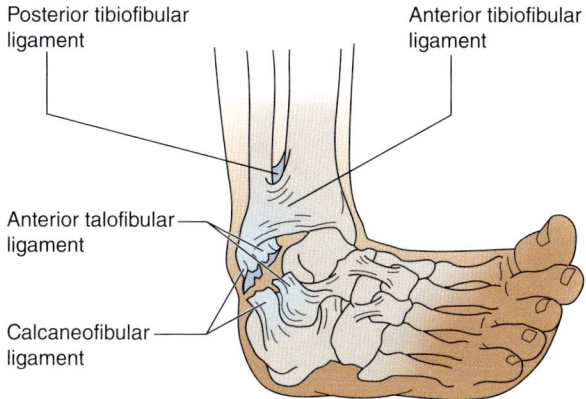

Posterior tibiofibular ligament

Anterior tibiofibular ligament

Anterior talofibular ligament

Calcaneofibular ligament

Figure 3.2 Inversion injury to the ankle.

A ligament sprain is usually classified into one of the following categories:

- **Grade 1.** Mild pain, minimal swelling, minor or microscopic ligament damage
- **Grade 2.** Moderate pain, moderate swelling, possible tearing of ligaments, some joint laxity
- **Grade 3.** Considerable pain, swelling, discoloration, complete ligament disruption, joint instability

Inflammation typically causes three to five days of swelling, discomfort, heat, discoloration, and limitation of motion. The discoloration results from damage to blood vessels and ligaments. The lack of mobility usually results from inflammation in the joint. The inflammatory process stimulates cells known as fibroblasts that work to build connective tissue and promote healing. Inflammation is a natural response to injury and contributes to the healing process.

Depending on the severity of the sprain, repair may take four to six weeks. Due to the fibroblast activity, adhesions will form. Gentle strengthening and controlled range-of-motion exercises can be used to align and strengthen the collagen fibers of the ligaments that support your ankle.

Remodeling may last six months. To improve function, the joint should be taken through a full range of motion as strength training continues. Scar tissue will be replaced by stronger connective tissue.

If you sustain an acute injury, find your mobility compromised, or otherwise feel unable to continue dancing, get yourself to a safe, calm place and assess your situation. Use the following guidelines (RICE):

- **Rest.** Get off your feet and take a moment to collect your thoughts. Minimizing stress on your injury now reduces the risk of incurring additional trauma in the affected tissue later.

- **Ice.** Ice therapy may help reduce pain; however, it should be used only minimally so that the inflammatory process can continue. The National Athletic Trainers' Association cautions athletes against overusing ice. The inflammatory process is important for healing, and icing may cause constriction of the cells that are needed to begin the healing process (Mirkin 2014).

- **Compression.** If your joint feels loose or unstable, you can support it with an elastic compression-type wrap. For example, if you have suffered an ankle injury, wrapping your ankle may help to reduce swelling and make it feel more secure. Begin wrapping your ankle at the metatarsal-phalangeal joints, keeping it snug as you continue to circle around your arch. Keep wrapping around the midfoot as you move up toward your ankle. As you pull upward from under the arch, circle around the ankle in a figure-eight pattern. Repeat the figure eight pattern under the arch and around the ankle, moving upward until you have reached the lower calf area. Secure the end of your wrap with tape if it doesn't come with a self-fastening end. Your wrap should feel secure but should not cut off circulation.

- **Elevation.** Keeping your injury elevated above your heart helps manage the pain.

It's also important to get an excellent diagnosis to help get you on the road to recovery. You do not have to dance through pain. If you are experiencing pain, swelling, joint instability, or difficulty bearing weight, please contact your physician. Once you have received a proper diagnosis, you can begin the rehabilitation process including strength training, balance training, and a safe return to dance.

Suggestions for a Dance Studio Injury Care Plan

- Staff are educated on the RICE procedure for acute injury care.
- Dancer immediately communicates with instructor and family.
- Studio provides a list of health care resources such as orthopedic doctors, physical therapists, athletic trainers, and nutritionists.
- Studio, dancer, and family maintain open communication lines about diagnosis, physical therapy, and recovery timeline.
- Dancer is responsible for performing home exercises and watching rehearsals to stay connected with dance team.
- Dancer should incorporate coping skills, if needed.
- All parties encourage quality nutrition, hydration, and adequate sleep.
- Studio instructors are supportive of the dancer's healing journey.

Healing Journey

Injuries can be caused by poor technique, muscular imbalance, weakness, poor alignment, fatigue, overuse, or growth. Early diagnosis is imperative for rehabilitation and recovery so that you can get back to what you love to do. Education, however, is the best policy: the more you know about how injuries can occur, the more you can take the steps needed to avoid them in the first place.

After suffering an injury, you can enhance and improve your healing journey in several ways:

- **Plan of care.** After assessing your injury, your health care provider may provide a plan of care. Your plan of care may include education on your diagnosis, advice on when to apply ice or heat, appropriate exercises or stretches, your rehabilitation progression, and a tentative timeline of when you can return to dance. Follow your health care provider's instructions.

- **Dance teacher communication.** Keep communication lines open with your instructors so they know your status and can support your healing journey.

- **Nutrition.** Good nutrition promotes healing. In fact, 80 percent of injury recovery comes from rest and good nutrition. Generally, to fuel your body properly, take in 55 to 60 percent of your nutrition in the form of carbohydrate, 20 to 30 percent in the form of fat, and 12 to 15 percent in the form of protein. With these proportions in mind, read food labels so that you know exactly what is in the foods you eat. Avoid foods that increase the inflammatory process, such as processed meats (hot dogs, bacon, lunch meats), refined grains (white bread, white pasta, white rice), fried foods, and foods high in sugar and fat. Good carbohydrate choices include oatmeal, whole-wheat pasta, sweet potatoes, vegetables, and fruit. Good fat choices include nuts, seeds, olive oil, avocados, and salmon. Good sources of protein include Greek yogurt, turkey, chicken, fish, milk, eggs, beans, and chickpeas. In addition, drink plenty of fluids. If you have questions or concerns about your nutritional needs, please consult with a nutritionist or registered dietician.

- **Hydration.** Water flushes out toxins; try to drink half of your body weight in ounces per day.

- **Emotional well-being.** If you notice an increase in anxiety, anger, sadness, or excessive worrying, do not be afraid to reach out for help. Practice positive self-talk and meditation, spend time with family and friends who support you, and stay involved at your dance studio.

- **Sleep.** Rest and recovery promote healing. Maintain a regular sleep routine.

To get more information, consider a screening program, which can help you learn about your strengths and weaknesses and identify any red flags for potential injury. Learn as much as you can about your body so that you understand proper alignment and know which muscles should participate in each movement. Begin strength training outside of your dance classes and engage in regular cardiovascular exercises. Don't work through pain; instead, when necessary, seek medical assistance so that you can get better, get stronger, and keep dancing.

CHAPTER 4

Spine

Your spine can create multidirectional movement in various dance styles with fluidity and ease. For contemporary combinations, your spine can portray a resilient and fluid look, and for ballet posture, it can display a majestic and elegant look. It all depends on placement, balance, and organization of muscle contractions. To perfect your body placement, you need healthy, balanced muscle action to support proper alignment of your spine. This chapter introduces the spinal regions and muscles associated with optimal placement and support of the spine. Dance can put enormous stress on your back, especially in the segments with the most mobility. If you learn to use your entire spine to balance stability, mobility, and flexibility, you can improve your performance skills and reduce your risk of injury.

In anatomy, the term *axial* refers to anatomical direction; regarding the skeletal system, it refers to your bones being aligned vertically along a longitudinal axis. Your axial skeleton consists of your skull, spinal column, ribs, and sacrum. Given this structure, you need to remember to move against the resistance of gravity; in other words, create length, or axial elongation, along your spine while incorporating stability for placement and support.

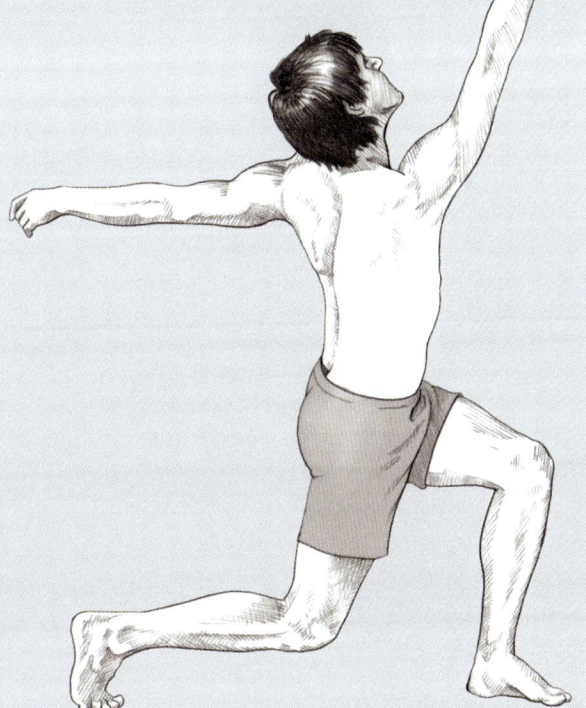

Spinal Column

Your spine is the center of your skeleton. It is a column of 33 strong bones—called *vertebrae*—that connect your skull, shoulders, ribs, hips, and legs. (Twenty-four vertebrae are movable vertebrae, while those in the sacrum and coccyx are fused.) The vertebrae surround and protect your spinal cord, which transmits the nerve impulses that control all voluntary and involuntary movements. Vertebrae are connected by small, fluid-filled sacks of tough, fibrous cartilage called *discs*, which allow for vertebral support as well as a small amount of cushioning. These discs help absorb shock, especially when you perform jumping and lifting movements.

Movement between the vertebrae creates flexibility throughout the spine. For instance, although a large cambré (backbend) can be gorgeous, the tendency of many dancers is to overextend in the neck and lower back without trying to incorporate effective movement through the thoracic spine (midback). Transmitting forces equally through the entire spine creates more efficient movement. This effect is particularly significant in the lower back; if forces such as gravity, hyperextension, and compression are transmitted only through the lower spine, you overwork that portion of the spine, thus putting yourself at serious risk for fracture, soft-tissue damage, or disc degeneration. Repetitive spine malalignment can also lead to overuse of various muscles, which can have a negative impact on your alignment and balance skills.

The vertebrae cannot stand upright without the support of an elaborate system of ligaments. The major connecting ligaments are the anterior and posterior longitudinal ligaments, which are continuous bands that run down the anterior and posterior areas of the spine. All vertebrae have a common basic structural pattern: a body, the vertebral foramen, a spinous process, and two transverse processes (figure 4.1). The body area of the vertebra bears the weight of the body above it, the foramen creates the space for the spinal cord, and the processes serve as sites for various muscle and ligament attachments.

The point at which each process meets the next process creates a gliding joint, or facet. At these small joints, the vertebral processes are

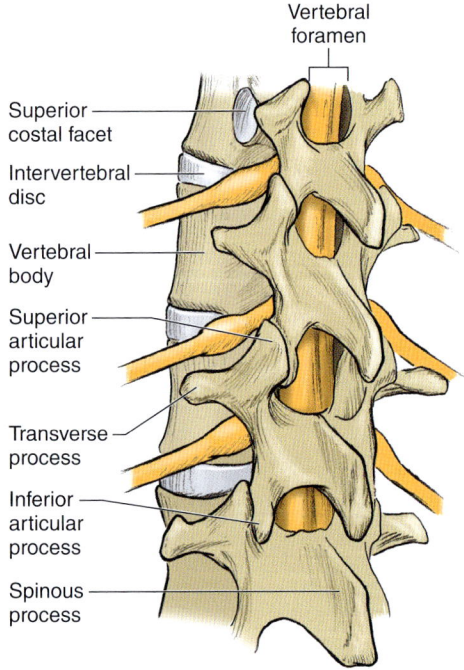

Vertebral foramen

Superior costal facet

Intervertebral disc

Vertebral body

Superior articular process

Transverse process

Inferior articular process

Spinous process

Figure 4.1   Structure of the vertebrae.

flat, and each surface must slide smoothly against the other when you twist or bend. Facet joints allow you to flex forward, extend back, and rotate from side to side. Injury to these small facet joints is usually produced by repetitive, uncontrolled movements that create asymmetry. When these joints do not glide smoothly, movement becomes limited and rigid, may cause pain, and may eventually lead to compensations. Visualize this smooth sliding effect between all vertebrae and incorporate it with control while executing the exercises presented in this chapter.

Spinal Regions

The spine consists of four main sections: the cervical area, the thoracic area, the lumbar area, and the sacrum. Take a moment to examine the regions of the spine depicted in figure 4.2 and how the vertebrae stack up neatly. The sacrum and coccyx are also illustrated in figure 4.2. Excellent spinal health depends on maintaining the natural curves designed to enable good balance and postural stability.

Cervical Spine

The neck, or cervical spine, contains seven vertebrae, along with ligaments, tendons, and muscles, and presents with an anterior curve. It supports your head, which can weigh 10 to 12 pounds (4.5-5.4 kg). The neck is both relatively flexible and fragile, because its vertebrae are slightly smaller than those of other spinal regions. The cervical vertebrae are labeled C1 through C7. Neck anatomy also includes cervical nerves, which are labeled C1 through C8. Cervical nerve impulses from the spinal cord control the following parts of the body:

- C1 and C2 control the head and neck.

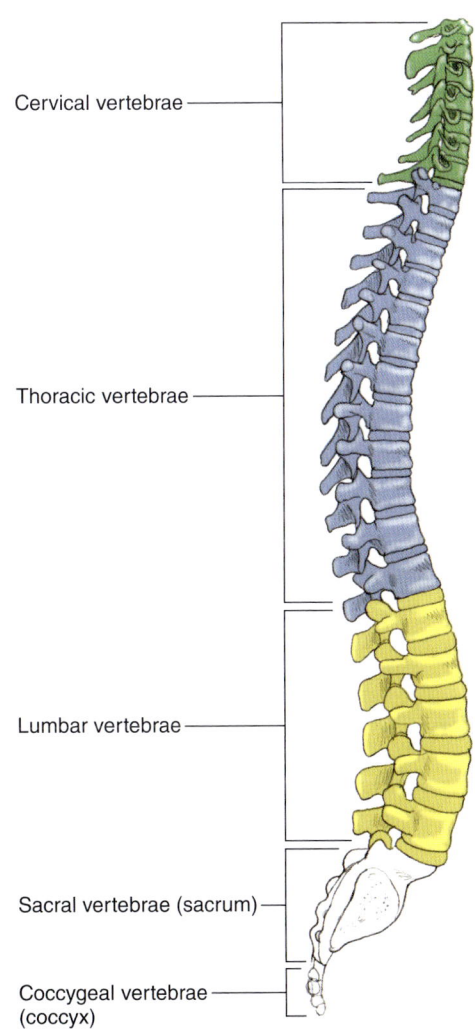

Figure 4.2 Main regions of the spine: cervical, thoracic, lumbar, sacral (sacrum), and coccygeal (coccyx).

- C3 controls the diaphragm.
- C4 controls some muscles in the upper body.
- C5 and C6 control some muscles in the wrist.
- C7 controls muscles in the back of the upper arms.
- C8 controls muscles in the hands.

Injury to the neck can be serious, especially if it causes damage to the discs or numbness or tingling in the body parts controlled by the nerves. One key to avoiding dangerous injuries is to develop balance and strength and maintain alignment.

The first two vertebrae are quite interesting. C1 is called the atlas, as in the Greek myth of Atlas supporting the heavens; in this case, the atlas carries the skull. C2, referred to as the axis, has a small, bony projection that rises into the ring of C1. This connection creates a pivot that allows rotation between the atlas and the axis, thus enabling movements such as nodding and rotating the head. Physical tension in the muscles supporting the cervical spine can limit efficient movement for spotting during turning.

Picture your head centered and balanced on C1 and C2. If your head is balanced, the neck muscles that control movement can work with ease. Any time your head moves outside of this balanced state, the opposing muscles of that movement overwork to maintain alignment. Beyond turning, head placement is an important element in the aesthetics of the upper-body poses and choreography of any dance genre.

Thoracic Spine

The vertebrae increase in size further down the spine. The thoracic spine contains 12 large vertebrae, T1 through T12, and presents with a posterior curve. The thoracic vertebrae connect to the ribs, which provide protection for some of your organs. The increasing size of the vertebrae and the presence of the rib attachments limit the flexibility and mobility of this area, which is known as the thoracic cage or rib cage. Thoracic nerve impulses control the following areas of the body:

- T1 and T2 control the shoulders and arms.
- T3 through T6 control part of the chest.
- T7 through T11 control part of the chest and abdomen.
- T12 controls the abdominal wall and buttocks.

Learning how to move through your entire spine will encourage mobility throughout your thoracic region. The exercises presented in this chapter focus on axial elongation in all movement planes, thus moving in the longest possible arc.

Lumbar Spine and Sacrum

The lumbar, or lower-back, region includes five vertebrae (L1 to L5) and is more flexible than the thoracic region. This region carries most of your weight, takes on the most stress, and presents with an anterior curve.

The lower segments of the spinal column can move more in extension than in rotation, which can create a shear force. In other words, the vertebrae can slide in an anterior and posterior pattern, thus creating an unnecessary and excessive sliding or shearing motion. This unsupported movement can eventually wear down the discs and cause weakness in the ligaments, which increases the risk of lower-back injury. You can reduce your risk of lower-back injury by understanding basic facts about the spine, gaining awareness of body placement, and developing a stronger core to anchor and stabilize your spine.

The spinal cord does not continue all the way through the lumbar spine. It ends at the first lumbar vertebrae, where the nerve roots branch off to control various body parts:

- L1 controls certain abdominal muscles.
- L2 and L3 control the area from the thighs to the knees.
- L2, L3, and L4 control the inner thighs, the hip flexors, and the tops of the thighs.
- L5 controls the outsides and backs of the thighs.

Greater movement is possible between the last lumbar vertebra (lumbosacral joint) and the sacrum. The sacrum is composed of five fused vertebrae (S1 to S5), is triangular in shape, and presents with a posterior curve. The sacrum together with the coccyx take on the load of the upper body, transferring it to the pelvic girdle. The base of the sacrum forms the coccyx, or tailbone. The coccyx is one of the sites for pelvic floor muscle attachment. A lot of force comes through the lumbosacral joint in various spine-extension dance movements. Because the lower spine vertebrae have more flexibility in extension and take on more of the load, it is important to strengthen your core and lower back to improve your lumbar spine stability and reduce your risk of injury. The importance of the core musculature for trunk stabilization is addressed in the following section; the principle of trunk stabilization is discussed in more detail in chapter 6.

Neutral Spine

The spine is capable of flexion, extension, side-bending, rotation, and various combinations of these movements, thus giving you the ability to perform any type of choreography. A significant role in body placement is played by the four spinal curves within the sagittal plane (figure 4.3). In the cervical and lumbar areas, the curve is concave or anterior (i.e., it moves in a forward motion), whereas the thoracic and sacral curves are convex or posterior (i.e., they move in a backward direction). The vertebrae within these curves are cushioned by the intervertebral discs. Changing the curves as a base for your placement causes undue stress on the discs; it also requires otherwise unnecessary muscle activity to maintain the misalignment.

Excellent body placement skills depend on creating strength and stability along the spine while keeping the natural curves intact. This is known as neutral posture, neutral spine, or neutral pelvis. When you maintain your natural, neutral curves while dancing with axial elongation, you create less stress on your discs and vertebrae. Axial elongation helps to decrease the load on various segments of the spine. Of course, challenging choreography requires movement of the spine in all directions and in combinations of all directions, but a strong dancer can control the spine throughout challenging movements. The locating neutral exercise presented in this chapter helps you locate the natural, neutral pelvis position.

In visualizing your body from the side, you should be able to draw an imaginary line from the middle of the ear down to the lateral malleolus, or ankle bone, without any deviation. This is called the plumb line (figure 4.3). As it

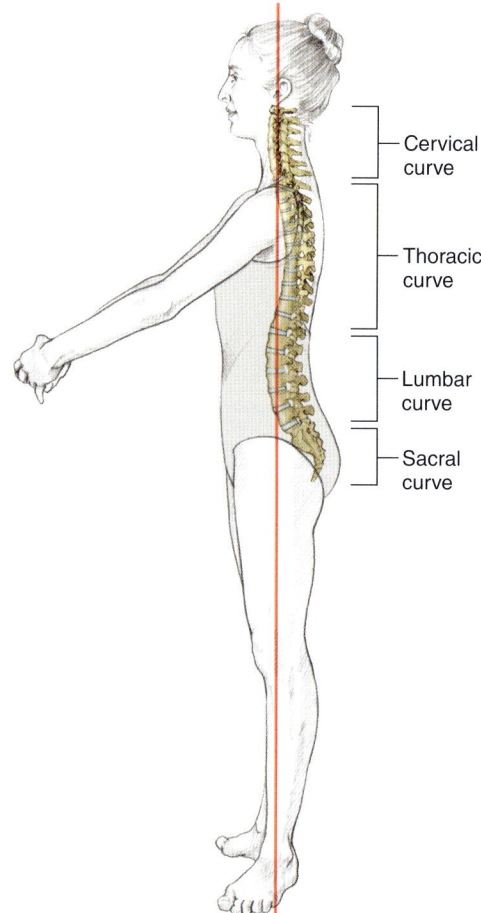

Cervical curve

Thoracic curve

Lumbar curve

Sacral curve

Figure 4.3 Four curves of the spine and the plumb line.

descends, the plumb line moves through the center of the shoulder, through the center of the greater trochanter at the hip, down to the knee, and on to the lateral malleolus—again, with no deviation. You should also be able to secure this alignment with your legs either parallel or turned out.

Unfortunately, some dancers have difficulty maintaining a neutral or natural position of the lower back. For example, the lumbar spine may extend slightly, thus creating lordosis, or excessive lumbar curve in extension. Lumbar lordosis can result from various causes. One possible cause is abdominal weakness, which leaves the lower spine unsupported, thus causing the lower back to arch. Another potential cause is that the posterior spinal muscles are tight and short, which pulls the lower spine into an arched position. Alternatively, the iliopsoas muscles may be tight and short, which also pulls the lumbar spine into a lordotic position.

Maintaining good alignment along your spine involves having your head in a neutral position as well. Specifically, your head must be aligned and balanced on top of your cervical vertebrae. If you look at yourself from the front, your neck should be nice and straight. Viewed from the side, your cervical vertebrae should have a small, natural anterior curve. Any malalignment of the vertebrae that pulls the chin forward causes undue stress along the neck and makes for unattractive body placement. Forward head posture seems to be more prevalent in dancers who spend a lot of time on media devices, promoting poor and incorrect posture. The forward head posture brings the ears forward of the plumb line, causing malalignment of the cervical vertebrae and unnecessary stress on the ligaments and joints. The head neutral exercise can help to strengthen the neck muscles to support a more neutral position of the cervical spine to avoid the forward head posture.

Muscular Balance

This section introduces the muscles that play a role in correct placement of the spine; these muscles are described in more detail throughout the rest of the book. The primary muscle along the anterior aspect of the spine is the rectus abdominis, which originates at the pubic bone and has insertions along the edges of the fifth, sixth, and seventh ribs and the xiphoid process of the sternum. Other muscles include the internal and external oblique muscles, which connect your ribs to your pelvis. The deepest of the abdominal muscles is the transversus abdominis, which is primarily a postural muscle and is very important for spinal stability. The deep transversus abdominis connects the lower ribs (7 through 12) with the pelvis; its fibers run horizontally.

Another muscle associated with spinal placement is the iliopsoas muscle (a combination of the psoas major and the iliacus), which has a direct connection with the lower spine, pelvis, and femur (thigh bone). Weakness or tightness in the iliopsoas can create instability in the lower region of the spine. This muscle is discussed further in chapter 8. To isolate the iliopsoas, you can use the standing hip flexor lift with a resistance band (chapter 8) to locate and contract it while working on lumbar stability as well. To stretch the iliopsoas, you can use the hip flexor stretch (chapter 8).

The posterior aspect of the spine is supported by the sacrospinalis (erector spinae) and deeper multifidus muscles, which run from the pelvis to the base of the skull. The deep multifidi are also extremely important for improving body placement; on contraction, they aid in trunk control and spinal stability by providing gentle compression along the spine. More information on the importance of the multifidi musculature is in chapter 6 on the core.

The pelvic floor provides a strong base of support for the lower spine and pelvis. It is discussed further in chapters 5, 6, and 8. For now, note that it attaches at the base of the pelvis and sacrum, which is located at the bottom of the spine. To learn about engaging these muscles and using them for improved placement, see the ischial squeeze exercise presented in this chapter.

Along either side of the trunk, the quadratus lumborum runs from the last rib to the iliac rim and the lower spine along its way. This muscle helps you side-bend and extend your lower back, but when it is tight, it can elevate the pelvis or cause hip hike, especially with high kicking movements. A healthy balance of strength and flexibility (along all sides of the spine) provides the needed support for attaining well-aligned body placement.

Several muscles that provide support for the head and neck are discussed in this chapter: the semispinalis cervicis, spinalis capitis, semispinalis capitis, and longissimus cervicis. Each has attachments along the base of the skull and various cervical vertebrae to provide support when you tip your head back or rotate your head. Again, the emphasis needs to be placed on axial elongation and on moving through your entire spine so as not to overuse the vertebrae in the cervical region.

Dance-Focused Exercise

As you execute the following series of exercises, remember to work with axial elongation and let your cervical spine be an extension of your thoracic spine. In the exercises that involve flexion of the spine, allow your cervical spine to finish the arch that your midback initiates. For example, in the trunk curl isometrics exercise, there should not be an excessive bend in the neck to try to force the upper back to move more. The same principle of axial elongation comes into play when the spine is required to move into extension. The neck should provide a beautiful continuation of the arch created by the spine.

Let's look at the start position of the locating neutral spine model to notice how the spine stacks up. The gentle curves of the spine are intact and supported, and the head balances on top of the cervical spine with ease. Also notice the balance between the muscles along the front and back of the spine. Think about how the deep multifidi under the erector spinae gently compress the spine to provide support. If you can also visualize activation of the iliopsoas, which connects the lower spine to the thighs and pelvic floor, to stabilize the base of the spine, then you have begun to develop improved placement. In fact, if you incorporate good muscular balance, you will need less overall muscle action and thus will create an excellent workplace in which your spine can function.

After the last of the following exercises, we'll examine the cambré derrière in detail and consider how the muscles work in this beautiful dance movement.

LOCATING NEUTRAL

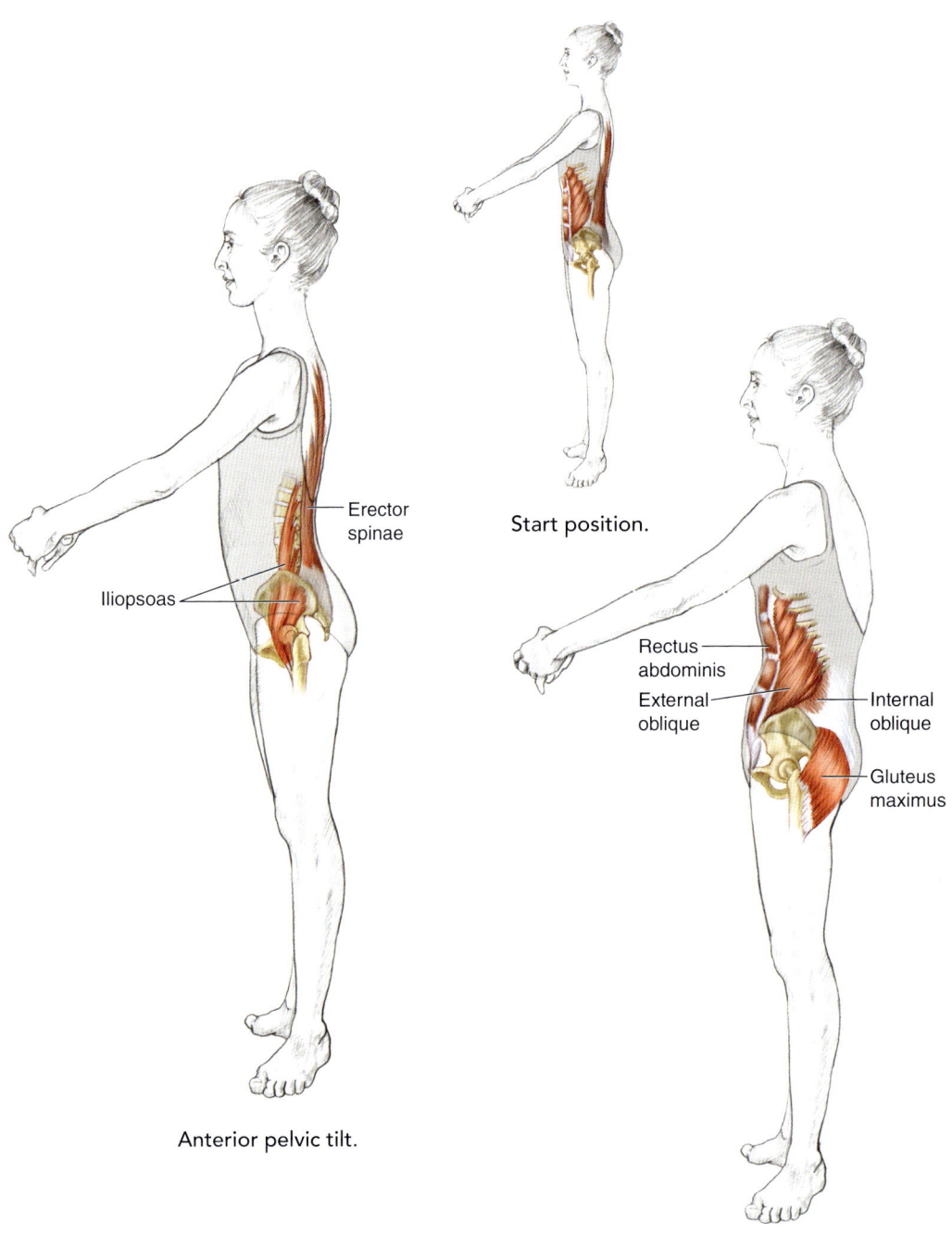

Erector
spinae

Start position.

Iliopsoas

Rectus
abdominis

External
oblique

Internal
oblique

Gluteus
maximus

Anterior pelvic tilt.

Posterior pelvic tilt.

EXECUTION

1. Stand with your legs and arms in first position. Create a lifted quality through your spine; gently engage the low abdomen and visualize the plumb line.

2. As you inhale, lift the ribs, release your abdominals, and gently rock the front of your pelvis forward, arching the lower back and moving into an anterior tilt. Notice the tightness in the upper and lower back and the looseness in your abdominals.

3. As you exhale, reverse the pelvic tilt and tighten through the abdominals; try to flatten the lower back and engage the gluteus maximus. Notice that the fronts of the hips tighten and the front of your chest drops.

4. Now return to a neutral position, visualizing the plumb line and gently lifting through your waist. There is a balance between the abdominals and the spinal muscles and a renewed lengthened feeling in the spine.

5. Next, as you inhale, move into your anterior pelvic tilt. As you exhale, move into your neutral position. Emphasize abdominal contraction and the external obliques to move into neutral. Perform 10 to 12 times.

MUSCLES INVOLVED

Anterior pelvic tilt: Iliopsoas, erector spinae (iliocostalis, longissimus, spinalis)

Posterior pelvic tilt: Rectus abdominis, internal oblique, external oblique, gluteus maximus, hamstrings (semitendinosus, semimembranosus, biceps femoris)

DANCE FOCUS

Let this exercise help you work through your center and notice the changes that occur along your spine. Knowing that the lower portion of your spine has more flexibility than the other portions, you must be aware of activating the abdominal muscles to control your pelvis and spine in a more natural, neutral position. Instructors can help cue students to engage their abdominals to support the curves of their spine. Visualize how the external oblique musculature connects the ribs and the pelvis. Keep that connection working when your leg needs to move to the back; doing so helps you keep your lumbar spine or lower back from overextending. All styles of dance require three-dimensional spine and hip movement, and control of these movements is one of the keys to technical improvement.

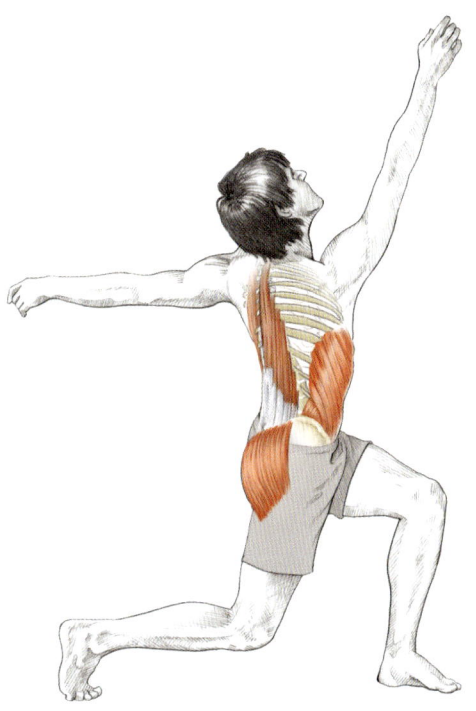

HEAD NEUTRAL

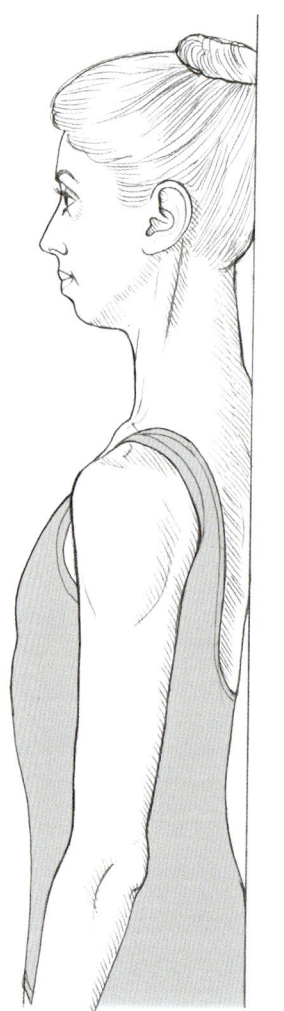

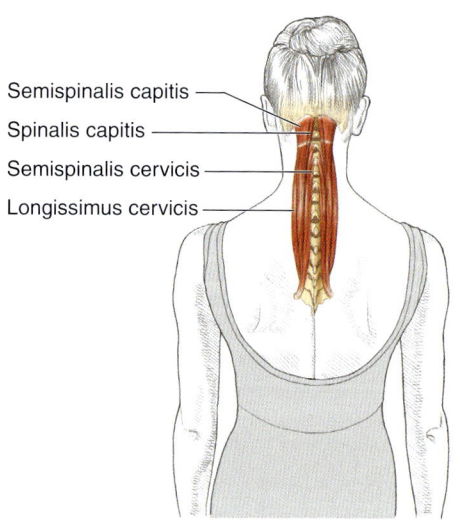

Semispinalis capitis
Spinalis capitis
Semispinalis cervicis
Longissimus cervicis

SAFETY TIP: Maintain alignment with your chin; do not drop or lift your chin.

EXECUTION

1. Stand with your back against a wall and maintain a neutral spine position. Visualize your plumb line. The back of your head should be against the wall as well.

2. Inhale to prepare. As you exhale, gently push your head back into the wall to engage the muscles along the back of your neck, creating an isometric contraction. For this exercise, focus on the four neck muscles that help support the head and bring your head and neck into extension.

3. As you contract the muscles along the back of your neck, hold for 6 to 8 counts and then release. Perform 10 times.

MUSCLES INVOLVED

Semispinalis cervicis, spinalis capitis, semispinalis capitis, longissimus cervicis

DANCE FOCUS

As you move, carry your upper body with elegance. Focus on balancing your head on your spine. Feel length along your cervical vertebrae and strength along the back of your neck. In performances, you may be required to wear headpieces, masks, wigs, crowns, or hats of various sizes and weights. Although a simple tiara might weigh only a couple of ounces, more dramatic headgear can weigh 15 pounds (7 kg) or even more. Dancing with a heavy headpiece alters your dynamic posture. Because your neck supports your head, you will need a strong neck to move efficiently with ease and avoid muscle-strain injuries. In addition, while executing any cambré type of movement or excessive spine extension, you must support your head and neck and move in the longest possible arc. Maintain control of your cervical spine-extension movements.

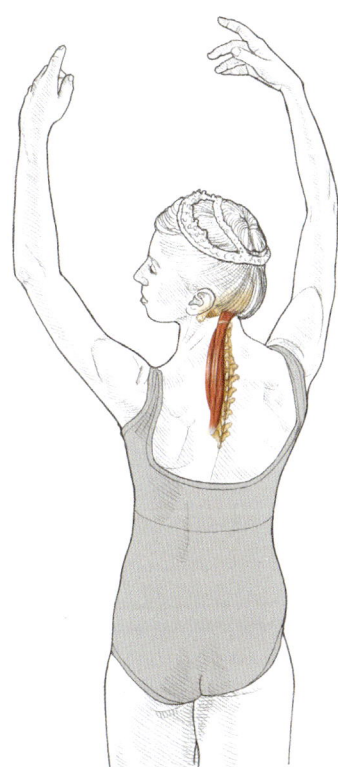

LEG GLIDE

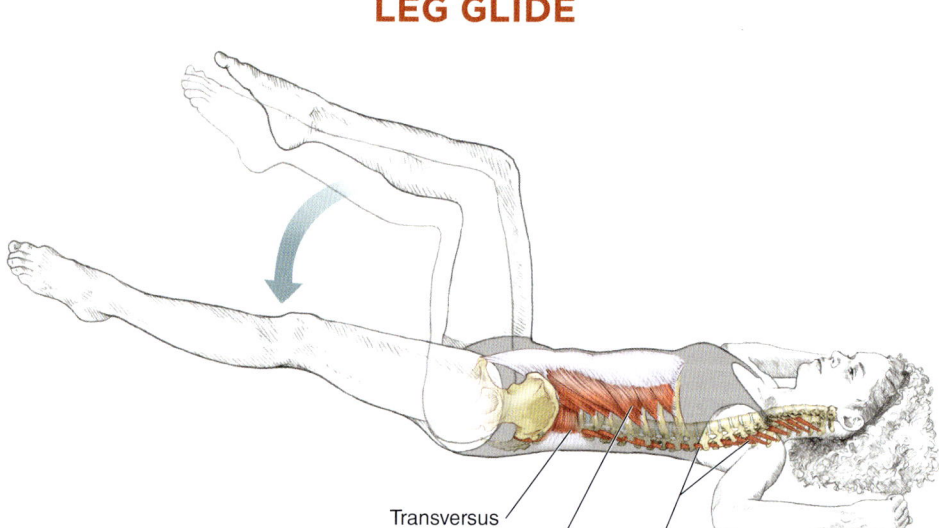

Transversus abdominis

External oblique

Multifidi

EXECUTION

1. Lie on your back with your arms in first position. (Note: In the illustration, the arms are placed in a position to make the abdominals more visible.) Locate your neutral pelvic alignment and bring one leg at a time to 90-degree hip flexion and 90-degree knee flexion (i.e., 90/90 position). Align your knees with your hips.

2. Inhale to prepare. As you breathe out, deepen the abdominal contraction, and let one leg glide away from you at approximately 60 degrees. Allow your knee to extend fully. Focus on anchoring your abdominals to your lower back; allow no movement of the pelvis. Feel the deep transversus abdominis and external oblique firing to help stabilize your pelvis.

3. Inhale to bring the leg back to the starting position. Repeat the sequence with the other leg. As you exhale, focus on flattening your abdomen to anchor your pelvis; emphasize deep contraction in the abdominal muscles, not the hip flexors. Perform 10 to 12 times with each leg.

4. As your leg moves away from your center to extend the knee, notice the movement of your leg occurring along the sagittal plane and actively increase the abdominal contraction to resist pelvic movement.

SAFETY TIP: Maintain stability in your lower back. If you find that it is too difficult to hold your lower back in a stable neutral position, do not take the leg as low; try it again with the leg extending higher. You may lower the leg when your back is stable.

MUSCLES INVOLVED

Transversus abdominis, external oblique, multifidus

DANCE FOCUS

This exercise emphasizes that what matters is not how many abdominal exercises you can perform but how well you can use abdominal strength to improve your technique. For example, Irish dance requires intense trunk control in a neutral position to maintain a stable position. Focus on the deep transversus abdominis along with the deep multifidus to give you double support. This cocontraction provides the anchor you need before performing any arm or leg motion. Remind yourself that only your legs are moving, not your pelvis or your spine!

The same principle applies to jumping combinations. Visualize your navel moving toward your spine for added support. Put energy into your abdominal muscles, not tension into your neck and shoulders. Take a moment to practice a few small jumps in place. Feel the core muscles bracing your spine and feel the external oblique working to connect your ribs and your pelvis. Relax and enjoy the ride! Teachers can use this tool to help students move from the center with less stress on the spine.

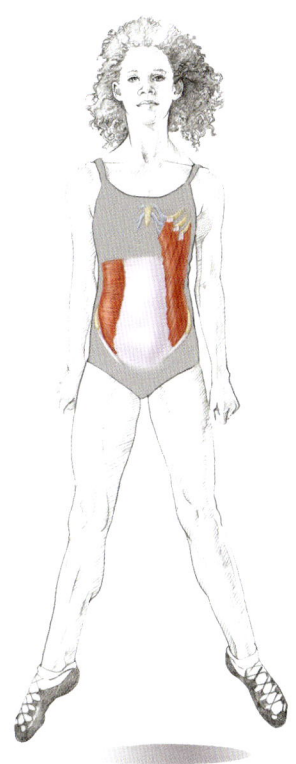

VARIATION

Rotated Leg Glide

Begin with your legs at 90/90 position and turn out both thighs. On exhalation, deepen the abdominal contraction and lower one leg to about 60 degrees as you extend your knee. Maintain turnout and emphasize leg movement only—not pelvis or spine movement. Inhale on the return and focus on deepening the abdominal contraction while maintaining hip turnout. Perform this exercise 10 to 12 times with each leg.

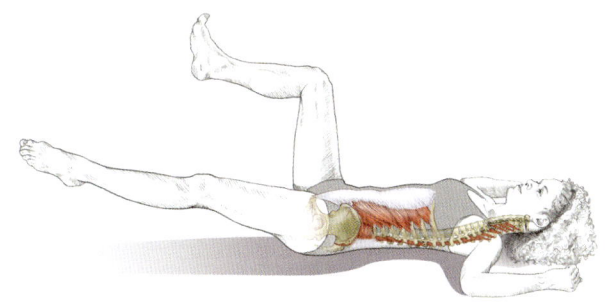

TRUNK CURL ISOMETRICS

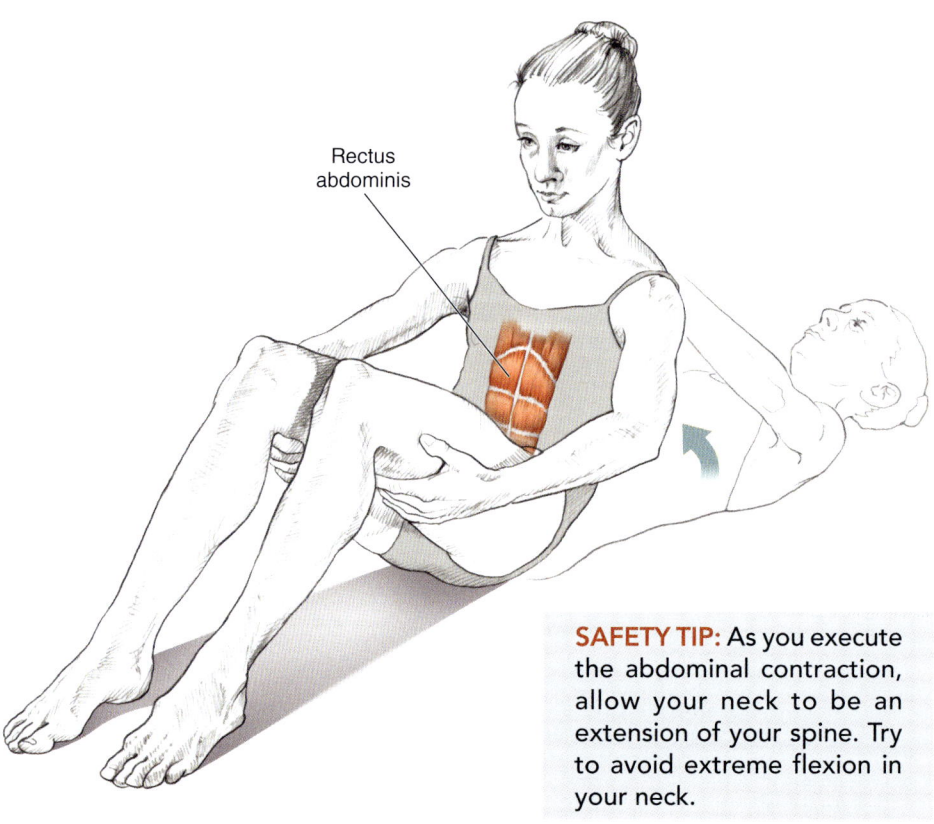

Rectus abdominis

SAFETY TIP: As you execute the abdominal contraction, allow your neck to be an extension of your spine. Try to avoid extreme flexion in your neck.

EXECUTION

1. Lie on your back with your knees bent, your feet flat on the floor, and your arms by your sides. Inhale on the preparation and exhale as you engage the rectus abdominis to curl your trunk until the bottom edges of the shoulder blades lift off the floor. Gently glide your chin toward your Adam's apple and allow your arms to reach for the backs of your thighs.

2. Place your hands behind your thighs and hold an isometric contraction. Emphasize moving the entire thoracic spine into a curl; allow the spinal muscles to support that curling effect. Keep the sacrum firm on the floor; do not use the hip flexors.

3. Hold this position and feel the strength of the abdomen. With control, breathe in and slowly return to the floor, emphasizing the eccentric contraction of the rectus abdominis. Move along the sagittal plane, curling as much as possible through your upper back on the way up and uncurling on the way down. Work with control. Count to 4 for each action; perform 10 to 12 times.

MUSCLES INVOLVED

Rectus abdominis

DANCE FOCUS

When thinking about using the rectus abdominis, attend not only to the look that this muscle provides ("six-pack abs") but also to its responsibilities. You know that this muscle flexes your trunk and therefore can also help with added mobility in your stiff thoracic spine. If you are executing a contraction in modern dance, visualize how the rectus abdominis connects your ribs to your pubic bone; maintain this visualization of it as the muscle that creates concentric contraction to curl your spine and focus on thoracic mobility. As you extend your spine by performing a cambré back or an arabesque, the rectus abdominis engages eccentrically to support and provide a lifting effect for your spine, which enhances your movements. Using the rectus abdominis effectively helps increase your core strength and decrease overuse of the hip flexor muscles. Because the abdominals make up your center, let all your movement radiate from this point; this is where you improve your body placement.

HIP FLEXOR ISOMETRICS

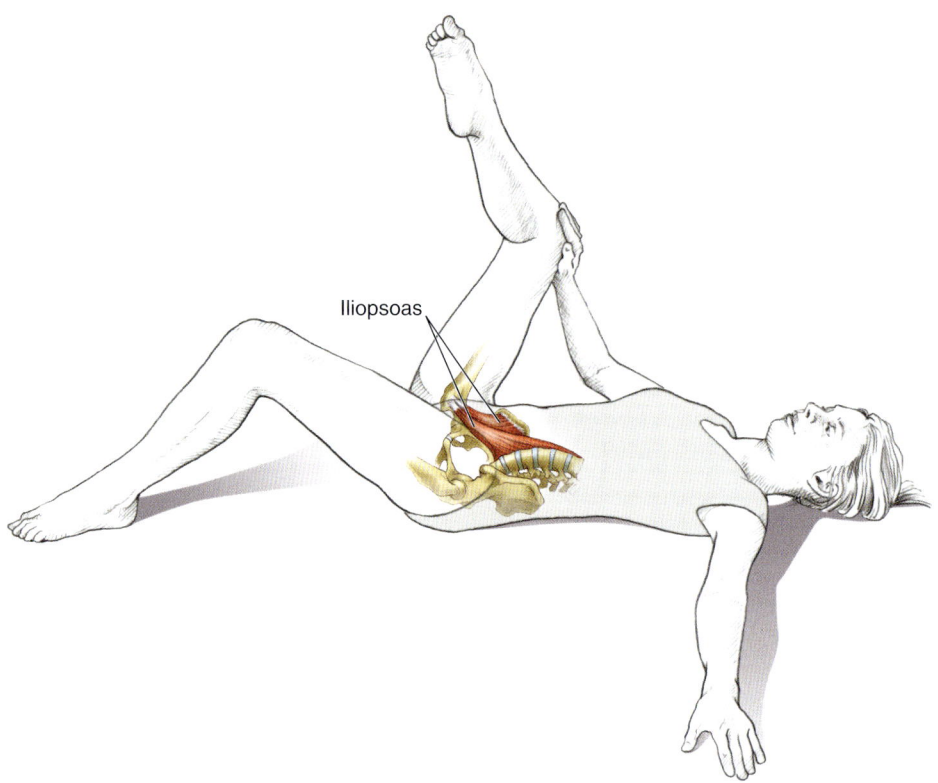

Iliopsoas

EXECUTION

1. Lie on your back with both legs bent and your feet on the floor. Coordinate a small pelvic tilt by engaging the lower abdominals and maintain that slight tilt throughout the exercise to help you locate the iliopsoas.

2. Focus on the deep iliopsoas to contract and elevate the thigh toward the same shoulder with slight turnout. Maintain leg height just above 90 degrees.

3. Press against the thigh with one hand to perform an isometric contraction of the iliopsoas. Hold for 4 to 6 counts, then relax. Perform 4 times, then repeat with the other leg to focus on the location of the muscle.

4. Continue to the hip flexor neutral variation.

MUSCLES INVOLVED

Iliopsoas

DANCE FOCUS

This exercise involves a simple isometric contraction to help you visualize, locate, and contract the iliopsoas. The contraction provides some of the assistance you need to lift your legs higher than 90 degrees. As the iliopsoas contracts, do not allow the activation of the muscle fibers to pull your lower back into an overextended or arched position; to keep the pelvis from tipping forward, allow the abdominals to contract as well. As the iliopsoas contracts, visualize the muscles that run vertically along the back of your spine lengthening and stretching. Release the tension in your upper body and send the energy down to the iliopsoas. If you need to, close your eyes and visualize the muscle origin and insertion. Knowing that this muscle connects your lower spine to your femur, imagine pulling your femur closer to your spine—not pulling your spine to your femur. This image creates more awareness of spinal alignment and helps you lift your legs higher.

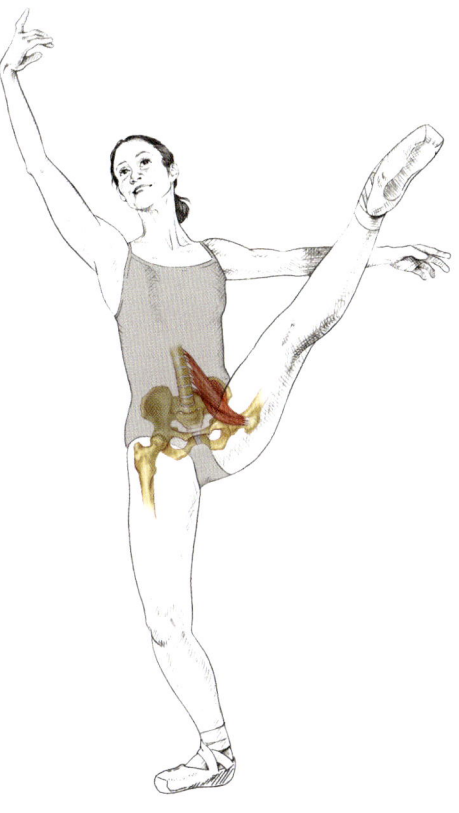

VARIATION

Hip Flexor Neutral

Repeat the main exercise. While maintaining the deep iliopsoas contraction, bring your pelvis into a more neutral position. This is challenging! While maintaining control, slowly lengthen the abdominals to slightly roll the pelvis back to neutral. Maintain the contraction of the iliopsoas. On reaching neutral and still feeling the contraction of the iliopsoas, relax and perform the exercise another 4 times.

SAFETY TIP: As you move into neutral, avoid overextending in the lower back; move with control.

BRIDGE

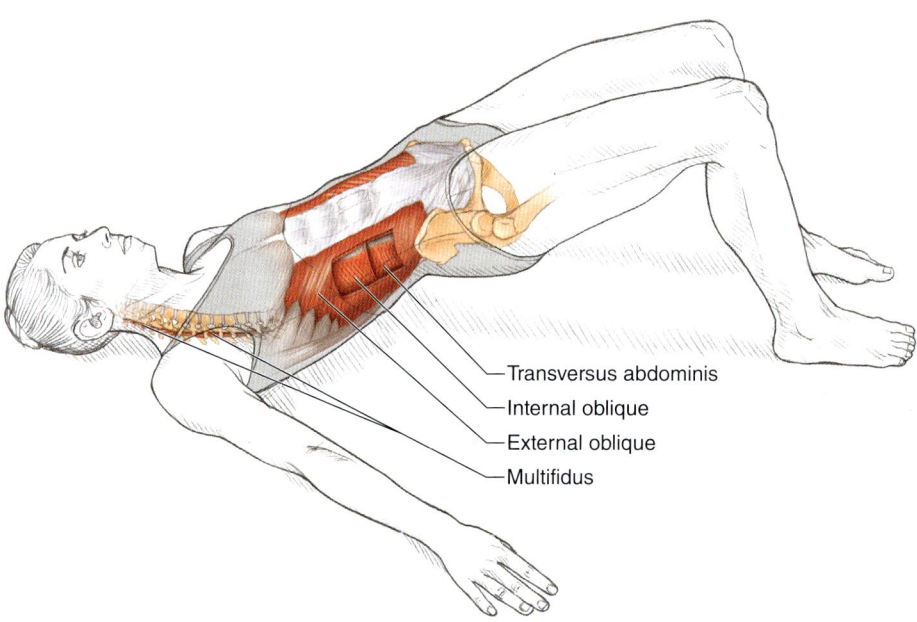

Transversus abdominis
Internal oblique
External oblique
Multifidus

EXECUTION

1. Lie on your back with your arms by your sides and locate your neutral spine position. Inhale to prepare.

2. As you exhale, engage your abdominals and begin to elevate your hips. Maintain the feeling of pulling your navel to your spine to support your lumbar spine. Relax your neck and shoulders.

3. Continue to elevate your hips until they are in line with your shoulders and knees. Notice the deep transversus abdominis supporting your spine.

4. Hold the position as you inhale. As you exhale, lower your hips to return to the starting position under control. Execute 2 sets of 10, resting in between each set if you need to.

MUSCLES INVOLVED

Transversus abdominis, internal oblique, external oblique, multifidus

DANCE FOCUS

The bridge is a great option for bracing the spine. Because this chapter focuses on support for the spine, it emphasizes the deep core muscles involved in providing that support. As you continue through the text, however, you will notice that the bridge is much more than a spinal bracing exercise; it involves more muscles. It will also help you learn to support your lower spine with your abdominals, which is a useful skill in all dance movements involving extension of the spine, especially arabesque and attitude derrière.

The emphasis on spine control begins with tendu derrière. From the tendu, as you move your leg into arabesque or attitude derrière, think about bringing your navel toward your spine to fully support the lower segments of your spine. Allow the arabesque or attitude to move freely through the hip. While executing a basic bridge, you will notice that your hips go into extension so you can move freely through your hips as well. Both lumbar spine support and efficient movement through the hip are required in any contemporary movement, floor work, or partnering work in which you move your leg to the back.

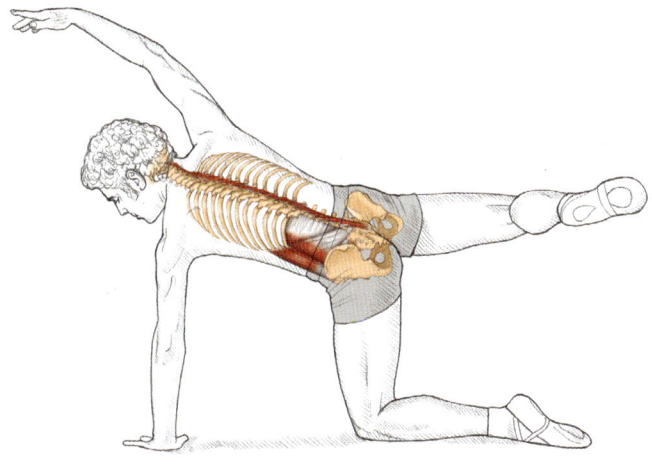

QUADRUPED MULTIFIDUS

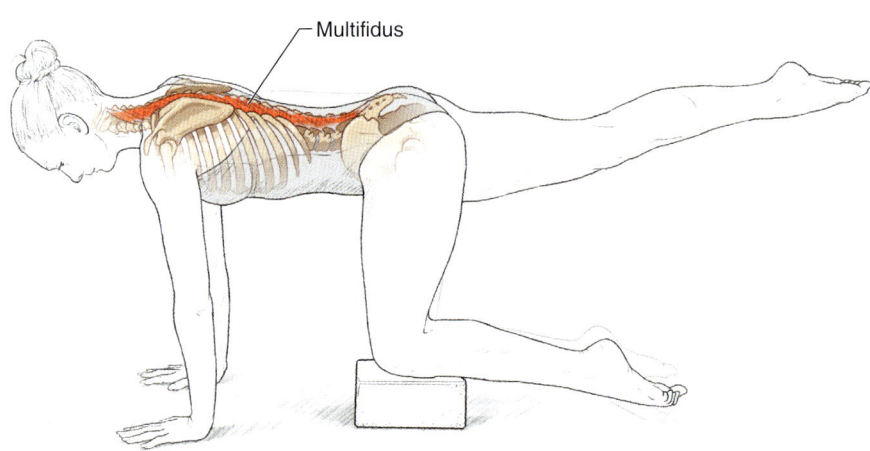

Multifidus

EXECUTION

1. Begin in a quadruped (hands and knees) position with a pillow or yoga block under the left knee to elevate that hip. Align the knees directly under the hips and the wrists directly under the shoulders.

2. Inhale to prepare. On exhalation, engage the abdominals and elevate the right hip and thigh to align with the left. The right knee and lower leg will now be hovering over the floor. Be mindful to maintain a neutral position of the spine.

3. Hold this position as you inhale. On exhalation extend the right hip and knee back as if moving into parallel arabesque without any movement along the spine. The right shoulder, hip, knee, and ankle should be in line as if the imaginary plumb line was assisting your alignment. Visualize the small deep multifidus muscles contracting and supporting the spine. Hold this position for 5 to 10 seconds, focusing on spine stabilization.

4. Inhale as you return to the starting position with control while maintaining a neutral spine position. Repeat 4 or 5 times before executing the exercise on the other leg.

SAFETY TIP: Avoid locking or hyperextending the elbows and remember to maintain a neutral spine position.

MUSCLES INVOLVED

Multifidus

DANCE FOCUS

Let this small, detailed exercise help you feel the power and strength of the spine to secure your placement. Visualize the small, deep multifidi gently compressing or hugging your spine as if to brace it. Although larger muscles create extension of the spine, use this exercise to emphasize a secure, braced spine. Without the power of the multifidi and coordinating efforts of the abdominals, your spine would collapse under the pressure created by dance motions. This fact is critical to understanding placement and stabilization of the spine before making any movements with the arms or legs. Strengthening the multifidus gives you excellent placement skills by creating segmental stability of your vertebrae. All movement of the arms and legs should be initiated by contraction of the deep transversus abdominis and the deep multifidus muscles.

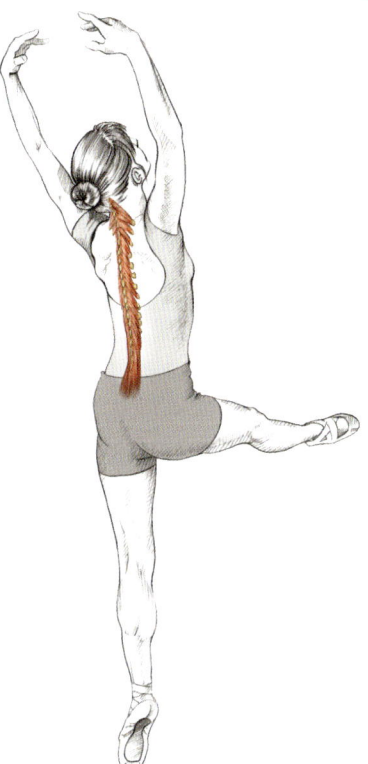

VARIATION

Multifidus Pointer

Repeat the main exercise. If you can maintain a stable and neutral position of the spine as the right hip and knee extend back, you can also release the left arm and bring it into an overhead position similar to a high fifth position. Balancing in this position will activate the multifidus, requiring more control and spine stability. Hold the balance for 5 to 10 seconds before returning to your starting position with control. You may repeat on each side 4 or 5 times.

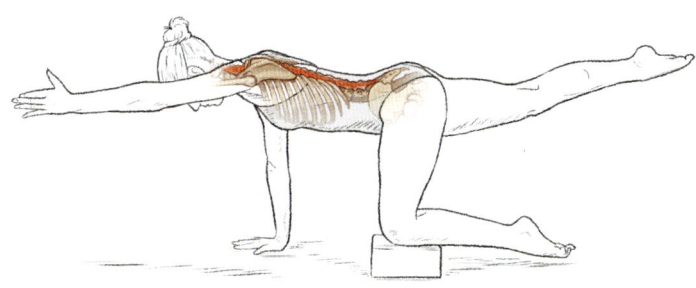

ISCHIAL SQUEEZE

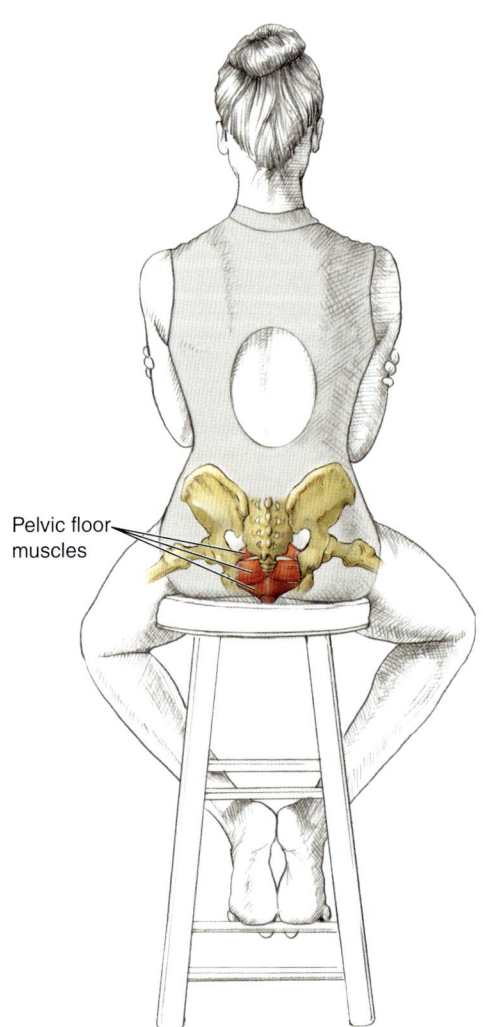

Pelvic floor muscles

EXECUTION

1. Sit on a stool or chair with your legs and hips slightly turned out. Rock the pelvis side to side to locate the ischial tuberosities (sit bones) along the lower portion of the pelvis. Return and locate your neutral alignment while seated. Check to make sure that you are not in a posterior pelvic tilt or overextending the lower back with an anterior tilt. Rest your crossed arms in front of your body and gently inhale.

2. As you exhale, engage the pelvic floor muscles, and pull the sit bones together. Try to organize this muscle contraction with your exhalation. Visualize the muscles of the pelvic floor shortening, allowing the sit bones to pull toward each other. Notice how your spine gently lifts with this supportive contraction.

3. Relax and feel the muscles lengthen. Repeat; as you begin to experience this contraction, visualize the pubic and coccyx bones pulling together as well. Perform 10 to 12 times.

MUSCLES INVOLVED

Pelvic floor muscles: Coccygeus, levator ani (pubococcygeus, puborectalis, iliococcygeus)

DANCE FOCUS

In all creative dance motions, you may never think about using the pelvic floor muscles. However, if you consider where the pelvic floor is located, you will understand the significance of its ability to form the basin of support for your pelvis. Even so, these muscles are hardly ever mentioned during technique classes, choreography work, or rehearsals. Therefore, let's take a moment to understand this exercise and its relationship to placement. This is an excellent exercise for emphasizing body awareness; if it doesn't come right away, zero in on the sit bones and visualize the basin shrinking. The movement is very small and fine, but small shifts can lead to large supportive changes. Pelvic floor musculature is addressed further in chapters 5, 6, and 8; in the meantime, you can use this exercise as an introduction and become familiar with the lift and support that it provides.

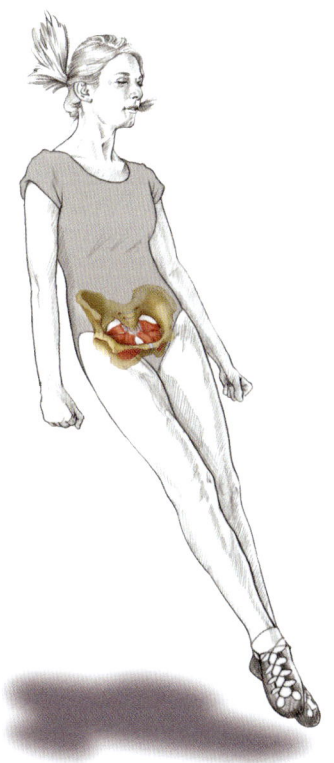

CAMBRÉ DERRIÈRE

Let's look at the safest way to execute a healthy cambré derrière. This chapter focuses on the spine, and the following dance movement execution expands the focus to incorporate various other muscles that are targeted in other chapters. Support for spinal extension involves several factors, one of which is intra-abdominal pressure (which is discussed in more detail in chapter 5). The dance movement execution presented here gives you an overall idea of the functionality of the cambré derrière.

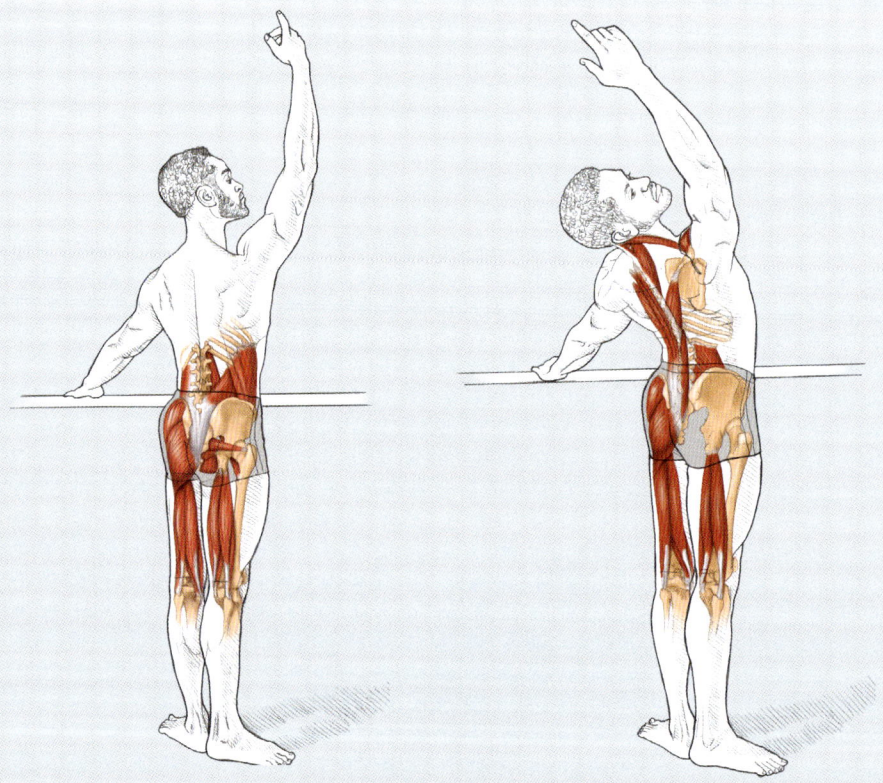

1. Begin in first position, with your left hand on the barre, and bring your right arm to high fifth, using your anterior deltoid and your pectoralis major. Check that you are placed in your neutral spine position. Your legs are turned out from the hips. Activate your quadriceps, hip adductors (inner thighs), pelvic floor muscles, and deep hip external rotators. Maintain muscle tone of the hamstrings, gastrocnemius, soleus, tibialis anterior, peroneal muscles, and intrinsic muscles of your arches.

2. As you inhale, your diaphragm contracts and begins to push down. Lengthen through your spine to help unload your spinal joints. As if to grow taller, feel the lengthening of your abdominal muscles and the intra-abdominal pressure supporting your lower spine. Try to feel the pelvic floor muscle tone supporting your low back and pelvis.

3. Initiate the cambré from your upper back or thoracic spine; allow your head and neck to follow the line of your upper back. Engage the cervical and spinal extensors while maintaining the abdominal tone. Glide your shoulder blades down away from your ears toward your hips by using your lower trapezius and serratus anterior.

4. As your spine begins to extend, continue providing support from the abdominals, pelvic floor muscles, and hip adductors. Incorporate movement through the thoracic spine; don't allow your hips to rock forward or your neck to collapse and overextend. Your right arm stays in high fifth position.

5. As the spine continues to extend, moving in a long arc, begin to turn your head to the right, maintaining support and control of your neck. Lift your sternum to assist with thoracic mobility.

6. On exhalation, reengage your abdominals to reverse the movement and slowly return to the starting position, maintaining length along your entire spine and moving in the longest possible arc.

Muscles Involved

Neutral spine placement: Transversus abdominis, internal oblique, external oblique, pelvic floor, iliopsoas

Hips and legs: Quadriceps (rectus femoris, vastus lateralis, vastus medialis, vastus intermedius), sartorius, hamstrings (semitendinosus, semimembranosus, biceps femoris), gluteus maximus, gluteus medius, deep hip external rotators, adductor brevis, adductor longus, gracilis

Spine extension: Diaphragm, abdominals (eccentric contraction), multifidus, erector spinae (iliocostalis, longissimus, spinalis), quadratus lumborum, spinalis thoracis, longissimus thoracis, iliocostalis lumborum

Cervical extension: Splenius capitis, semispinalis capitis, splenius cervicis, sternocleidomastoid (as the head begins to rotate)

Arm: Anterior deltoid, lower trapezius, pectoralis major, serratus anterior

While executing your cambré derrière, focus on axial elongation to take the load off the vertebrae; lengthen and grow tall through your spine. Incorporate movement through your thoracic spine as you use your abdominals to support your spine. Use your entire spine, not just the lower segments, to execute the cambré. Poor technique can create repetitive microtrauma and overuse, which contribute to lower back injuries. To advance and grow as a dancer, focus on maintaining stability, using your abdominals, and minimizing stress on the lower spine.

CHAPTER 5
Ribs and Breath

Even though breathing is the natural process of bringing oxygen into the lungs, many dancers are unclear about exactly how to breathe. Granted, you know *how* to breathe—indeed, you take some 17,000 breaths per day—but can you use your breath efficiently to reduce tension and improve core strength? Quality breathing also supplies the muscles with needed oxygen to enable your dance movement. As mentioned in chapter 2 on brain health, since deep breathing can help reduce performance anxiety, breathing has a positive influence on the mind–body connection. Rhythmic, organized breathing can also reduce stress and improve mood. Breathing is really an important link to your brain, as well as to your movement patterns.

During class, how many times do you receive cues to pull your tummy in and up? Typically, you suck your belly inward and throw your ribs and chest upward, elevating your shoulders. In doing so, you increase the tension in your upper body and make it more difficult to breathe! How, then, can you possibly move with ease and grace? The cue to "pull up" can be used more efficiently by incorporating axial elongation or lengthening through your spine. Axial elongation takes stress off your spinal joints without involving shoulder elevation.

Breathing is part of all movement, including dance; in fact, your diaphragm and high-quality breathing skills are critical to spinal stability. Instructors could consider adding breathing exercises to dance combinations. For example, breathing can be choreographed into exercises with music so that dancers become more aware of their breathing patterns. This organized, rhythmic breathing can be a great tool for instilling better breathing habits.

Physiological Benefits of Quality Breathing

Quality breathing provides many physiological benefits besides helping improve your dance skills. Chapter 2 discussed how breathing can help manage stress. Deep, effective breathing stimulates the vagus nerve, which is responsible for providing a relaxation response. The vagus nerve runs from the brain through the neck and down to the abdomen. It helps to regulate the relaxation response, heart rate, blood pressure, and digestion.

Why is this important? The vagus nerve is responsible for both sensory and motor activities in your body. It can help reduce inflammation, lower your heart rate if you are under stress, and communicate with your gut. Deep breathing and vagus nerve stimulation are also associated with reducing and controlling pain. Endorphins, hormones that play a role in pain perception, are released during deep breathing. Stimulating the vagus nerve truly connects mind to body.

Good breathing habits have a direct correlation with improved posture, too. Diaphragm movement supports thoracic mobility, reducing pressure on the spinal discs and reinforcing the natural curves of your spine. Sleep can also be improved by practicing deep breathing, which will allow your heart rate to slow down and reduce any anxious feelings that keep you awake. Deep breathing also has been associated with melatonin production, which regulates sleep–wake cycles.

Quality breathing even helps your immune system, which protects you against disease and bacteria. Oxygen carries nutrients to your cells. When your body is fully oxygenated, your immune system can function appropriately. Breathing also helps regulate the lymphatic system, which is responsible for immune responses and detoxifying the body.

Be mindful of the benefits of deep and purposeful breathing and how it can help you improve many areas of your life.

Breath Anatomy

Breathing, or respiration, consists of two phases: inspiration, the period when oxygen flows into the lungs, and expiration, the period when carbon dioxide leaves the lungs. At rest, the lungs move about 0.5 liters of air with every breath; during exercise, they move as much as 3 liters per breath. Every part of your body needs oxygen, which allows cells to release needed energy for muscular work, including that of dancing.

Both phases of breathing can be either passive or forced. While reading this book, you are most likely unaware of your breathing. This is probably also the case during the warm-up at the beginning of a technique class, where you are focused on organizing your body. These are examples of quiet, passive breathing. This type of breathing is also required when you hold a beautiful balance in relevé.

In contrast, active inhalation and exhalation involve deeper breathing that uses more musculature for inspiration and expiration. For instance, you may find yourself breathing more deeply while executing jumping combinations or when the choreography requires other kinds of more challenging muscle work. Organizing the process of breathing can reduce tension in your upper body, improve oxygen flow to your muscles, and engage your core muscles. With these benefits in mind, the exercises presented in this chapter will help you organize your breathing.

Your lungs are soft, spongy, elastic organs that provide a passageway for air. They are surrounded and structurally supported by your ribs. Lung function is aided by the respiratory muscles, and a basic understanding of how these muscles function can help you become a better dancer. Key respiratory muscles include the diaphragm, the transversus abdominis, and the pelvic floor muscles.

As you inhale, air comes in through your nose or mouth, goes into your trachea (windpipe), and travels deep into very small tubes in the lungs called bronchioles. From there, the air moves into microscopic sacs called alveoli (figure 5.1), which are surrounded by small capillaries; this is where red blood cells distribute carbon dioxide and pick up oxygen. In other words, oxygen is added to the red blood cells from the air, and carbon dioxide is taken out of the cells and released into the air. The blood carrying the oxygen gets delivered to your heart via the pulmonary vein, and your heart then pumps the oxygenated blood to your body.

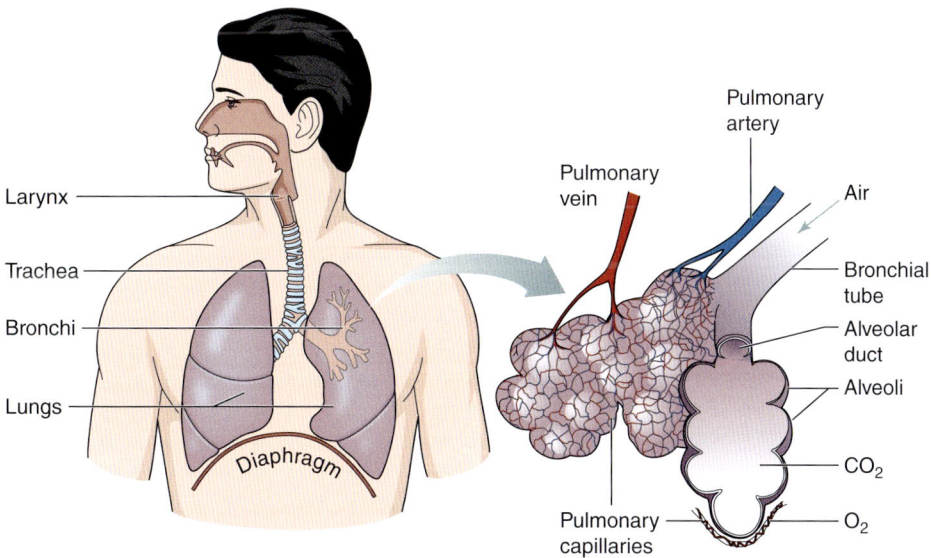

Figure 5.1 In the alveoli, oxygen moves into the blood while carbon dioxide is removed from it.

So, as you progress through a warm-up in class, move to the center, and begin to perform more jumping and strenuous combinations, your carbon dioxide levels go up, which triggers your breathing rate to increase. As your breathing rate goes up, your body gets rid of the carbon dioxide and provides you with more oxygen. The more efficient your breathing patterns are, the more oxygen your muscles can receive to help you perform your best.

The most important muscle of the respiratory system is the diaphragm, which is the primary mover. It is a large, dome-shaped muscle that lies within the rib cage and just above the abdominal organs (figure 5.2); visualize an open parachute inside your rib cage. All its muscle fibers run up and down, which determines how it contracts. The diaphragm is attached to the lower end of the sternum (chest bone), the lowest six ribs, and the spine. It creates the three-dimensional changes in shape in the thoracic and abdominal cavity. As you inhale, the diaphragm contracts, moves downward, and flattens out. This contraction allows the lungs and ribs to expand a small amount in all planes, which increases the volume of the thoracic cavity. This expansion moves your ribs in a three-dimensional pattern.

The abdominal wall is made up of four layers, the deepest of which is the transversus abdominis muscle, which supports your trunk like a corset; visualize the tight bodice of a costume. This muscle's fibers run horizontally, and the diaphragm weaves into them. The diaphragm and the transversus abdominis work together for postural control and spinal stability. For more details, see Hodges and Gandevia (2000), who address the connection between these two muscles and their important role in providing intra-abdominal pressure to stabilize the spine.

On forced exhalation, the transversus abdominis begins to contract, thus increasing abdominal pressure. Typically, forced exhalation can help you on the downward phase of some movements by enhancing your control of the landing. For example, you can try this approach when performing a slow grand battement (high kick). Inhale on the preparation and into the leg lift, then actively exhale on the way down. Notice how the exhalation supports the downward phase—you have more control over your leg. Thus, although the abdominal wall was discussed in chapter 4 as an important source of support

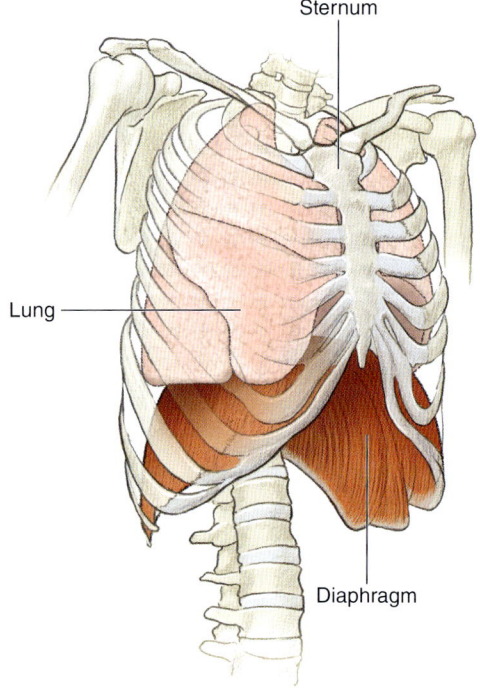

Sternum

Lung

Diaphragm

Figure 5.2   The diaphragm.

for the spine and core, remember that contraction of the deep transversus abdominis muscle also relates directly to forced exhalation.

Forced exhalation also involves several layers of muscles that support the pelvis. Known as the pelvic floor muscles, they connect between the ischia (sit bones), the pubic bone, and the coccyx (tailbone). Visualize a horizontal diamond shape with the sit bones along the side points of the diamond and the pubic and coccyx bones along the front and back points. During forced exhalation, the muscles that align and attach along the points of the diamond engage, pull together, and provide support for the position of the pelvis. This muscular contraction becomes more apparent while practicing the breathing plié described later in this chapter. Specifically, the upward phase of the plié coordinates exhalation with engagement of the deep core and the pelvic floor.

Diaphragm Movement

Do you ever wonder why you're so fatigued after performing challenging choreography? Do you conclude that you must keep practicing to build stamina? How can you build stamina if you're not getting adequate oxygen? Very simply, on inhalation, the lungs and ribs widen, the diaphragm moves downward, and the abdominal muscles lengthen (it's okay to let your belly relax a little). Try not to get caught up in your belly always having to be flat; your belly needs to be able to expand as you inhale. On exhalation, the diaphragm moves upward, the ribs return, and the abdominal muscles contract or shorten. There is more emphasis on three-dimensional movement of the lungs and ribs to provide adequate space for oxygen to enter.

If your stamina is in question, you have probably been rehearsing with upper chest breathing or shallow breathing while trying to hold your belly in. With upper chest breathing, air enters only the top part of the lungs, which raises your center of gravity. If your chest is too high, you will find it harder to balance and will have difficulty feeling freedom in your shoulders. In addition, since the lower portion of your lungs doesn't get adequate oxygen, you create a leaner look (for the moment) but reduce the ability of your diaphragm and lungs to work properly—thus limiting your oxygen intake! Your diaphragm takes care of about 80 percent of your breathing; therefore, strengthening your diaphragm can help you improve your endurance.

The diaphragm also has muscle attachments to the iliopsoas, which is the powerful hip flexor muscle. Thus, when you aggressively suck in your belly, you limit the efficient movement not only of the diaphragm but also of the iliopsoas, which can create unwanted tension in the hip joint. The iliopsoas is composed of two muscles: the iliacus and the psoas major.

- The iliacus originates along the iliac crest and inserts into the femur.
- The psoas major originates along the lumbar vertebrae and the 12th thoracic vertebra and inserts into the femur.

The balance of these two muscles is extremely important for dancers. The iliopsoas connects the spine and pelvis to the legs. Therefore, healthy balance between strength and flexibility in this muscle helps you achieve leg height above 90 degrees; it can also decrease lower-back pain. When elevating your leg, allow the inhalation to create a lengthening feeling through the spine and allow the exhalation to create a deep contraction of the abdomen so that the hip joint is free to move with ease.

With any forward cambré position, the flexion action deep in the front of the hips compresses the abdomen and brings the diaphragm toward your head. Therefore, efficient breathing must occur more in the back of the rib cage. Feel as though you are breathing more into the back of the lower ribs to provide adequate space for taking in oxygen. In contrast, tension in the hip joint causes labored breathing, which limits oxygen flow.

Muscle Action

Respiration also involves work by other muscles (figure 5.3). For instance, the external intercostals, which lie between the ribs, contract on inhalation to open the ribs and bring the sternum forward. Because of the shape of the ribs, they move laterally, anteriorly, and posteriorly to widen the chest; visualize how a curved bucket handle is raised. The ribs can be raised even more by the scalene and sternocleidomastoid muscles in the neck, along with the pectoralis major muscle of the chest. In addition to their other jobs, these muscles activate to elevate the ribs during inspiration.

- The scalene muscles originate along the cervical vertebrae and insert into the first two ribs.
- The sternocleidomastoid muscle originates on the sternum and clavicle and inserts into the temporal bone (jaw).
- The pectoralis major originates at the clavicle, the sternum, the cartilage of ribs 1 through 6, and the external oblique and inserts into the humerus bone.

Given that these muscles are so involved in inspiration, can you see how overactivation can create tension in the upper body? You can use this understanding to your advantage when lifting your arms overhead in any dance position. Rather than elevating your rib cage, think about elongating axially on inhalation and widening with your rib cage. By emphasizing the lateral rib movement, you create mobility throughout your thoracic spine and freedom in your shoulders.

Thus far, we have seen that during the active process of forced expiration, the deep abdominal wall contracts along with the pelvic floor. In addition, the ribs are depressed by engagement of the intercostal muscles within the ribs, the latissimus dorsi in the back, and the quadratus lumborum. Get into the habit of using your exhalation to release superficial tension but increase deep abdominal tension. You certainly don't want your audience to see you

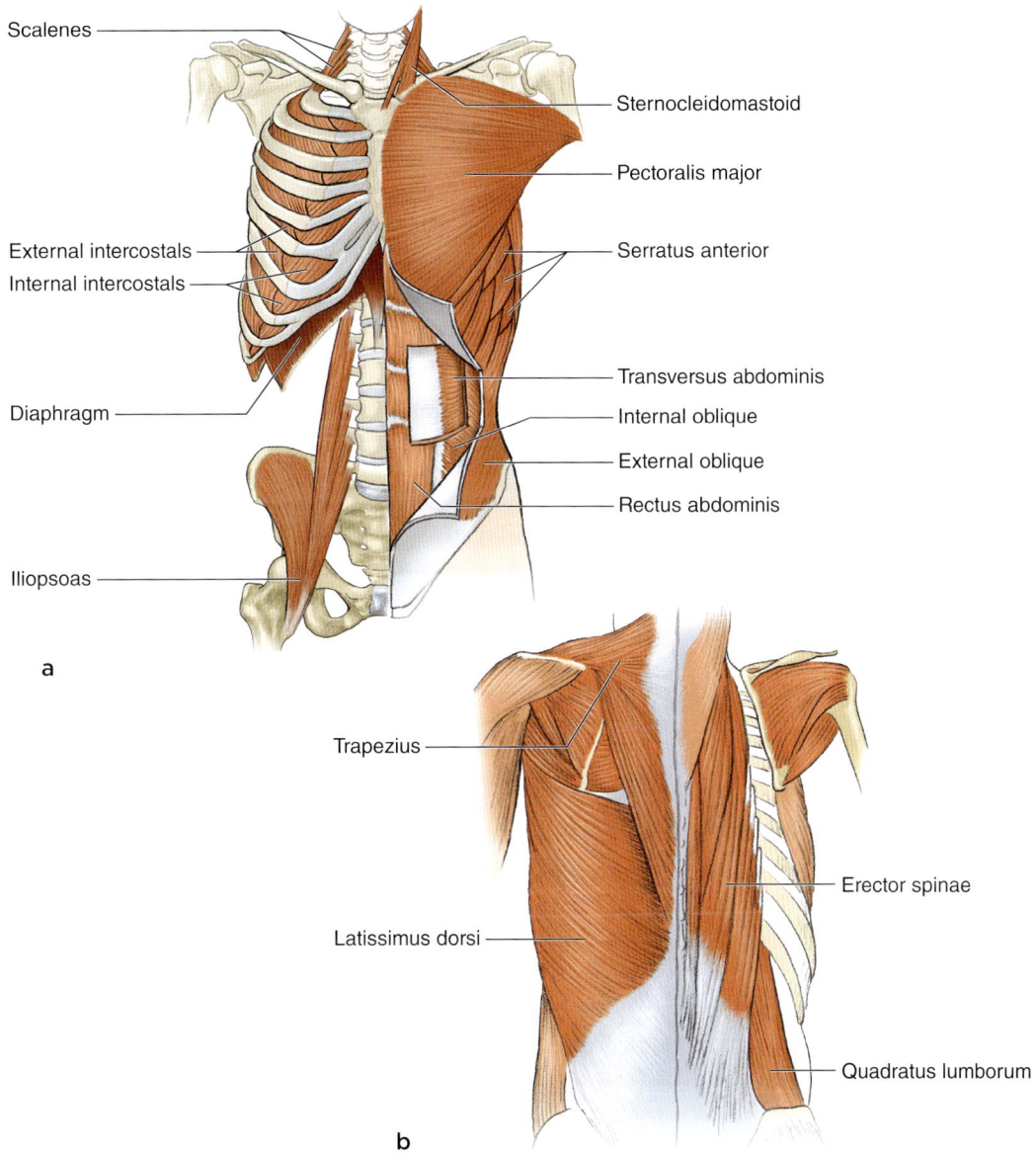

Figure 5.3 Muscles that work during respiration: *(a)* front; *(b)* back.

fighting with tension, panting heavily, or gasping for air. Your audience wants to see incredible skill exhibited without physical exhaustion.

To enable this kind of performance, think of your diaphragm floating up and down within the movement of the ribs—not creating tension in your jaw, neck, and shoulders. Visualize the lungs moving softly so that the ribs can be flexible. When exhaling during the exercises presented in this chapter, focus on relaxing your neck and shoulders but increasing the abdominal pressure by bringing your navel toward your spine as if you are trying to flatten your waist.

Two other layers of the abdominals are the internal and external oblique muscles. As you will learn in chapter 6, the oblique muscles play a role in supporting your trunk and improving your basic body placement in dance. The internal oblique muscles have fibrous attachments to the internal intercostals, and the external oblique muscles have fibrous attachments within the external intercostals, thus emphasizing, once again, the relationship between breathing and the core.

The oblique muscles are involved in twisting dance movements. Specifically, the upper body rotates in one direction against the resistance of the lower body, which is holding in the opposing direction. To make this twisting of the torso more effective, you must maintain freedom in the shoulders and hips; otherwise, it will be too difficult for the diaphragm, abdomen, and ribs to move for breathing. Although it is next to impossible to choreograph breathing into every dance step, you can practice using active (forced) exhalation when you need control. Inhale on the preparation and exhale on the movement.

Remember the discussion about the gliding joints? They involve the ribs and their attachments to the spine. There is typically very little movement along the midspine or thoracic spine; thus, you want to improve the mobility at these joints to release tension. Use the inhalation phase to help you lengthen through your spine through all planes of movement. The lengthening effect gives you more space between vertebrae and incorporates a small amount of movement along the rib attachments. Let the exhalation phase occur deep within the abdomen and pelvic floor to ground your pelvis and support your spine.

Nasal Breathing

Nasal breathing consists of inhalation and exhalation through the nose. It is emphasized, for example, in many yoga exercises. In addition, some Pilates exercises are based on inhaling through the nose and exhaling through the mouth. A combination of nasal and mouth breathing is also used in the Alexander technique, especially for training singers. Inhaling through the nose helps filter the air and warms it before it enters your lungs. Exhaling through the nose helps control the amount of carbon dioxide leaving your body, whereas exhaling through the mouth may help you focus on deep abdominal contraction and may therefore help when you are short of breath. Holding your breath is not efficient. In fact, it can hamper the return of blood to your heart and can increase your blood pressure.

For the purpose of this text, some exercises use both nasal and mouth breathing patterns. Excellent breathing techniques can help you execute your dance movements and provide a pleasing, visual quality to your upper body. You can train your lungs and ribs to move more efficiently and limit tension in various joints by practicing the following exercises. Use these exercises as part of your daily warm-up and cool-down.

Dance-Focused Exercise

Before proceeding to the exercises, take a moment to practice breathing. On inhalation, widen through the ribs laterally; on exhalation, feel the ribs return-

Healthy Lungs

The American Lung Association encourages breathing exercises to maintain healthy lungs. Two important breathing exercises include pursed lip breathing and deep belly breathing. Since the diaphragm does approximately 80 percent of the work for efficient breathing, it needs to be healthy and strong. If the diaphragm is not working at full capacity, your lungs will take in less oxygen.

Pursed lip breathing will aid in keeping airways open longer so you can take in more oxygen for your lungs and your muscles. Practice breathing in through your nose for 2 or 3 seconds and exhaling with pursed lips for 4 to 6 seconds. This technique will help slow down your breathing, giving you more control.

Deep belly breathing is another technique for improving oxygen–carbon dioxide exchange. As you practice deep breathing, place your hand on your belly as you breathe in through your nose, feeling your belly expand. Exhale through your mouth and feel your belly deflate as you exhale. This technique also helps to fully engage the diaphragm.

Practice these two techniques several times a day to help improve your oxygen levels and strengthen the diaphragm.

The following are other important factors associated with healthy lungs:

- Avoid smoking, breathing secondhand smoke, and vaping.
- Participate in cardiovascular exercise to increase lung capacity.
- Avoid pollutants, dust, mold, and pet dander.

ing, along with a deep abdominal supportive contraction. Each time you inhale, expand through the ribs and lungs with minimal movement in the upper chest. When you exhale, feel tension leaving your neck and shoulders and flatten your waist by pulling your navel toward your spine. You can place a hand on your chest to check that you are not elevating your chest as you breathe in and place your other hand on your belly to feel your belly expanding with inhalation. Practice various breathing styles while lying down, sitting, and standing, just to change your base of support. You can also try it in front of a mirror and focus on your neck and the tops of your shoulders. Are they rising—that is, are you adding muscle tension? The goal is minimal upper-chest movement and maximal freedom in the neck and shoulders. Look in the mirror to see your ribs moving laterally.

Let your chest feel weightless and your neck be long and free. Try a few arm movements; inhale as your arms go up, and exhale as they return. Think of the smooth movement in the shoulders as separate from the widening movement of the lungs and ribs.

In the exercises presented here, you will practice breathing while lying down, sitting, standing, jumping, and with spine extension. How do you put it all together when you're moving and trying to focus on choreography, placement, music, and rhythm? In the final exercise, breathing sauté, you will practice vertical jumping while incorporating lateral breathing. This final exercise is a functional breathing exercise that advances the lateral breathing. The chapter concludes with a detailed look at a breathing plié.

LATERAL BREATHING

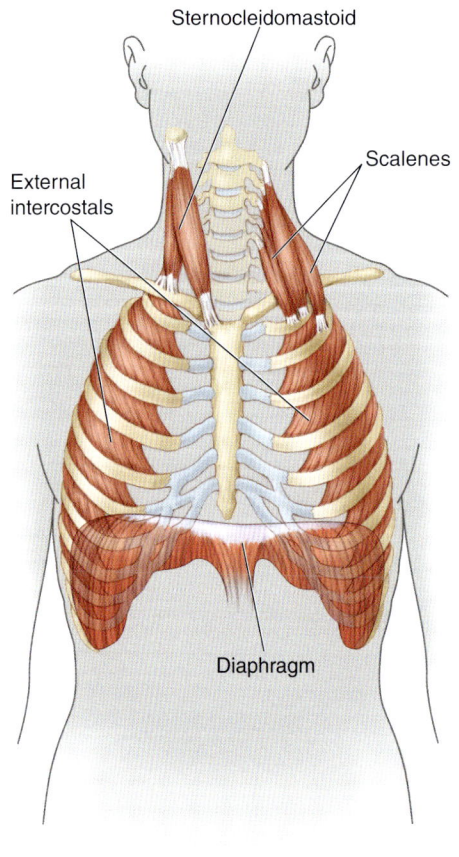

Sternocleidomastoid

Scalenes

External intercostals

Diaphragm

Inhale.

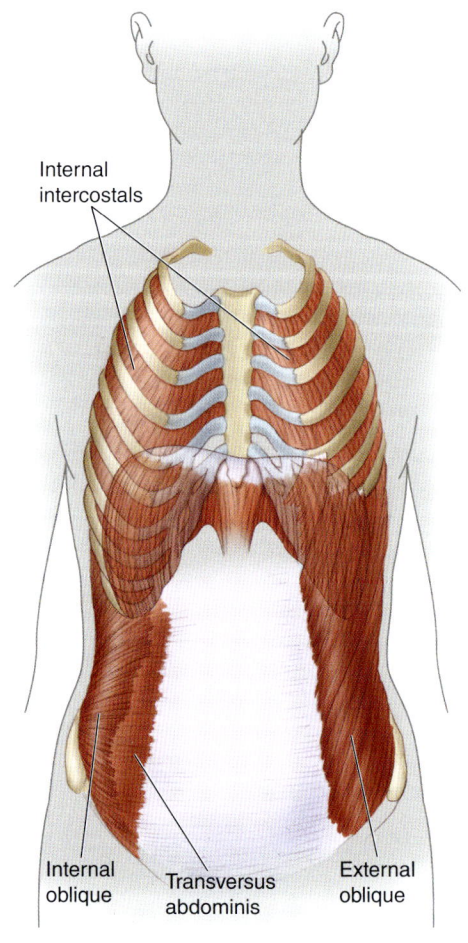

Internal intercostals

Internal oblique

Transversus abdominis

External oblique

Exhale.

EXECUTION

1. Lie faceup with your knees bent, your feet on the floor, and your arms at your sides with your palms up. Locate your neutral position. Place one hand on your upper chest (sternum) and the other hand on your belly. On inhalation through the nose, relax the belly and allow it to expand, feel the ribs opening and widening, and visualize the diaphragm moving downward. Continue to expand through the middle of the chest and the back of the ribs. Inhale for a slow count of 3; hold on the count of 4. Do not allow your upper chest to lift or your spine to extend.

2. With forced exhalation through the mouth, feel the ribs returning and the chest relaxing, and visualize the diaphragm elevating. Feel your deep abdominal muscles contracting and release tension in the back

of your neck. Feel as though you are gliding your shoulders down toward your hips. Exhale for a count of 4. Perform the exercise 6 times.

3. You can also try this with your hands on your ribs. Focus on lateral rib movement without upper-chest movement; continue to relax the neck, jaw, and throat.

MUSCLES INVOLVED

Inhalation: Diaphragm, external intercostals, scalenes, sternocleido-mastoid

Exhalation: External oblique, internal oblique, transversus abdominis, internal intercostals, latissimus dorsi, quadratus lumborum

DANCE FOCUS

For visual help with moving the ribs in a more lateral direction, you can try this exercise while sitting or standing in front of a mirror. You can also try the following with a partner: Place your hands on the back of your partner's ribs. When your partner inhales, feel the ribs moving into your hands; on your partner's exhalation, gently press into the ribs to assist with the returning of the ribs.

As you dance, feel less restriction in your neck and chest. Let your spine move in response to the antagonistic efforts of the diaphragm and the abdominals. Use your breathing with jumping combinations as well (more on this later with the breathing sauté exercise). Notice how the coordinated lateral breathing makes you feel lighter. Remember your breath and use it to move with fluidity and depth.

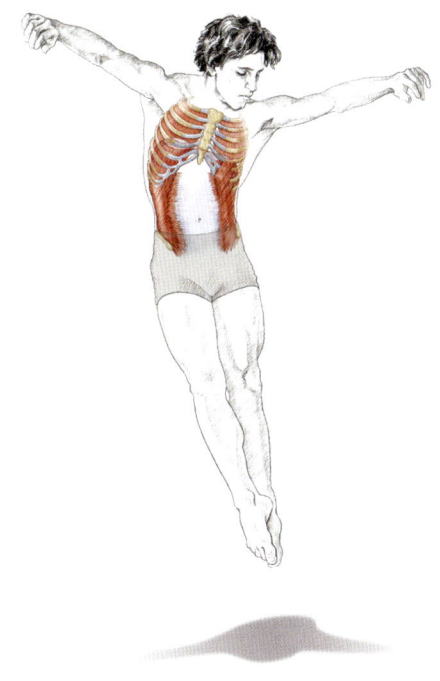

LATERAL BREATHING WITH RESISTANCE

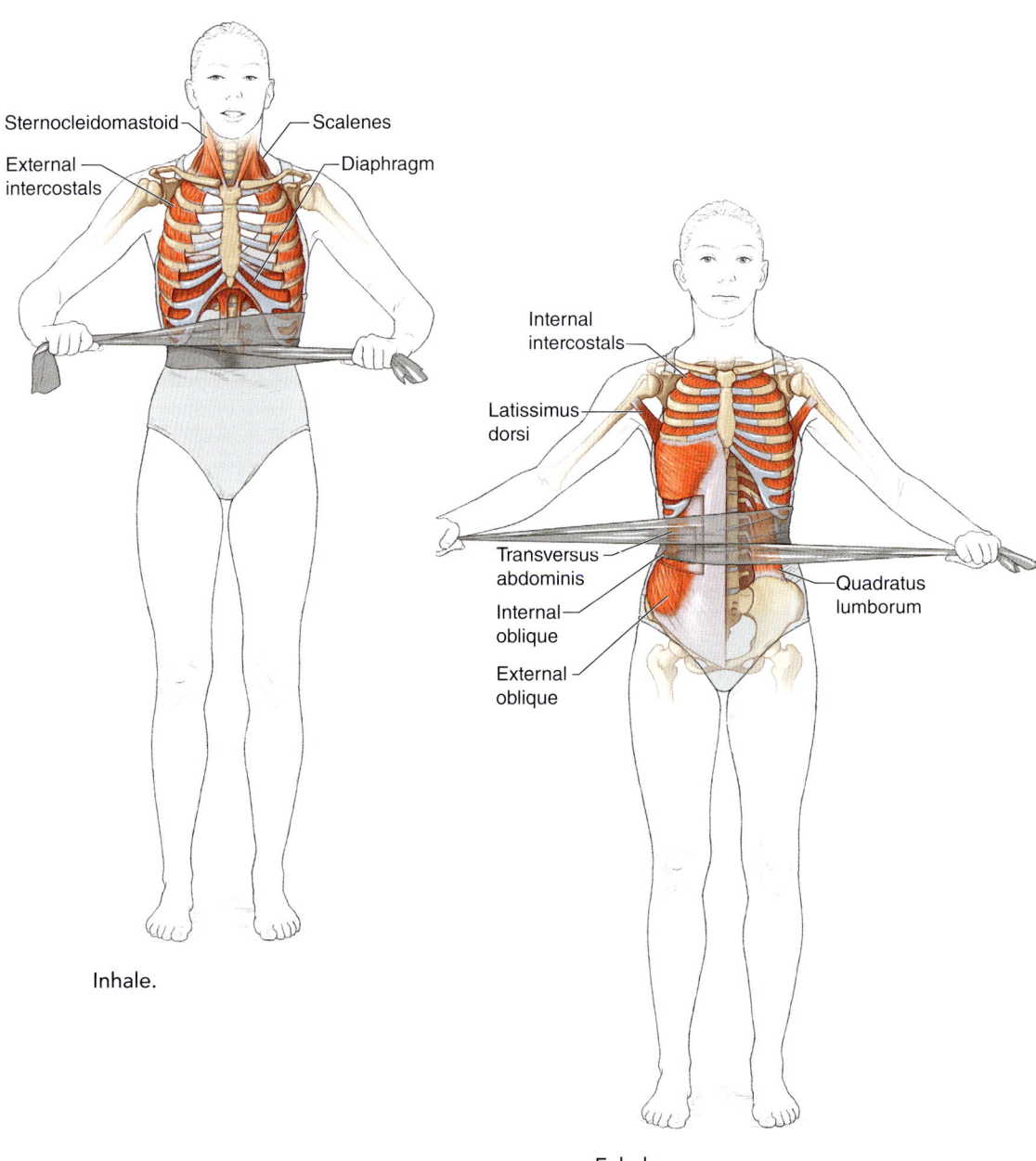

Sternocleidomastoid — Scalenes

External — Diaphragm
intercostals

Internal
intercostals —

Latissimus —
dorsi

Transversus — Quadratus
abdominis lumborum

Internal —
oblique

External —
oblique

Inhale.

Exhale.

EXECUTION

1. Wrap a resistance band around your ribs from the back; cross it in the front and hold the ends with your hands. You may try this while seated or standing.

2. On inhalation, widen the rib cage into the resistance of the band. Feel as if you are pushing the ribs into the resistance of the band.

3. With forced exhalation, actively pull the band to help the rib cage retract. Working with the resistance band allows you to advance the inhalation technique to improve lung capacity. Focus on deep breaths, lateral rib movement, diaphragm movement, and activation of your deep abdominals. Perform 6 to 8 times.

MUSCLES INVOLVED

Inhalation: Diaphragm, external intercostals, scalenes, sternocleido-mastoid

Exhalation: External oblique, internal oblique, transversus abdominis, internal intercostals, latissimus dorsi, quadratus lumborum

DANCE FOCUS

Learning to focus on breathing patterns while dancing will enhance your movement. Allowing the ribs to move in a more three-dimensional pattern will help produce quality movement through the entire spine. Holding your breath or shallow breathing creates tension in the neck and shoulders, creating stiffness in your dancing. Deeper breathing from your diaphragm will give your movement more fluidity, flow, and dynamics. Use the lateral breathing with resistance exercise to emphasize more three-dimensional movement from your torso. Modern dancers, for example, have learned how to incorporate deep breathing and their abdominal muscles to create the signature modern contraction, encouraging the relationship of breath and movement. Dance sequences can be choreographed using breath to tune in to movements of the torso in various planes of motion. Using efficient breathing can also give you a deeper emotional connection to your dancing.

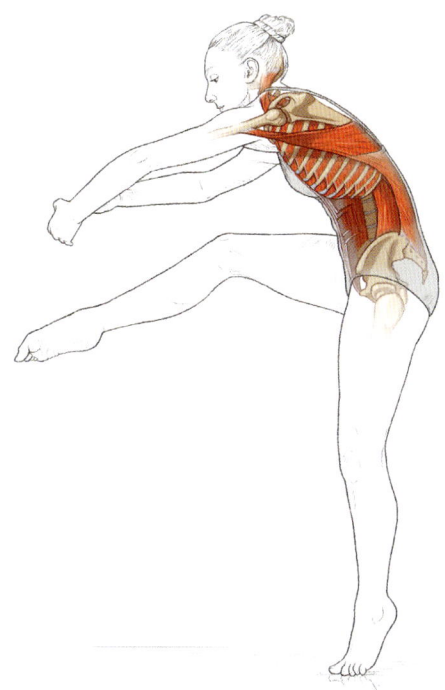

BREATHING WITH SIDE-BEND

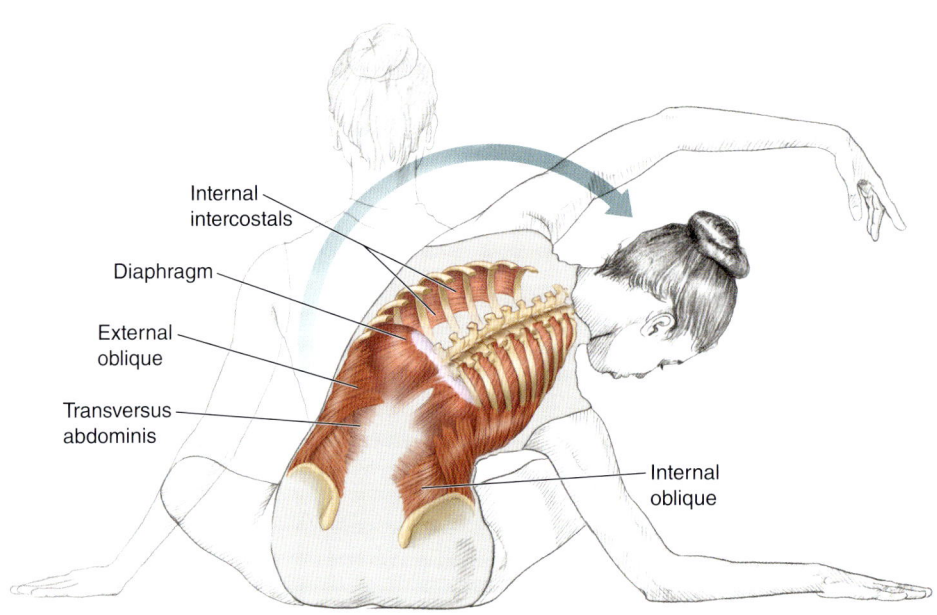

Internal intercostals

Diaphragm

External oblique

Transversus abdominis

Internal oblique

EXECUTION

1. Begin in a seated position with your legs comfortably crossed in front and your hands placed by your sides. Locate your neutral position and inhale through the nose. On exhalation through the nose, lengthen through your spine. Engage the core musculature; gently slide your right hand along the floor and bend laterally, directly along your frontal plane. Keep both ischia firmly on the floor. Allow your left arm to lift overhead while you maintain width through the chest. Your head may remain facing the front or may be gently turned toward the direction of the bend.

2. Gently rest your right elbow on the floor while continuing to lift through your center. Do not collapse into the elbow. Hold for a breath cycle. As you inhale, feel the lower ribs of the left rib cage opening wide. Be aware of the difference between the left rib cage's expansion and the right's compression.

3. With forced exhalation, feel the left rib cage pulling together and the diaphragm lifting. Engage the deep transversus abdominis and oblique muscles while moving in the longest possible arc. Return to your seated starting position. Perform 2 to 4 times on each side.

SAFETY TIP: Try to avoid letting the neck collapse. Maintain axial length and spine support.

MUSCLES INVOLVED

Exhalation: Diaphragm, internal intercostals, transversus abdominis, external oblique, internal oblique

DANCE FOCUS

Allow yourself the privilege of moving smoothly through various planes, trusting the flexibility and stability that your respiratory system provides for you. As you bend your trunk to the side, notice how the top of the lung lifts and the bottom slides downward. Let this principle of internal elasticity give you more fluidity throughout your upper body and more mobility through your thoracic spine. Moving from your center will have more meaning for you when you can feel freedom in your movements. Each time you inhale, allow the air to fill the entire portion of the lungs. As you continue to increase your lung capacity and breathe more comfortably, you will find that you have more mobility in your lateral bends. With each exhalation, notice how the abdominals can anchor your pelvis and support your spine. Remember to emphasize moving along the longest arc through the entire range of the side-bend.

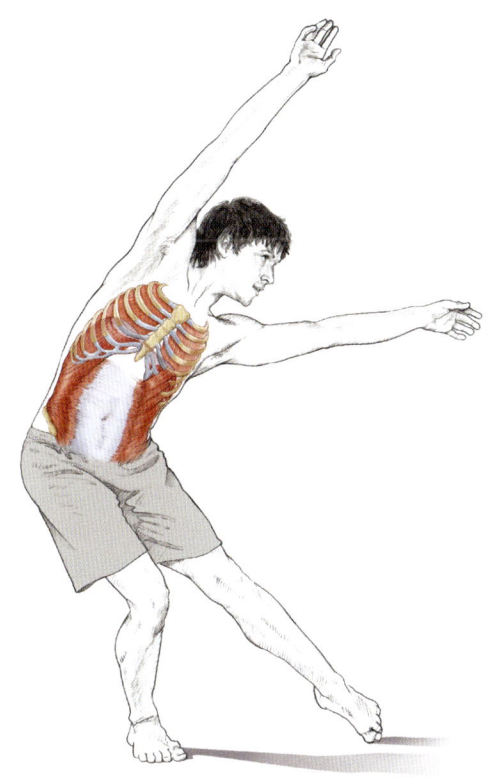

BREATHING WITH PORT DE BRAS

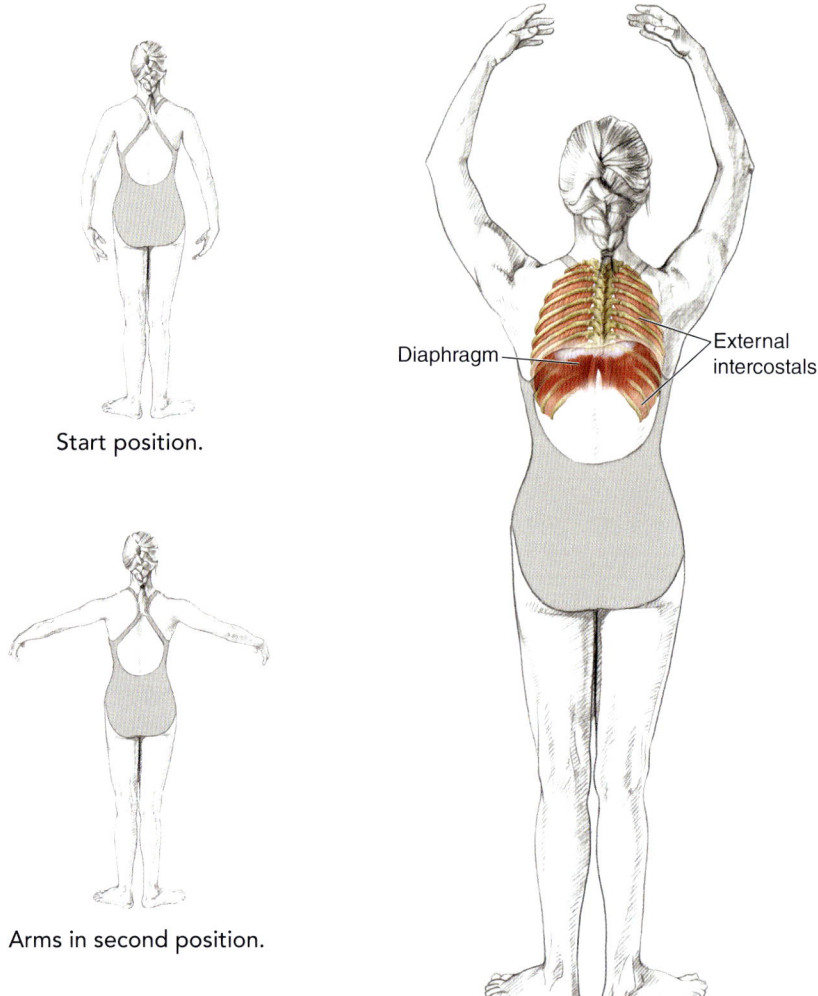

Start position.

Arms in second position.

Diaphragm

External intercostals

Arms overhead.

EXECUTION

1. Stand comfortably in a narrow second position. Focus on neutral alignment and provide a firm base for your balance. Before beginning, feel your arms lengthening by your sides and a sense of relaxation around your neck and shoulders. Neatly stack your spinal curves on top of each other and feel a gentle lift through your waist.

2. As you begin to inhale through your nose, open your arms to second and continue to lift them to high fifth. Feel your rib cage expanding with air. When your arms are overhead, focus on the weight of the arms moving down the spine to release tension in the neck and shoulders.

3. Hold that position briefly and notice the relaxation in the back of your neck. Exhale through your nose as you bring your arms down by your sides, allowing the lungs and ribs to return. Perform 4 to 6 times; inhale for 4 counts and exhale for 8 counts.

SAFETY TIP: Avoid overextension in the neck, which compresses the cervical discs; just let your neck be an extension of your spine, maintaining its natural anterior curve.

MUSCLES INVOLVED

Inhalation: Diaphragm, external intercostals

DANCE FOCUS

The key to this basic breathing exercise is to coordinate the lifting of your arms with effective inspiration. This coordination gives your upper body a light, lifted feel without tension surrounding your neck and shoulders. When you breathe in, fill the lungs with oxygen and feel the rib cage expanding and the diaphragm gliding downward, which allows the lungs to move with ease and flexibility. Visualize the external intercostals contracting to expand your ribs so that your upper chest does not elevate. Notice a gentle mobilization where your ribs meet your spine; this action improves your thoracic mobility and spinal alignment. Inhale through your nose and imagine your arms floating up as your ribs swell. Exhale through your mouth when your arms float back down. Do not allow your spine to move into extension; doing so will cause your chest to elevate, your tension to increase, and your alignment to suffer. As it becomes easier to breathe and lift the arms without tension, add relevé and then repeat with jumps and leaps.

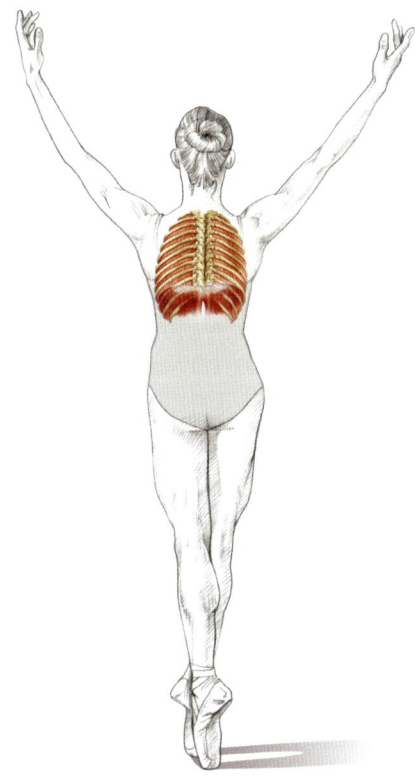

KNEELING SPINE EXTENSION

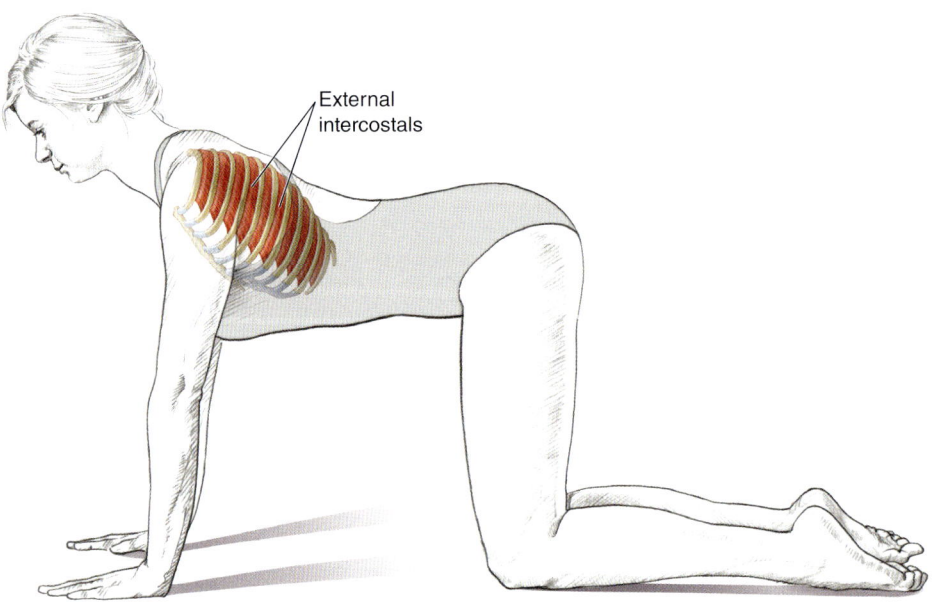

External
intercostals

EXECUTION

1. Center yourself comfortably on your hands and knees. Align your shoulders over your wrists and your hips over your knees. Practice this exercise with pursed lip breathing. Remember to release tension in the neck.

2. As you begin to inhale through the nose, extend through your entire spine. Think about lengthening and moving in a long arc. Just allow your head to follow; don't overextend in the neck. Feel the abdominals being stretched as you expand through the lower ribs. Inhale through the nose for 4 counts.

3. On exhalation through pursed lips for 8 counts, reverse the arch and move into your starting position. Perform 6 times and focus on lengthening through the spine and widening through the ribs. Let all segments of your spine move equally.

SAFETY TIP: Avoid overextension in the lower segments of the spine.

MUSCLES INVOLVED

Inhalation: External intercostals, rectus abdominis (eccentric contraction), internal oblique, external oblique, erector spinae (concentric contraction of iliocostalis, longissimus, spinalis)

DANCE FOCUS

Most of the time, inhalation should occur with back-bend positions. However, as you may have noticed, it is difficult to breathe if you are trying to hold your abdominals tight. So, to keep your lower back safe, you must use the entire spine. Allow the abdominals to stretch and the chest to expand laterally. Let your inhalation help you extend your spine in that long arch position. Remind yourself to glide your shoulder blades down toward your hips and relax your neck and shoulders. Allow the tightness of the abdominals to create pressure against the abdominal cavity to give your spine needed support. Remember to feel your weight settling into your legs; secure yourself by feeling stable in the lower spine and pelvis. Feel as though you are breathing into the spaces between your ribs, allowing them to expand. You will notice that you have more three-dimensional range of motion in your chest.

VARIATION

Trunk Extension

Perform an upper-back extension from a standing position with one hand on the barre for balance and the other in a low fifth position. As you inhale, widen through your chest. Lengthen and extend the spine, creating a long arch and moving evenly. Do not raise your shoulders or create tension in your neck; continue to feel the abdominals stretching. Your pelvis remains directly over your legs and feet. On exhalation, control the movement on the return and again feel lengthening through the spine. Perform 4 to 6 times.

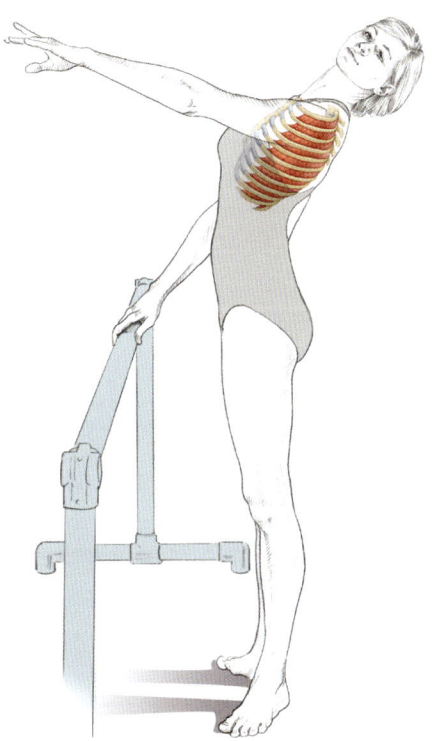

BREATHING SAUTÉ

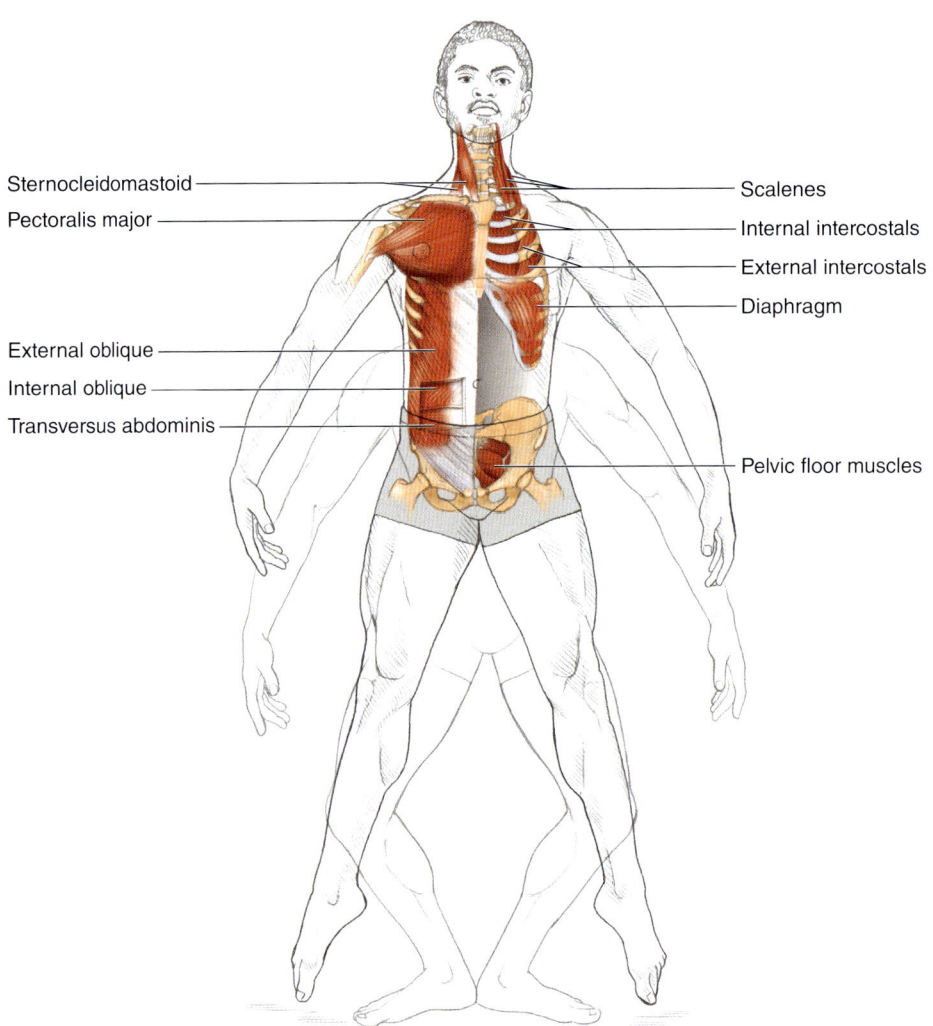

Sternocleidomastoid

Pectoralis major

External oblique

Internal oblique

Transversus abdominis

Scalenes

Internal intercostals

External intercostals

Diaphragm

Pelvic floor muscles

EXECUTION

1. Begin in first position with arms en bas. Find your neutral and stable placement as you lengthen your center, providing axial elongation.

2. Inhale and exhale along with your demi-plié. Execute eight sautés in first position. Inhale for 2 jumps and exhale for 2 jumps. Begin in a slow and controlled manner, especially on the landing. Remember to maintain turnout in the air and on landing.

3. Hold and rest after 8 sautés. Notice how smooth your jumps were and how easy you felt in your neck and shoulders.

4. Repeat the same breathing pattern while you execute 16 sautés. Inhale for 2 jumps and exhale for 2 jumps. You will find that by coordinating your breathing and oxygenating your muscles, you accomplish fewer rigid landings, your jumps feel lighter, and you experience less fatigue.

MUSCLES INVOLVED

Inhalation: Diaphragm, external intercostals, scalenes, sternocleidomastoid, pectoralis major

Exhalation: Diaphragm, internal intercostals, transversus abdominis, external oblique, internal oblique, pelvic floor muscles (coccygeus, levator ani), latissimus dorsi, quadratus lumborum

DANCE FOCUS

Once you are comfortable using organized breathing while dancing, your diaphragm will get stronger, and your breathing will help you move better. The increased oxygenation in your muscles will help your endurance, and the strong exhalation will fire up your deep transverse muscles to stabilize your spine. Keep practicing breathing sautés until you feel comfortable adding more petit allegro combinations with coordinated breathing. As your breathing patterns get stronger, try 4 sautés as you inhale and 4 as you exhale. You will feel less tension in your neck and shoulders and more stability in your lower back and pelvis. You will be able to dance longer with less fatigue.

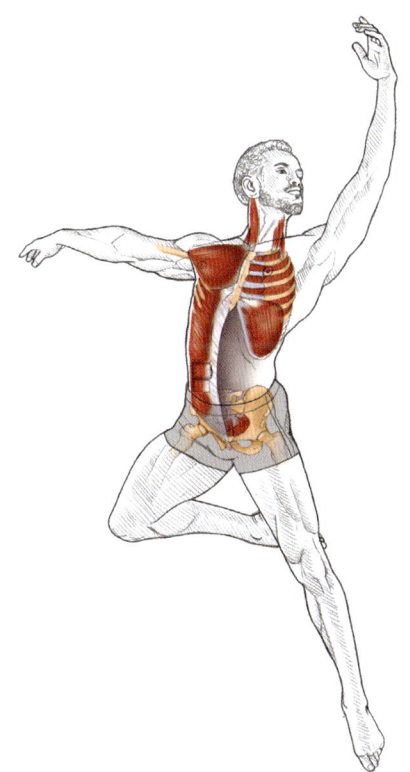

BREATHING PLIÉ

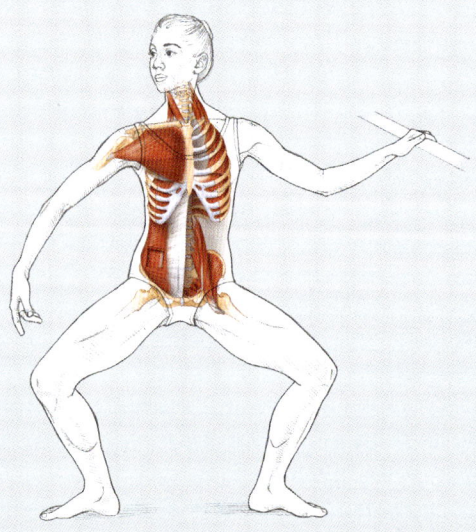

Plié is the basis for almost all movement in dance, whether is it parallel, turned in, or turned out. Instructors are responsible for teaching their students how to plié, which prepares them for relevé, pointe work, turns, and jumps; it also prepares them to absorb the forces involved in landing. For dancers, proper execution is imperative for making technical gains, enhancing performance, and reducing injury risk. Key elements include equal weight placement through your first and fifth metatarsals and your heel, as well as maintenance of a neutral lumbar and pelvic position through the entire range of the plié. Throughout your plié, feel your equal weight placement and try to engage your hamstrings.

Now, let's examine proper and safe execution of demi-plié. We will focus on breath and on spinal and pelvic placement.

1. Stand in a firm, turned-out second position with arms in second position. Locate your neutral spine and pelvic position. On inhalation through the nose, widen through the ribs and lungs as you move into demi-plié. Continue focusing on axial elongation. The hips and knees will flex.

2. Balance your weight evenly throughout all five toes and your heel on each foot. The pelvis and lumbar spine remain in a neutral position. As the hips flex, the thighs are turned out directly over the toes along the frontal plane.

3. The hips are turning out with an eccentric contraction of the deep external rotators and the hip adductors. The knees are flexing with an eccentric contraction of the quadriceps.

4. The ankles move into dorsiflexion, contracting the anterior tibialis as the calves begin to gently lengthen. The intrinsic muscles of the arches are toned to support the foot. The heels maintain contact with the floor. The anterior lower leg muscles contract to assist with balance.

5. With forced exhalation through the mouth, begin the upward phase. As the lungs and ribs return, engage the abdominals, and feel the pelvic floor contracting. Visualize the ischia pulling together as well as the coccyx and pubic bone pulling together.

6. Press firmly into the floor through your first and fifth toes as well as your heels. Feel the deep hip rotators helping to maintain turnout.

7. The quadriceps, hip extensors, hip adductors, and hip external rotators create a concentric contraction as the knees and hips begin to extend. Continue to maintain a neutral lumbar spine and pelvis and work along the frontal plane through the entire range of the plié.

8. At the top (completion) of the plié, hold for 3 seconds and focus on neutral spine placement, hip external rotation, adductor contraction, and contraction of the pelvic floor. Allow the rhythm of the breathing with the plié to energize your body.

Muscles Involved

Inhalation: Diaphragm, external intercostals, scalenes, sternocleidomastoid, pectoralis major

Exhalation: Diaphragm, transversus abdominis, pelvic floor muscles (coccygeus, levator ani)

Spine and pelvic stabilizers: Transversus abdominis, internal oblique, iliopsoas, multifidus, erector spinae (iliocostalis, longissimus, spinalis), quadratus lumborum

Plié Descent

Gravity allows the plié to occur. Even so, the tibialis anterior contracts concentrically to dorsiflex the ankle, and the following muscles contract eccentrically for control.

Deep hip external rotators: Obturator externus, obturator internus, piriformis, quadratus femoris, gemellus superior, gemellus inferior, gluteus maximus, posterior fibers of gluteus medius

Hip adductors: Adductor longus, adductor brevis, adductor magnus

Knee extensors: Quadriceps (rectus femoris, vastus lateralis, vastus medialis, vastus intermedius), sartorius

Hip extensors: Hamstrings (semimembranosus, semitendinosus, biceps femoris), gluteus maximus

Posterior lower leg: Gastrocnemius, soleus

Plié Ascent

The following muscles contract concentrically.

Deep hip external rotators: Obturator externus, obturator internus, piriformis, quadratus femoris, gemellus superior, gemellus inferior, gluteus maximus, posterior fibers of gluteus medius

Hip adductors: Adductor longus, adductor brevis, adductor magnus

Knee extensors: Quadriceps (rectus femoris, vastus lateralis, vastus medialis, vastus intermedius), sartorius

Hip extensors: Biceps femoris, semimembranosus, semitendinosus, gluteus maximus

Posterior lower leg: Gastrocnemius, soleus

Anterior lower leg: anterior tibialis contracting eccentrically

Plié may be the most overlooked motion that you perform daily. Indeed, it is commonly used in all dance styles. It prepares you for relevé and for jumps and serves as the transition movement between steps. Without a pliable plié, you are left with choppy, rigid traveling steps.

As you begin your plié, visualize your three-dimensional movement of the thoracic spine with inhalation. Maintain axial length along your spine and organize your breath. Allow the inhalation to prepare your body and the exhalation to anchor your lungs, abdominals, and pelvic floor.

The upward phase engages the pelvic floor, abdominals, hamstrings, and hip rotators. On the upward phase, try to coordinate contraction of the inner thigh muscles as the legs begin moving together; this contraction adds support for your pelvis and prepares you for jumping combinations with a smooth and controlled takeoff. A firm plié performed with smart breathing secures your pelvis and lower spine and frees your hips to turn out without constriction. Thus, you improve the quality of all your movement.

CHAPTER 6

Core

All movements in dance are generated from your torso, which serves as your foundation. A firm foundation promotes postural awareness, spinal stability, and gorgeous placement. Your goal is to move through space while creating the most challenging and interesting dance skills with ease, right? Achieving this goal requires strong trunk muscles. The trunk refers to your center (your deep core), the muscles that provide stability for your spine and pelvis.

Core strength promotes a healthy core stabilizing system. The core stabilizing system can be divided into two different levels of muscles—local and global—that provide support for your spine and pelvis. The local stabilizing muscles are primarily slow-twitch postural muscles. They are the deep transverse abdominis, internal obliques, pelvic floor, and multifidi. The global system is made up of the rectus abdominis, external obliques, erector spinae, and quadratus lumborum, which provide strength for larger, more powerful movements. The gluteal muscles are also included as part of the core musculature. The iliopsoas muscle is considered part of the core due to its attachments along the spine but will be discussed further in chapter 8. Weakness or an inability to effectively contract the core muscles prior to movement can be a precursor to injury.

One of the most basic movements in dance is the plié, which—whether it is performed parallel, turned in, or turned out—requires coordination of breath and core strength. Core muscle activity is important to help maintain neutral pelvic alignment during plié, which is the preparation for relevé, turning, and jumping. When choreography requires your torso to move off balance, the strength of your core keeps your spine from collapsing. In addition, whenever you need to extend your spine,

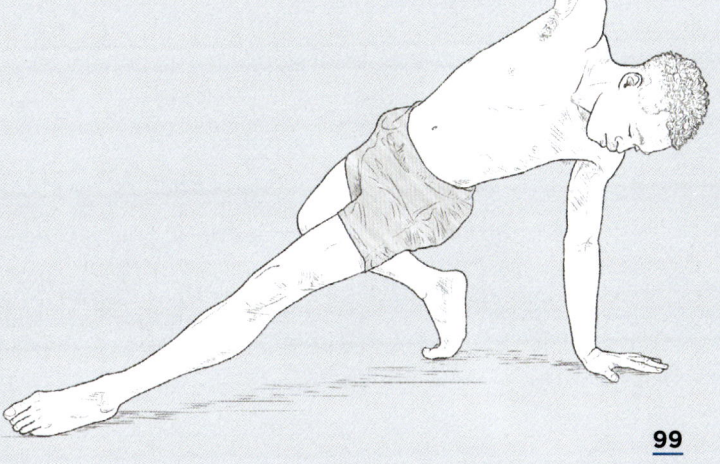

particularly while jumping, your core musculature must brace your spine for protection. In fact, all aspects of dance can challenge your spine. Fortunately, when you prepare effectively for movement, core activation allows you to exercise more control over all your movements, from a basic plié to any advanced jumping skill.

Abdominal training continues to be popular in gyms, physical therapy clinics, and dance studios, but as a dancer, do you really know how to use your abdominal muscles to improve your technique? It's not just about performing daily crunches; it's about understanding the anatomy of your trunk and coordinating the action of the muscles that make up your core to stabilize and support the segments of your spine.

The core muscles that contract to stabilize your spine receive a lot of attention regarding injury awareness and care of the spine. Numerous medical studies prove the connection between a cocontraction of the trunk muscles and a reduction in back injuries. For example, Hodges (2003) noted the importance of the deep transversus abdominis and the deep multifidus muscles in producing control and stability of the spine. In another study, Gildea, Hides, and Hodges (2013) investigated dancers with and without back pain. The dancers without back pain had larger multifidus muscles, which provide stability for the spine.

A research study published in 2017 in *The International Journal of Sports Physical Therapy* found improvements in pirouette ability, balance, and muscle performance after a nine-week core training program in college dancers (Watson et al. 2017). By performing specific exercises targeting the core muscles three times a week for nine weeks, the dancers saw significant improvement in core strength. A combination of local and global muscle strengthening will provide improvements in spine stability and has been included in the exercises in this chapter.

For spinal stability, you need to create an effective cocontraction of the local trunk muscles, which involves engaging the transversus abdominis, oblique, pelvic floor, and multifidus muscles. The core musculature has been described in various ways and has been given multiple names: center, trunk, abdomen, midline, powerhouse, spine stabilizers, torso, and abdominal wall. Remember, it's not just about doing multiple crunches; it's about how you apply abdominal strength to your functional dance work. Therefore, this chapter includes local, global, and functional exercises that use your entire core to help improve your strength.

Core Anatomy

The abdominal wall is made up of the following muscles, beginning with the deepest: transversus abdominis, internal and external obliques, and rectus abdominis (figure 6.1). When these muscles contract, they provide security for your spine and spinal curves.

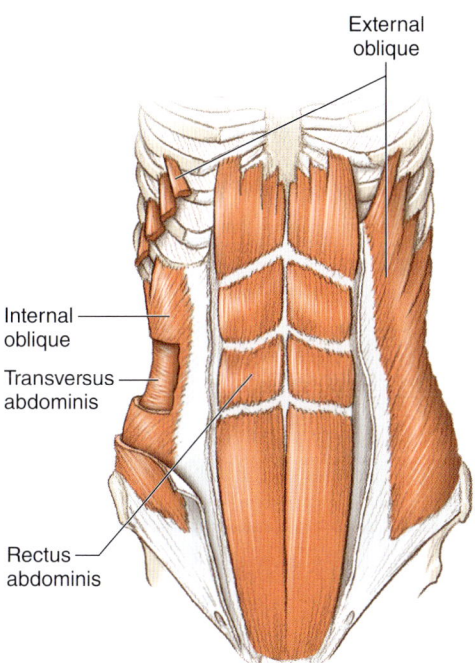

External
oblique

Internal
oblique

Transversus
abdominis

Rectus
abdominis

Figure 6.1   The four layers of abdominal muscles.

The deep, wide transversus abdominis contains fibers that run horizontally from the lowest six ribs and connect to the spine by means of a thick sheet of fascia, called the thoracolumbar fascia. The transversus abdominis is made up of type I fibers and stabilizes your spine before you move your arms and legs. To learn more about the importance of deep core musculature in supporting your spine and pelvis, see Richardson, Hodges, and Hides (2004). This publication explains the importance of training the core muscles to provide segmental stabilization for the lumbar spine and pelvis. It also addresses the role of core training in preventing injuries.

The internal oblique is a thin layer of muscle located along the side of the trunk, originating on the iliac crest and thoracolumbar fascia. The internal oblique muscle inserts along ribs 9 through 12 and the linea alba, a long, thin band of connective tissue that separates the right and left rectus abdominis. When the internal oblique muscle contracts, it provides spinal segmental support, and it pulls you into side-bend or rotation on that same side. This action accentuates side cambré movements, twisting motions, and jazz pelvic isolations. The external oblique, on the other hand, is the more superficial and larger of the two obliques; its fibers run in the opposite direction of the internal oblique. The external oblique muscle originates along ribs 5 through 12 and inserts into the anterior iliac crest and the linea alba. When the external oblique muscle is contracted, its primary action is to flex the spine and bend to the side, but it also contracts in spinal rotation from the opposite side. Your oblique muscles help your ribs feel connected to your pelvis. If you feel like you're dancing with your ribs elevated, think more about the diagonal fibers of the oblique muscles shortening to funnel the ribs in and downward.

The superficial rectus abdominis is a long, flat muscle that is divided into four sections and is sometimes referred to as the six-pack. The right and left rectus abdominis are separated by the linea alba along the anterior abdominal wall. The rectus abdominis originates at the pubic bone and inserts along ribs 5 through 7 and the sternum. The rectus abdominis is an important trunk flexor; it plays a significant role in modern contractions and posterior pelvic tilts and when rolling up from cambré forward. Although the walls of the abdomen have no bony reinforcements, the layering and directional changes of the fibers combine to create great strength.

The quadratus lumborum is also part of the posterior aspect of the core muscles. It originates off the iliac crest and inserts into the 12th rib as well as the transverse processes of the first four lumbar vertebrae. The quadratus lumborum helps to stabilize the lumbar spine and contracts during spine extension and lateral side-bends. Due to its attachment at the 12th rib, it also helps stabilize the 12th rib during deep breathing. Weakness in the quadratus lumborum can contribute to low back pain and overcompensation of other posterior muscles. Choreographic demand along with partnering and lifting sequences can also cause muscle imbalances in the lower back. Therefore, maintaining a healthy balance of exercises for the local stabilizing muscles as well as the global muscles can help reduce risk of injury.

The deep multifidus muscles run along the posterior aspect of the spine and provide spinal support for each vertebra, whereas the more superficial erector spinae muscles provide support when the spine extends. The multifidus has a high percentage of type I muscle fibers, which makes it very effective for stabilization and posture control. Both the multifidus and the erector spinae have numerous attachments along the entire spine, some of the ribs, and the sacrum, thus creating a detailed arrangement of intertwining soft-tissue structures that provide spinal security. These posterior core muscles can provide stability with fine coordination movements as well as large, forceful movements.

When combined, these muscles primarily make up the core. In figure 6.2, the gluteus medius and gluteus minimus have also been included due to their importance in helping you maintain a stable pelvis, which contributes to placement and dance skills. This topic is discussed further in chapter 8.

The muscles lining the deep pelvic region play a role in centering, pelvic stability, and postural awareness. This area is made up of several muscles, which for our purposes are combined and described as the pelvic floor (figure 6.3). The pelvic floor consists of several strong muscles located within the pelvis; visualize a basin as we continue exploring the anatomy of this area. The pelvic floor muscles work when you practice Kegel exercises or contract muscles to try to stop the flow of urine.

The pelvic bones are made up of two strong hip bones, each of which consists of an ilium, ischium, and pubic bone. This basin of bones is enclosed in the front by the pubic symphysis, or joint, and in the back by the sacrum, down to the tailbone (coccyx). When performing various seated floor exercises, notice the two bones that you are sitting on; these two sit bones are located at the base of the ischium.

Think of the bones of the basin as forming a diamond shape—the pubic bone in the front, the sit bones along the sides, and the tailbone in the back. The muscles that line this basin are layered for added strength and can tighten or stretch. Return for a moment to the first exercise (locating neutral) presented in chapter 4. Extend your lower back and move your pelvis into that forward (anterior) tilted position. Then begin to move your pelvis into your neutral position and visualize the diamond shape shrinking. Continue to practice this action, engaging the pelvic floor; feel the deep security of the lower portion of your trunk.

Awareness of the trunk should also address fascia, which is the superficial tissue lying just under the skin. It anchors the skin to the underlying organs and allows the skin to move freely. Fascia functions as a shock absorber and can help protect deeper body tissues from heat loss.

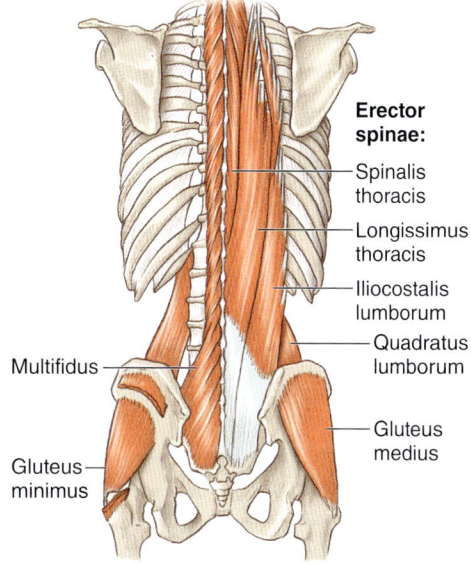

Figure 6.2   Posterior core muscles.

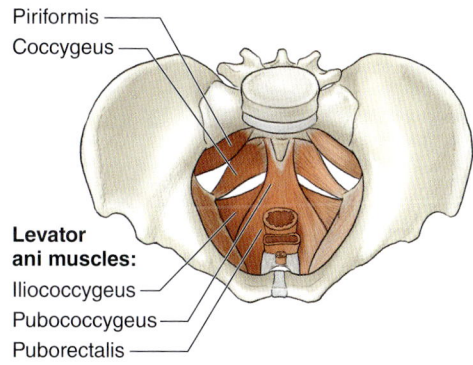

Figure 6.3   Muscles of the pelvic floor.

The thoracolumbar fascia is the membrane of fibers that covers the muscles of the back; it has connections to the core muscles, ribs, vertebrae, and sacrum. The anatomy of the core musculature and fascia has been detailed in a study by Willard et al. (2012), which shows how contractions of the core muscles provide lumbopelvic stability. However, if the abdominal muscles are weak and inactive, this tension of the fascia pulls the lower spine into extension, thus leaving it vulnerable to injury. Unsupported extension of the spine, along with tension in the thoracolumbar fascia, can also create tightness in the lower back.

Role of the Core in Dance Techniques

Every dance technique requires intense control, which is provided by core strength. Consider the technique of Irish dance. These dancers must hold their spine firm throughout their sets. Their trunk placement must be intensely secure so that they can emphasize their legs and feet with incredible speed. In this and all other forms of dance, technique is demanding, and injury can keep you from training and competing. Fortunately, you can improve your body placement and reduce your risk of injury by including core conditioning in your dance training.

We know that ballroom, social, or partner dance is fluid and beautiful to watch, but it is also quick and powerful. To achieve this effect, each partner must be acutely tuned in to each other's center and alignment to effectively lead and follow. Indeed, the swing, waltz, and salsa (to name a few) require extreme coordination; both dancers must hold their waists firm to provide stability for the pelvis and enable quick footwork. Strong core musculature also provides for safe and efficient support when the spine needs to be flexible and move in extension with rotation.

Ballroom encompasses all forms of social dancing: folk, Latin, and vintage dancing. This field is highly competitive. Contestants are judged not only on footwork and style but also on posture, body alignment, timing, and speed. Knowing what we know about deep core strength, wouldn't a series of exercises designed to improve posture and body alignment help improve rehearsal and performance efficiency? Even noncompetitive recreational social dancers will benefit from core training to improve their skills. Being centered and maintaining postural control provide long-term benefits for anyone who enjoys social dancing.

Modern and contemporary choreography require tricky, creative jump combinations and movement patterns that pose challenges for the spine. In dancers who lack the ability to contract the local stabilizers to provide dynamic segmental stability, their movements will be sloppy and weak. Moreover, if the spine and pelvis are unprepared, then landing from nontraditional jumping steps will create injury risks. Graham, Horton, Cunningham, and Taylor techniques require dancers to move against gravity with emotional conviction. There are movement sequences on the floor; while seated, kneeling, moving through space, jumping, and traveling; while deep breathing; and variations of all the above. Many of the movements include falling to the floor, partnering, and traveling–turning combinations. Some of the movements are lyrical and some are grounded; much of the choreography calls for excessive movement through the spine and hip, including spinal flexion and contraction, layouts, tilts, side-bends with rotation, extensions with rotation, and spiraling. Regardless of the style of modern dance, appropriate training of your core muscles can enhance your performance.

Specific exercises presented in this chapter can help you engage your core musculature while putting your spine in more nontraditional lines. Consider, for example, the functional obliques in second position and functional trunk-twist exercises. Both focus on nontraditional movement with muscular support for the spine—specifically, abdominal bracing while working in various planes and patterns.

Even if you are not interested in a career in professional ballet, you might be required to take classes in ballet technique as part of your training. Indeed, even if you just enjoy watching ballet and take a couple of beginner ballet classes per week, you need control of your spine. Whereas other styles of dance are more grounded, classical ballet gives the illusion of a lifted, light, and airy quality.

Ballet can be based on various styles—for instance, Vaganova, Cecchetti, Balanchine, and Bournonville—but the foundation always rests on five basic positions of the feet with the legs turned out. This technique alone requires centering and spine control. In addition, for dancers of all ages who perform ballet, a strong center is extremely important for placement, turns, jumps, landing from jumps, and, of course, pointe work. (We have Marie Taglioni to thank for being one of the pioneers in creating ballet movement en pointe)! Moreover, ballet calls for extreme joint motion and torso control. Therefore, proper alignment is crucial for both spinal control and injury reduction (recall the plumb line discussed in chapter 4). Once you have developed good alignment, you can emphasize strengthening.

In all dance styles, movement can be divided into phases: preparatory, ascending, flight, descending, and landing. The ascending phase usually engages muscles in concentric contraction. The flight phase should have a "lift, hold, and hover" look, which requires extreme core strength and isometric contraction. The descending phase requires eccentric contraction; some of the muscles lengthen but still support the movement while landing. This eccentric contraction, which is associated with control of the descending phase, is important for reducing injuries on landing. Some studies show, for instance, that landing from jumps can create a force up to 12 times that of your body weight! Improvements in trunk control and balance plus emphasizing strength and alignment can help reduce the risk of injury on landings.

Breathing With the Core

Remember that breathing plays a considerable role in strengthening the torso. When forcing air out of the lungs and gently pulling your navel toward your spine, you begin applying intra-abdominal pressure, or pressure in the abdominal cavity. Intra-abdominal pressure can play a role in stabilizing your spine, which in turn supports your trunk. It is advantageous to use forced exhalation when you execute a difficult task because it increases intra-abdominal pressure and activates the deep local muscles that provide you with spine stability. Each time you perform a high kick (grand battement), exhale and engage the core. When practicing a turning combination, inhale on the preparation and exhale on the turn; you will feel more secure along your spine. When executing a series of small jumping exercises, breathe comfortably and use the rhythm of the combination to balance inhalation and exhalation.

When you execute the exercises presented in this chapter, notice the breathing cues. With each exercise, the deeper you breathe, the more your abdominal muscles will work. Remember to inhale through your nose and use the forced exhalation principle to engage the deep stabilizers; doing so promotes security for your spine. Try exhaling through your nose for most of the exercises, but if you seem to be fatigued and need to exhale through your mouth, that's fine, too.

Dance-Focused Exercise

You can perform the following exercises in the progression given. Abdominal bracing, the first exercise, is meant to be used as a deep abdominal warm-up while you visualize the bracing effects of your core. Table 6.1 provides a summary of the core muscles highlighted in the exercises that follow. Use this exercise to prepare for the rest of the series presented here. The next four exercises focus on local stabilizing muscles and are performed in a non–weight-bearing position. The exercises then progress to plank variations to challenge the core and incorporate more global muscles. The chapter finishes with more functional exercises that use multiple muscles with multijoint movement.

Notice the details in the accompanying anatomical drawings and visualize the muscle fiber arrangements to help you understand the bracing effect that your core provides for your spine. Think about where the muscles attach and how those areas provide steady support for your placement. You want to build up strength to withstand any force that dancing may exert on your spine. Contract your core muscles with deep intensity. Many of the exercises presented in this chapter use other muscles as well, but the focus is on spine stability and helping you learn about and connect with the core muscles that provide that stability.

Table 6.1   Core Muscles

Muscle	Origination	Insertion	Action
Transversus abdominis	Ribs 7 to 12, thoracolumbar fascia, iliac crest	Linea alba, pubic bone	Abdominal wall tension, intra-abdominal pressure
Internal oblique	Iliac crest, thoracolumbar fascia	Ribs 9 to 12, linea alba, pubic bone	Lateral flexion, trunk rotation on same side
External oblique	Ribs 5 to 12	Iliac crest, pubic bone, linea alba	Flexion, lateral flexion to same side, trunk rotation to opposite side
Rectus abdominis	Pubic bone	Ribs 4 to 7, xiphoid process	Flexion
Multifidus	Sacrum, posterior iliac spine, lumbar vertebrae, vertebrae T1 to T3, vertebrae C4 to C7	All vertebrae except C1	Stabilize vertebrae
Quadratus lumborum	Posterior iliac crest	Rib 12 and lumbar vertebrae	Extension, stabilization, lateral tilt of pelvis
Pelvic floor	Ischial spine, pubic bone	Coccyx bone, ischial spine, sacrum, greater trochanter	Support bladder and bowels, pelvic stabilization
Iliacus	Iliac fossa	Lesser trochanter	Hip flexion, anterior pelvic tilt
Psoas major	Vertebrae T12 and all lumbar vertebrae	Lesser trochanter	Hip flexion, trunk flexion, anterior pelvic tilt

ABDOMINAL BRACING

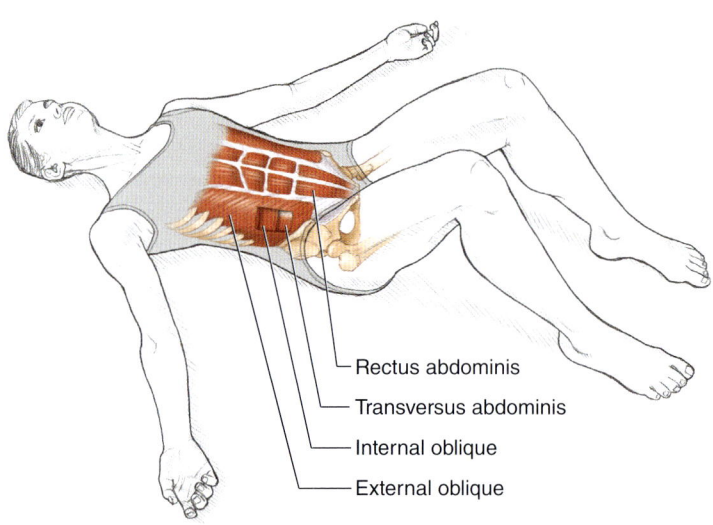

Rectus abdominis
Transversus abdominis
Internal oblique
External oblique

EXECUTION

1. Lie supine on the floor with your knees bent and your feet placed comfortably on the floor in parallel position. Your arms can be placed by your sides.

2. Feel axial elongation through your spine while relaxing through the base of your neck. Locate your neutral position. Inhale through the nose to prepare while widening through the ribs and lungs.

3. On forced exhalation, begin to contract the deep abdominals by drawing in and flattening your waist as if you were tightening a corset, but maintain neutral position. Gently pull your navel towards your spine without elevating your ribs or moving your lumbar spine.

4. Practice this exercise several times, then repeat it while seated on a ball or other unstable surface and while standing. Think about abdominal bracing prior to the exercises in this chapter, to assist in providing spine stability.

SAFETY TIP: Remember that the spine and pelvis do not move while you learn to isolate your transverse abdominis. This is a basic isometric contraction; the muscles tighten but do not create any flexion of the spine. Visualize the horizontal fibers tightening as you draw inward and engage your transversus abdominis (refer to figure 6.1). Remember to tighten the corset without elevating your ribs and chest.

MUSCLES INVOLVED

Rectus abdominis, transversus abdominis, internal obliques, external obliques

DANCE FOCUS

Terms such as *stability*, *cocontraction*, and *bracing* can be misleading if you associate them with a stiffening feeling along your spine. The word *stiff* is not one a dancer wants to associate with! Fortunately, this exercise is quite the opposite, because improving your deep core strength will enhance controlled movement of your spine. Spine stability is important for all movements in dance. For example, when you perform a cambré or spine-extension movement, think about bracing your spine with the deep local transversus abdominis to help stabilize each segment of your spine. The same principle applies when practicing side-bends, spirals, or contractions. Any chronic instability of spinal segments can increase the occurrence of back injury. It is important for the local muscles to contract to stabilize your spine so you can move safely with improved mobility and flexibility.

After practicing this abdominal bracing exercise, stand and perform a series of cambré movements, side-bends, or contractions, first slowly and then quickly. Think about drawing in and the corset effect as you move. Notice how controlled your movement is and how stable your spine is.

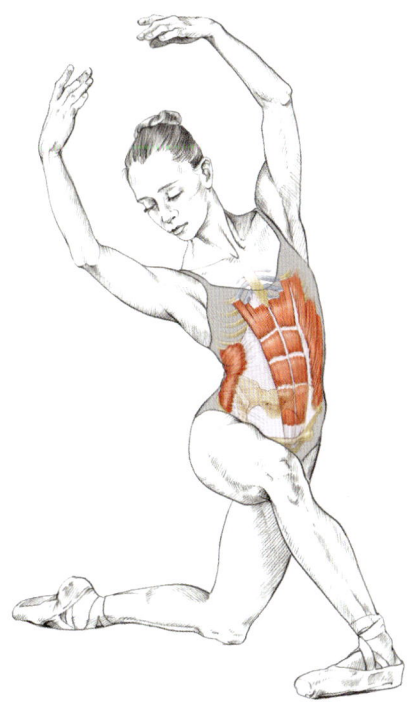

TRUNK CURL MARCHING

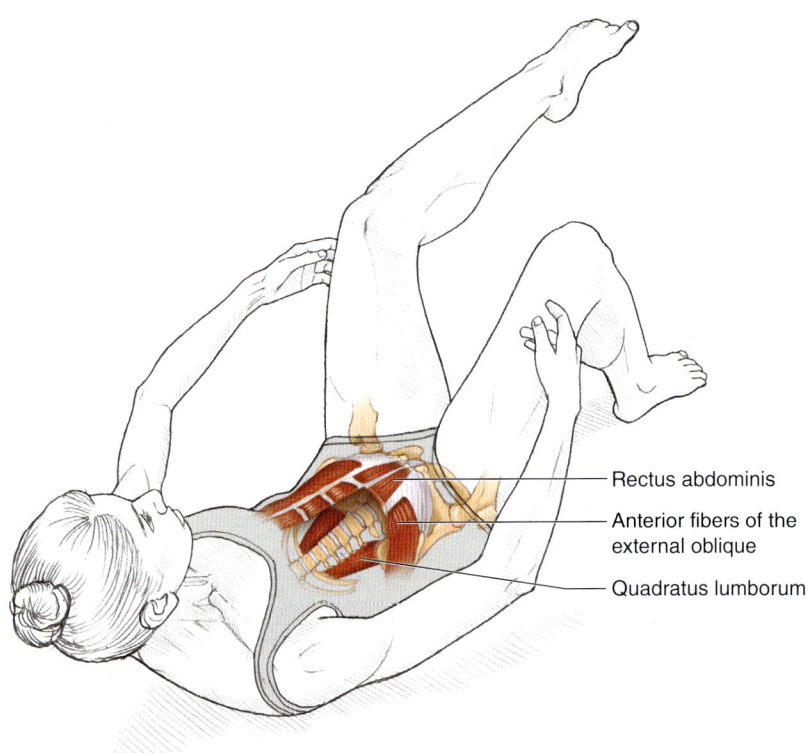

Rectus abdominis

Anterior fibers of the external oblique

Quadratus lumborum

EXECUTION

1. Lie on the floor with your knees bent and your feet placed hip-width apart on the floor. Position your arms in ballet first position. Inhale to prepare. (The exercise can also be done with your arms across your chest, by your sides, or on your shoulders.)

2. On exhalation, contract your rectus abdominis muscle to curl your trunk 45 degrees off the floor. Stabilize your pelvis; your sacrum must stay on the floor. Focus on the upper body moving into flexion; your chin gently comes toward your Adam's apple.

3. Hold this position for about 8 counts and visualize your rectus abdominis muscle fibers shortening. Elevate yourself enough that your shoulder blades are off the floor. Think about curling or flexing throughout your upper or thoracic spine.

4. Take small marching steps with one foot and then the other, trying to maintain stability. Incorporating this leg movement with the abdominal contraction provides dynamic stability.

5. Inhale on the controlled return, resisting gravity as your trunk returns to the floor. Reorganize and repeat; perform 8 to 10 times. Remember to feel the trunk moving toward a stable pelvis. As you get stronger, perform 6 sets of 10 repetitions each.

> **SAFETY TIP:** Avoid pulling with the neck or overusing the hip flexors when you flex the trunk. The deep hip flexors will be activated, and the abdominal contraction reduced, if you try to lift your trunk higher than 45 degrees. During the marches, keep your feet and legs low to avoid overusing the hip flexors. Do not increase your repetitions unless you are able to maintain control and alignment.

MUSCLES INVOLVED

Rectus abdominis, anterior fibers of the external oblique, quadratus lumborum

DANCE FOCUS

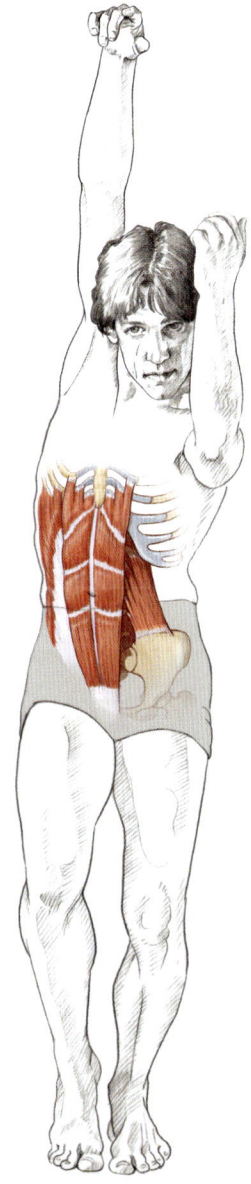

The firm center emphasized in this chapter not only helps reduce injury risk but also is quite visually appealing. Remember, however, that you are doing this work to improve your dance *technique*. Specifically, strength in the rectus abdominis can provide you with more thoracic mobility. The stronger this portion of your trunk is, the greater your range of motion will be in your upper body.

If you perform choreography that requires you to flex your trunk but maintain a neutral position of the pelvis, you will need to visualize the attachments of the rectus abdominis along the fifth, sixth, and seventh ribs and the sternum as they pull and shorten vertically to flex your spine. If you are required to hold this position (and, possibly, lift a prop or partner), then you will need even more strength and muscle tone. Allow the rectus abdominis to provide power for trunk flexion as well as eccentric length for the back of your spine. Don't allow the movement to compress your spine; think more about your spinal muscles lengthening.

MODIFIED SWAN

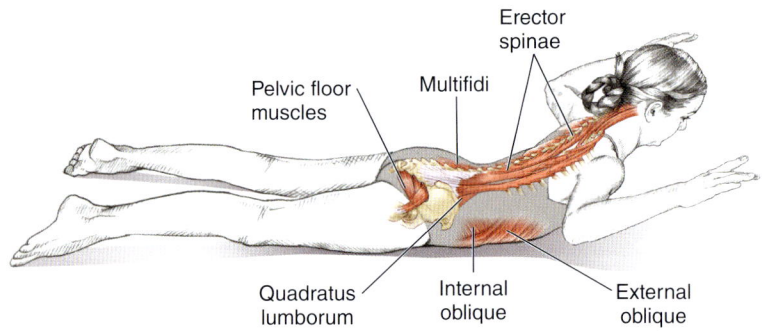

Start position.

EXECUTION

1. Lie facedown with your arms resting on the floor and your shoulders and elbows at 90 degrees. Extend your legs along the floor, slightly turned out and just farther than hip-width apart. Lengthen through your spine; gently squeeze the lower buttocks and sit bones.

2. As you inhale, begin to lift your upper body along your sagittal plane while keeping your arms along the frontal plane and maintaining the 90/90 position. Feel extension equally throughout the entire spine. Try to lift the sternum off the floor. Hold for 4 counts.

3. On exhalation, continue the axial elongation and return to start position with control. Reemphasize abdominal support and contraction of the pelvic floor. Perform 8 times.

SAFETY TIP: Lengthen through the back of the neck to avoid overextending the neck and possibly causing strain. Remember to allow movement to occur along all spinal segments, not just the lower back and neck.

MUSCLES INVOLVED

Erector spinae (iliocostalis, longissimus, spinalis), multifidi, quadratus lumborum, external obliques, internal obliques, pelvic floor muscles (levator ani, coccygeus)

DANCE FOCUS

Spinal extension is seen in all dance styles. The Swan Queen shows off her effortless spinal flexibility, as does the advanced jazz dancer with the signature layout or stag leap. The keys are timing, core strength, and axial elongation. Before you move into any type of spinal extension, remember to elongate through your entire spine; feel as though you are growing taller. Visualize the long, deep multifidi firing for deep control and the erector spinae muscles firing to help you extend your spine. The strength in your abdominals will also brace and support your spine along the front of your body. This is a beautiful preparation for arabesque in ballet or stag leaps in contemporary dance. Visualize the upper segments of your spine having individual movement in extension and a beautiful lift in your chest to create this long arch.

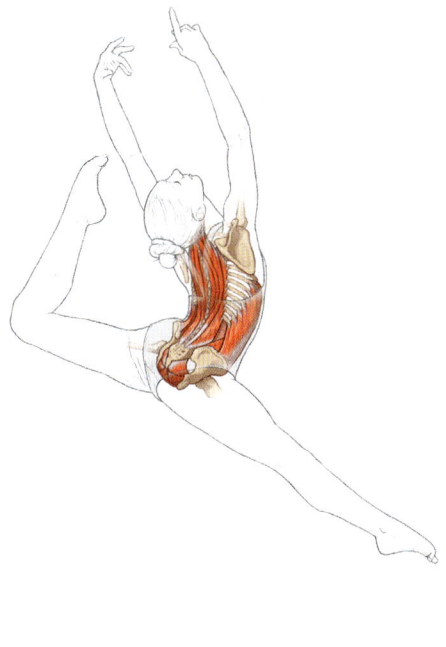

Always keep in mind that breathing will help you. Inhale as you extend, feeling the abdomen lengthening and the diaphragm moving down. You will be surprised at how much more range of motion you can attain. Let the exhalation help you hold the position and return by reemphasizing abdominal anchoring and support from the pelvic floor. You are set, your foundation is secure, and you are ready to show off your effortless spinal flexibility.

VARIATION

Modified Swan on Ball

Lie prone on a stability ball, with the ball supporting your lumbar spine. You may need to place your feet on the floor at the wall to get your balance. Place your hands behind your head to start. As you inhale, slowly lift your upper body to your neutral position and hold. On forced exhalation, engage the transversus abdominis, multifidus, and pelvic floor and continue to hold for 3 to 5 counts. As you inhale, lift your upper body a little higher into spinal extension, emphasizing thoracic mobility and avoiding overextension of the neck. Focus on stabilizing the lumbar spine while moving through the thoracic spine. As you exhale, hold for another 3 to 5 counts and emphasize lumbar spine stability. Slowly return to the starting position before repeating 10 times.

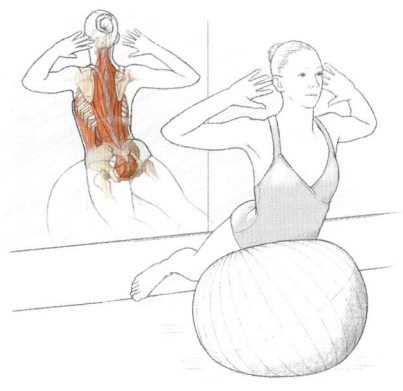

REVERSE LIFT

SAFETY TIP: Lift your hips until you are balanced along your midspine—not your neck. Throughout the exercise, maintain an abdominal contraction to support your spine.

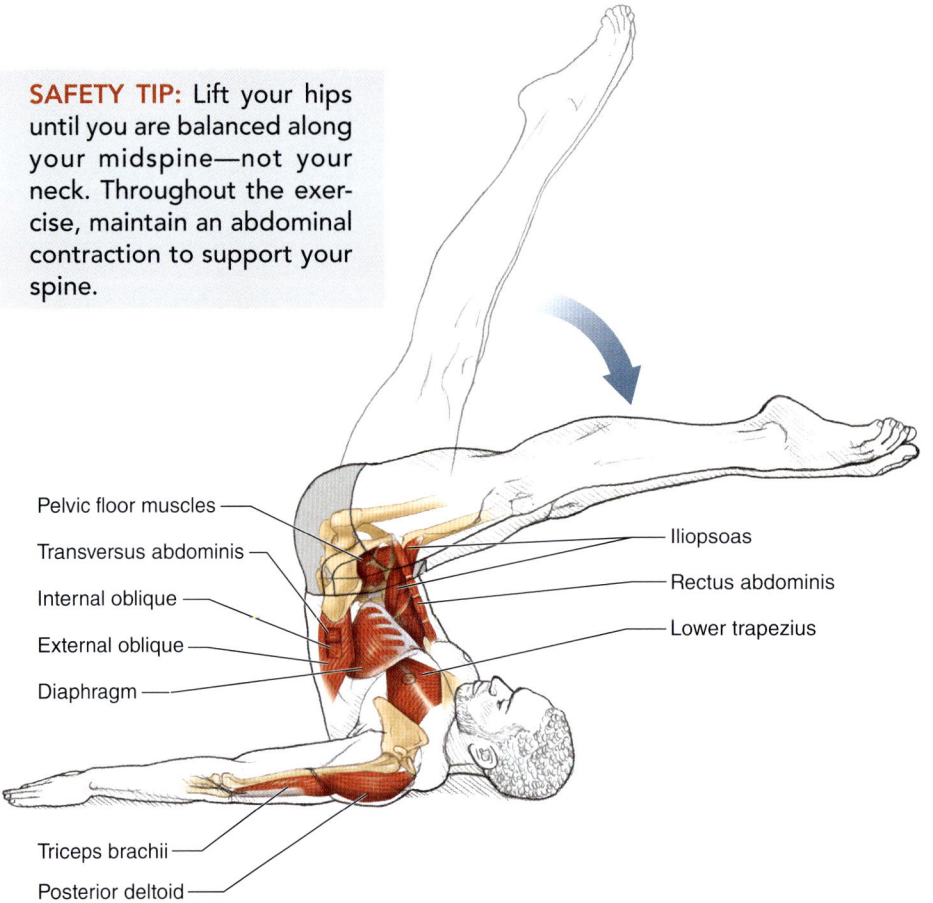

Pelvic floor muscles

Transversus abdominis

Internal oblique

External oblique

Diaphragm

Iliopsoas

Rectus abdominis

Lower trapezius

Triceps brachii

Posterior deltoid

EXECUTION

1. While supine, locate your neutral spine position. On exhalation, bring your knees to a tabletop or 90/90 position. Position your arms by your sides. Relax your neck and shoulders. Inhale to prepare.

2. On exhalation, engage your abdominals, extend, and straighten your knees with your hips remaining at 90 degrees. Begin to roll your pelvis up off the floor, emphasizing your abdominal contraction.

3. Elevate your hips until you have lifted to your midback or thoracic spine. Begin to extend your hips and lift both legs toward the ceiling. Try not to use your momentum or to press down onto the floor with your arms.

4. Inhale and, with control, gently begin to flex your hips, keeping your knees straight and allowing your legs to move along your sagittal plane until they are horizontal with the floor. Hold and stabilize your spine.

Feel a nice stretch through the lower back and hips. On exhalation, feel your abdominals contracting and hold a firm balance along your upper thoracic spine.

5. Inhale to prepare. On exhalation, roll back down to the starting position. Move slowly and with control while maintaining a strong abdominal contraction. Return to your neutral spine position and bring the legs to the original 90/90 position. Perform 10 repetitions.

MUSCLES INVOLVED

Transversus abdominis, rectus abdominis, internal obliques, external obliques, diaphragm, iliopsoas, pelvic floor muscles (levator ani, coccygeus), lower trapezius, posterior deltoid, triceps brachii

DANCE FOCUS

You never know what a choreographer will want you to do! Contemporary dance combines movement from many genres, including street dance, hip-hop, jazz, modern dance, and ballet. Contemporary choreography may include dancing off balance, in which case you need strong trunk-stabilizing skills. Contemporary work also uses creative positions for the spine, as well as different bases of support. You might be dancing on your knee, hand, or spine!

This exercise is a wonderful way to engage your core musculature for spine stability while practicing the use of a different base of support. It is also very effective for lengthening and stretching through your spinal and hip extensors, and it is an excellent lesson in control and balance.

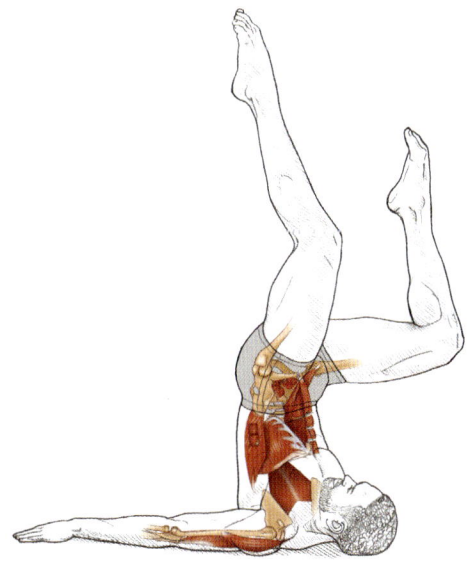

SIDE PLANK ON BALL

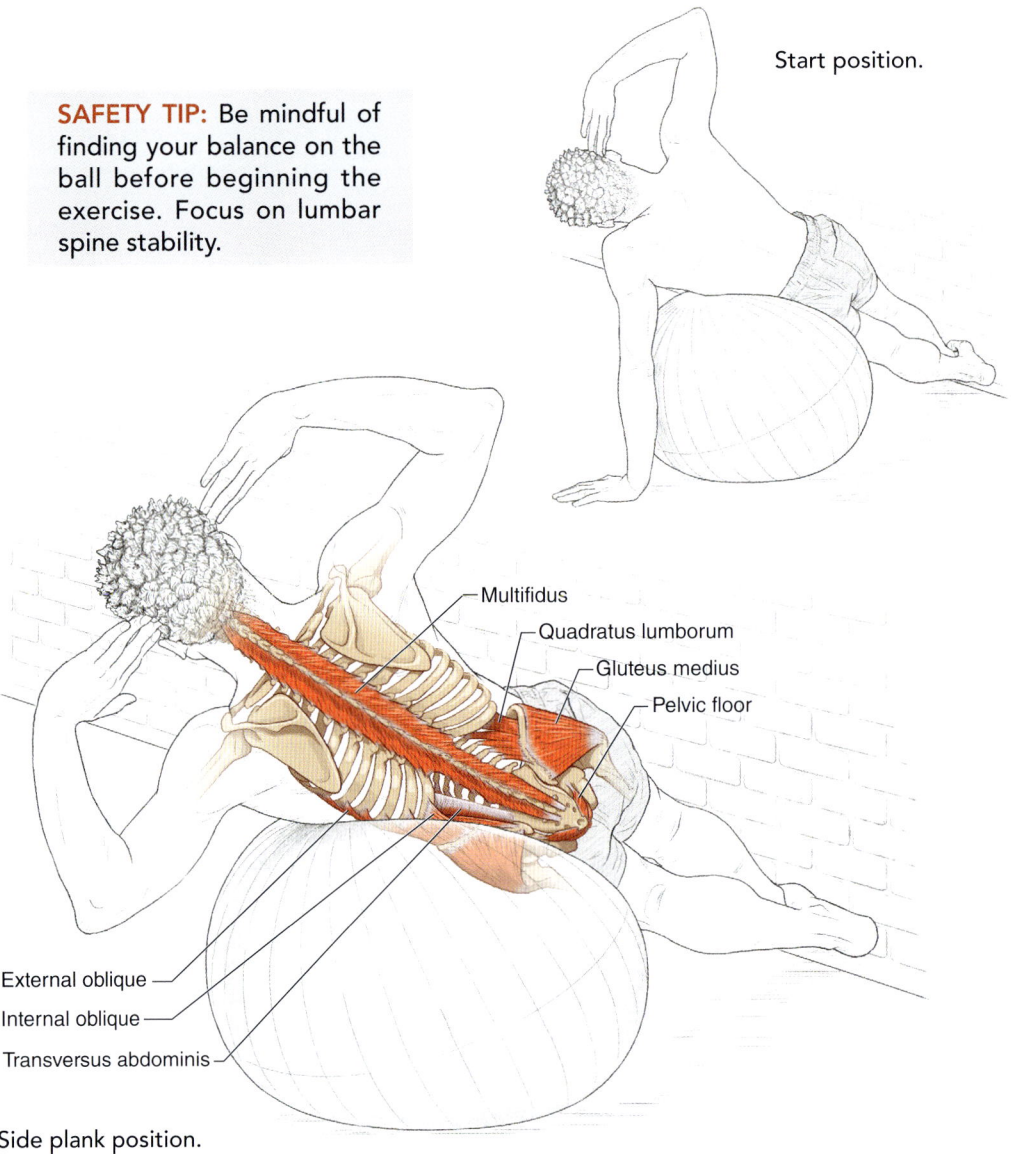

Start position.

SAFETY TIP: Be mindful of finding your balance on the ball before beginning the exercise. Focus on lumbar spine stability.

Multifidus
Quadratus lumborum
Gluteus medius
Pelvic floor

External oblique
Internal oblique
Transversus abdominis

Side plank position.

EXECUTION

1. Lie on an exercise ball on your left side. You can place your feet against the floor and wall to help stabilize you. Feel as if your hips and waist are supported by the ball.

2. Once you have found your placement, place your fingertips behind your ears. You should already begin to feel your core working to stabilize your spine on the ball.

3. Inhale as you begin to move into a right lateral side-bend, focusing on the transversus abdominis and the left internal oblique. You will also be using your multifidi and left quadratus lumborum.

4. Hold and exhale for an isometric contraction of 3 to 5 counts before slowly returning to your starting position. Practice this 8 to 10 times before going to the other side. As your skills improve, try holding for longer than 3 to 5 seconds to challenge your core and stabilizing ability.

MUSCLES INVOLVED

Transversus abdominis, internal obliques, external obliques, multifidi, quadratus lumborum, pelvic floor, gluteus medius

DANCE FOCUS

A strong core is important for all dance genres, but let's focus on break dancing. Break dancing—also called b-boying, b-girling, or breaking—requires agility, strength, and balance skills. There is an artistic and athletic element to break dancing. Criteria judges at break-dancing competitions and championships are looking for clean and sharp foundational skills, musicality, style, and execution of high-level power moves, to name a few. Break dancing has an element of danger to it, which makes it exciting to watch. Core exercises can help to prepare the body for the challenging skills required for break dancing. Your muscles and nerves make up your neuromuscular system; exercising using a stability ball utilizes your neuro-muscular system, engages multiple muscles, and improves balance skills. All the tricks, flips, or head spins in break dancing require neuromuscular control, core strength, and exceptional balance skills.

BEAR TO PLANK

Start position.

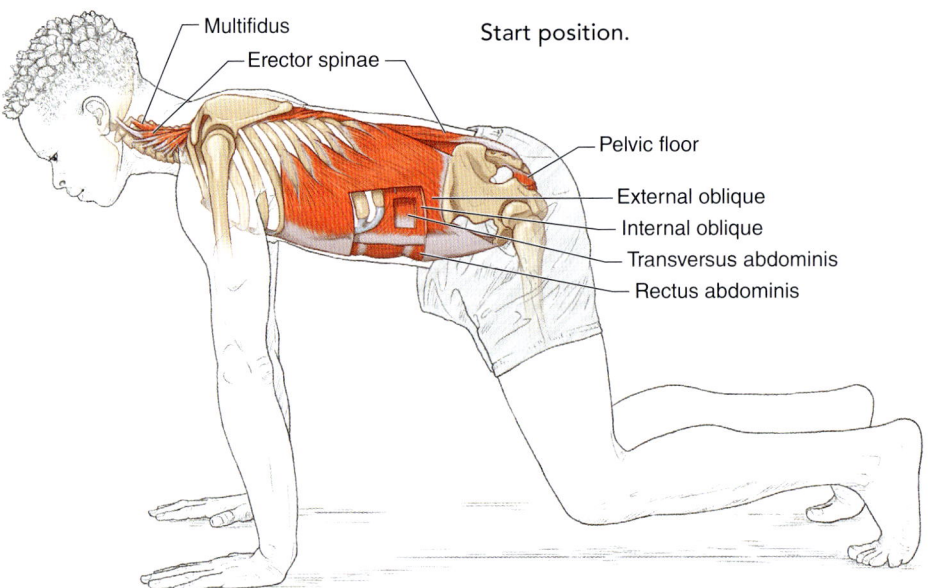

Multifidus

Erector spinae

Pelvic floor

External oblique

Internal oblique

Transversus abdominis

Rectus abdominis

Bear.

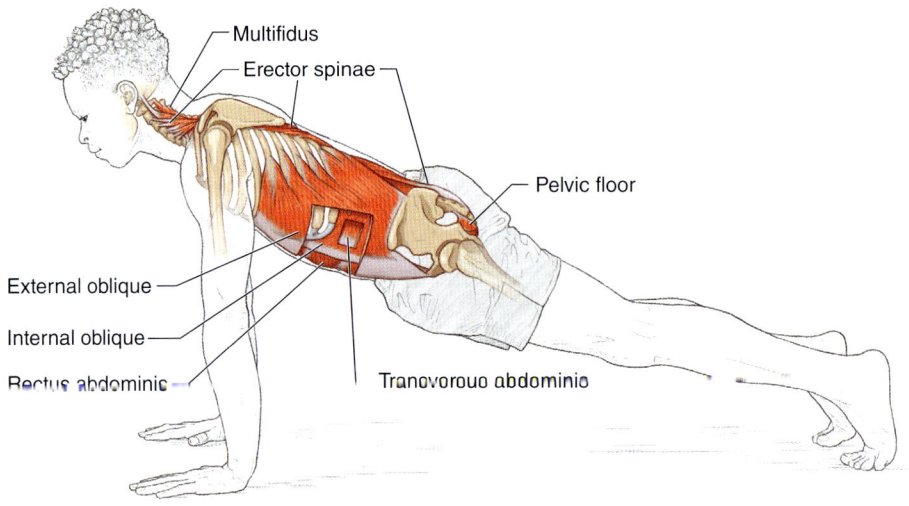

Multifidus

Erector spinae

Pelvic floor

External oblique

Internal oblique

Rectus abdominis

Transversus abdominis

Plank.

EXECUTION

1. Begin in a traditional hands-and-knees (quadruped) position, being mindful to align your shoulders over your wrists and your hips over your knees. Inhale to prepare. On forced exhalation, begin to tighten your transversus abdominis, and elevate your knees off the floor to the bear position. Allow the knees to hover and hold the position for 2 or 3 counts.

2. Inhale to prepare. On exhalation, move your legs one at a time into a plank position, maintaining strength in your core. Focus on supporting the lumbar spine to avoid spine extension. Hold for another 2 or 3 counts.

3. Inhale to prepare. On exhalation, alternate lifting your arms one at a time overhead to increase the challenge of the plank. Focus on supporting the spine. Breathe in and move the legs back into the bear position to begin the exercise again. Repeat 8 to 10 times. You can also increase the hold times as long as you can maintain alignment.

MUSCLES INVOLVED

Transversus abdominis, internal obliques, external obliques, rectus abdominis, multifidi, erector spinae (iliocostalis, longissimus, spinalis), pelvic floor

DANCE FOCUS

Moving from the bear to the plank position provides a more dynamic challenge to maintaining your neutral spinal alignment while adding some creativity. Many street-style movements require creative acrobatic skills interspersed between standing and floor work. Many contemporary styles use drops to the floor, in which it is challenging to maintain control of the spine. Strength in the core is important for reducing injury risk to the spine during all acrobatic-type dance movements. The bear to plank exercise is a bodyweight exercise, which can simulate some challenging choreography that you might be asked to perform. Multiple muscles are targeted while performing this exercise, which is beneficial for spine stability.

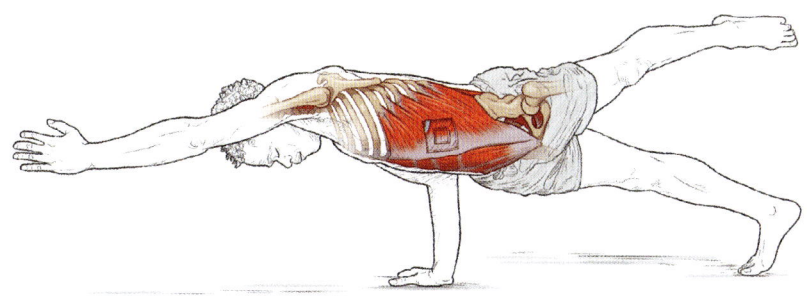

SIDE PLANK WITH PASSÉ

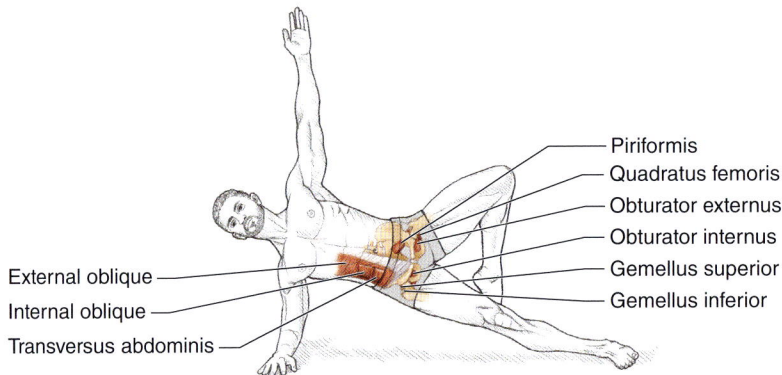

Piriformis
Quadratus femoris
Obturator externus
Obturator internus
Gemellus superior
Gemellus inferior

External oblique
Internal oblique
Transversus abdominis

Side plank position.

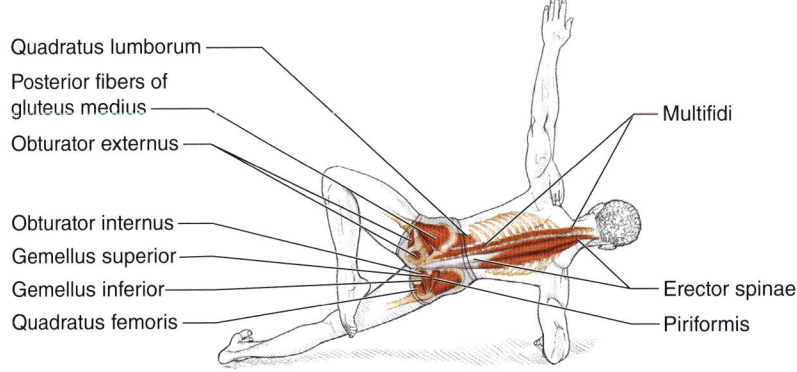

Quadratus lumborum
Posterior fibers of gluteus medius
Obturator externus

Obturator internus
Gemellus superior
Gemellus inferior
Quadratus femoris

Multifidi

Erector spinae
Piriformis

With leg in passé.

SAFETY TIP: Avoid sinking into the shoulder joint of the supporting arm. Maintain a lifted feeling in your trunk as you press down the supporting shoulder blade.

EXECUTION

1. Lie on your right side with your legs extended and one leg resting on the other. Your upper body is raised but supported by your right elbow. Allow your right forearm to be placed forward. Focus on your core; feel your center and your balance. Keep your shoulders down.

2. Inhale to prepare. As you exhale, pull your right shoulder blade down; activate your trunk muscles and elevate your hips. Focus on your center and work within your frontal plane to locate your balance. Maintain this position for 2 to 4 counts.

3. Once you balance and have secured the shoulder down and away from your ears, draw your top leg into passé while controlling your spinal movement. Engage in comfortable breathing during the passé lift. Slowly bring your leg back down and repeat. Perform the passé lift 3 to 5 times. Execute on the other side.

4. As you inhale, return to the start position with control; do not allow gravity to simply drop you to the floor. Feel your navel pulling in toward your spine for security. Visualize your internal and external obliques providing support along your rib cage.

MUSCLES INVOLVED

Core: Transversus abdominis, external obliques, internal obliques, quadratus lumborum, erector spinae (iliocostalis, longissimus, spinalis), multifidi

Hip external rotators with passé execution: Obturator externus, obturator internus, piriformis, quadratus femoris, gemellus superior, gemellus inferior, posterior fibers of gluteus medius

DANCE FOCUS

Planks and variations of planks continue to be an effective and versatile exercise for targeting multiple muscle groups. Again, for our purpose in this chapter, we are focusing on the core. Planks can be done anywhere; you don't need extra equipment or a gym, because you are relying on your own body weight. Variations of the plank can be executed to simulate various dance movements. For example, lifting one leg at a time can be like executing an arabesque. Your goal is to stabilize the lumbar spine as the leg moves into hip extension. You can also move one leg into a passé position to challenge pelvic stability and the turn-out muscles.

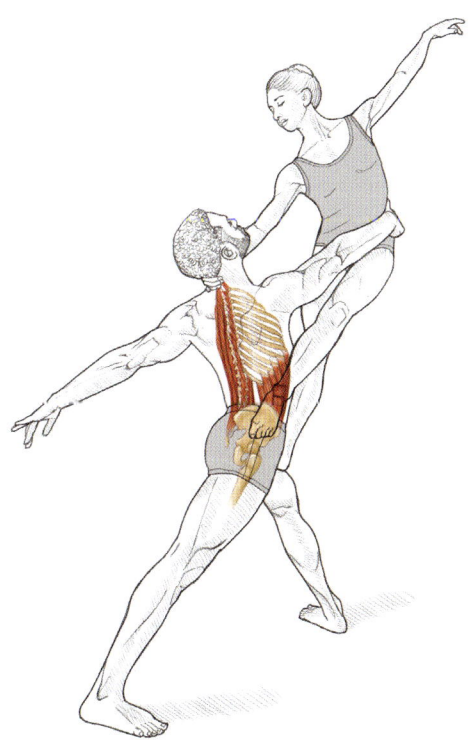

Side plank variations also increase the challenge by changing the base of support and continuing the focus on your core muscles. Dancers who perform floor work in contemporary dance styles, hand stands in acro dance, and groundwork in various street dance styles will benefit from performing plank exercises.

LUNGE CHALLENGE

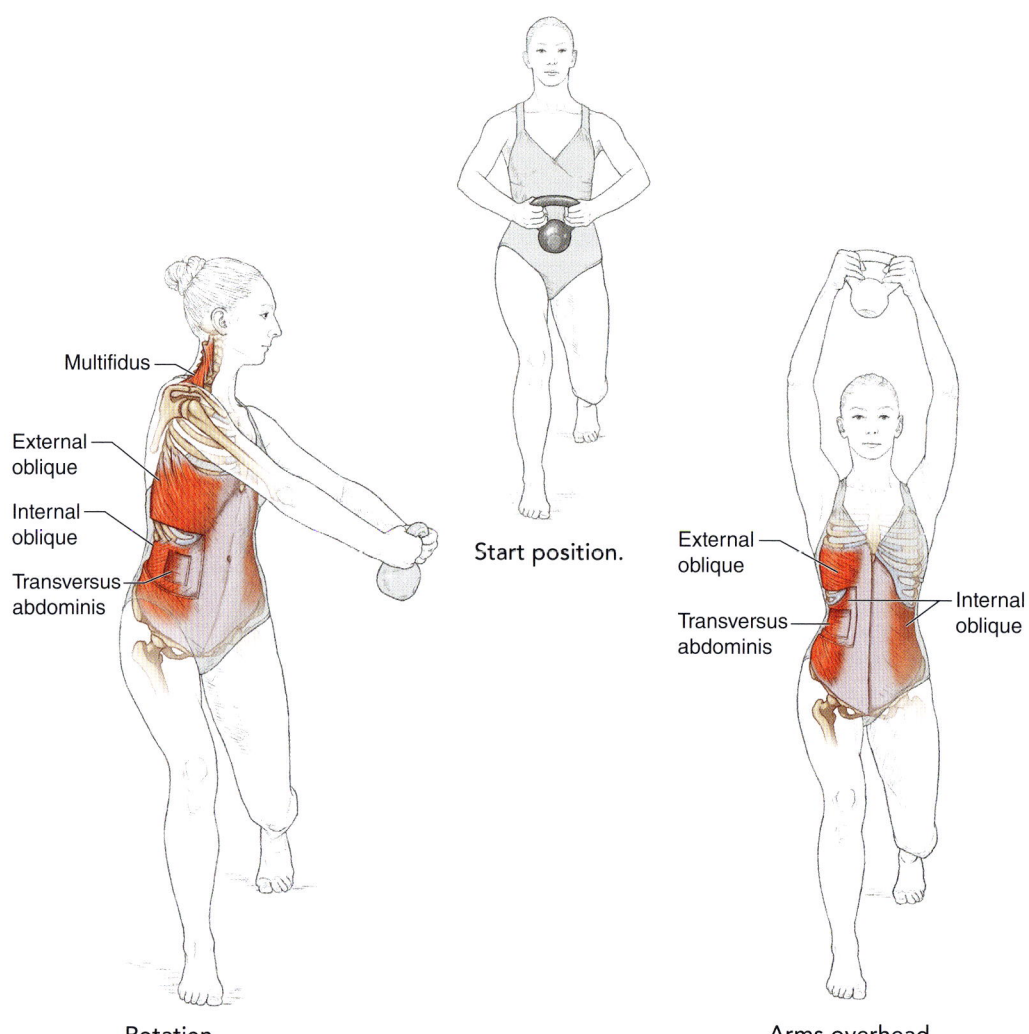

Start position.

Rotation.

Arms overhead.

EXECUTION

1. For this exercise, hold a 3- to 5-pound (1.4-2.3 kg) free weight or weighted ball in both hands at approximately chest height. Inhale to prepare. As you exhale, step forward into a lunge with your right foot, holding your arms forward along the sagittal plane.

2. Hold the lunge as you inhale. On exhalation, rotate your trunk with your arms over your left leg, working on lumbar spine stabilization and a strong contraction through your oblique muscles. Inhale as you return your arms to the front.

3. Stay in your lunge. On exhalation, lift both arms overhead, being mindful to stabilize your spine by contracting the deep transversus abdominis. Move your arms back to your start position as you return to standing. Repeat on the other side. Alternate trunk rotation with the arms and elevation of the arms while in the lunge to challenge your core. If you can maintain alignment, you may increase the weight of the free weight or ball.

MUSCLES INVOLVED

Core: Transversus abdominis, external obliques, internal obliques, multifidi

DANCE FOCUS

Lunge exercises can improve power, agility, and speed. For this chapter, our focus is the core. As you add the rotation of the trunk during the lunge, focus on the obliques. Turning combinations can be improved by learning to engage and strengthen your obliques to aid in balance and posture. As you begin a turning movement, think about the fibers of the oblique muscles firing to help brace your spine and hold you stable in your turns. The oblique muscles are also important for executing modern spiraling movements. The same principle applies if you are jumping in the air, like a tour jeté or calypso leap. Visualize the oblique muscles contracting to support your spine. The lunge challenge exercise emphasizes multijoint movement while challenging your core.

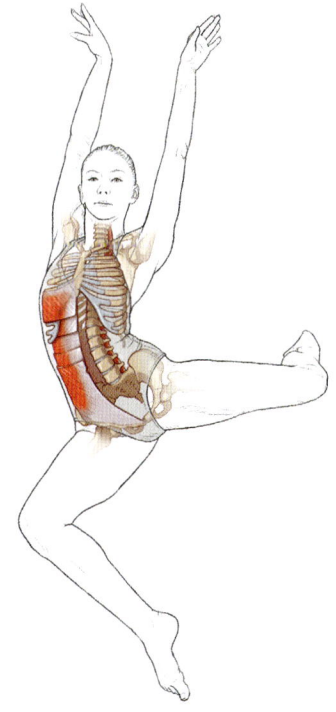

FUNCTIONAL OBLIQUES IN SECOND POSITION

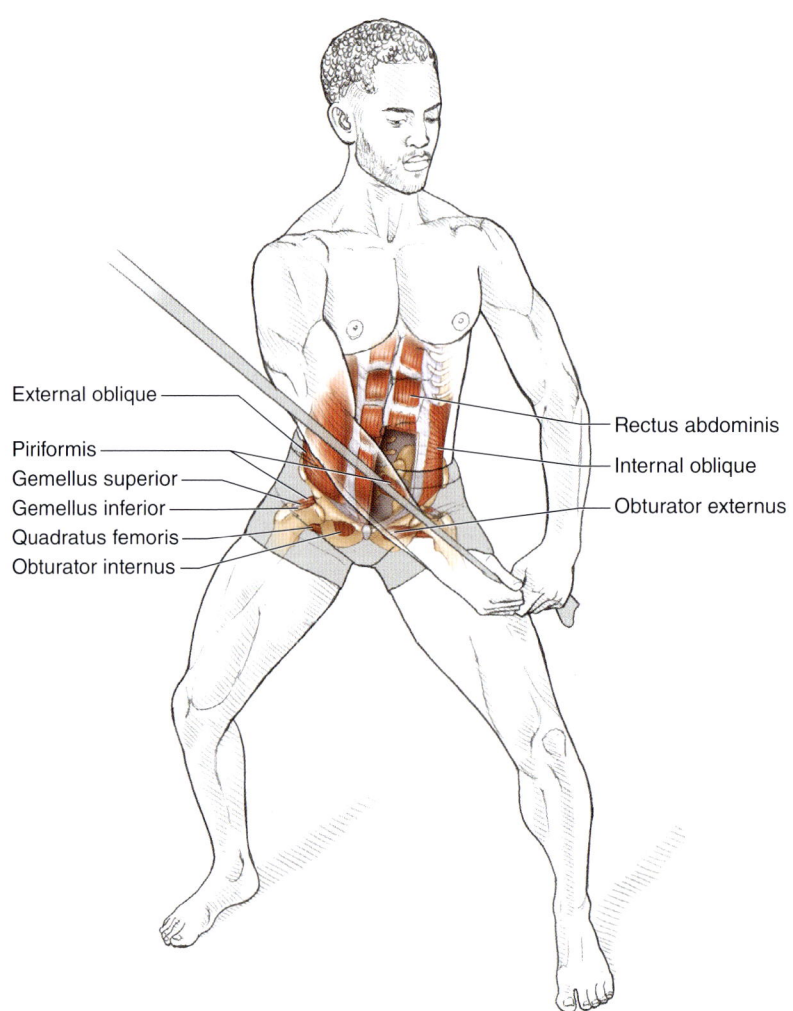

External oblique

Piriformis

Gemellus superior

Gemellus inferior

Quadratus femoris

Obturator internus

Rectus abdominis

Internal oblique

Obturator externus

EXECUTION

1. Stand in a firm second position. In both hands, hold a band resisted from above your right side, a weighted ball, or a hand weight. Locate your neutral spinal position and place your feet more than hip-width apart. Feel axial elongation through your center.

2. As you inhale, perform a demi-plié or move into a squat position, opening your thighs directly over your toes. Hold for 2 or 3 counts. On forced exhalation, come up from the squat and pull the band diagonally from above and across your body to your low left side as you slightly flex your trunk with rotation to the left. To help with the rotation, contract your oblique muscles. Keep pulling your shoulders down and away from your ears and keep your legs firmly turned out directly over your toes.

3. Allow for controlled movement along the entire spine. Hold for 3 counts. Emphasize the oblique contraction. Maintain your firm balance in second position and hold your pelvis stable.

4. On inhalation, begin the plié or squat and bring your arms back to the start position to begin again. Work up to 10 repetitions on this side, then repeat on the other side.

MUSCLES INVOLVED

Core: Transversus abdominis, internal obliques, external obliques, rectus abdominis, pelvic floor

Hips during plié: Deep hip rotators (obturator externus, obturator internus, piriformis, quadratus femoris, gemellus superior, gemellus inferior, posterior fibers of gluteus medius)

DANCE FOCUS

All turning movements require power through your torso; therefore, strength throughout the obliques helps you accomplish more refined turns. Modern choreography involves a lot of floor work with lateral and rotational motions; fall-and-rise techniques also need support from the oblique muscles. In addition, strong obliques increase the effectiveness of jazz warm-ups that focus on isolations. In addition to the internal and external obliques and the rectus abdominis, this exercise also uses the transversus abdominis, diaphragm, and pelvic floor muscles of the core and the deep hip rotators (obturator externus and internus, piriformis, quadratus femoris, gemellus superior and inferior, and posterior fibers of gluteus medius).

Each time you perform this exercise, focus on the navel-to-spine and abdominal bracing principle, which gives you added support for the lower segments of your spine, as well as a toned waist.

Think back to your plumb line posture; the oblique muscle fibers are positioned very well to help with proper alignment between the thoracic region and the pelvis. They tend to be overlooked because the rectus abdominis is given so much attention. To remedy this oversight, balance your exercise program to challenge all the core muscles.

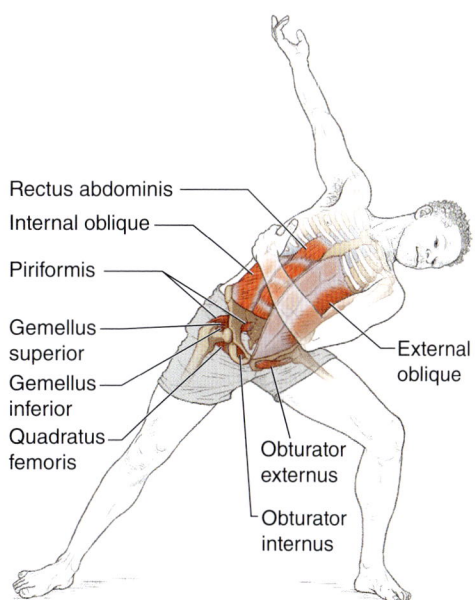

Rectus abdominis

Internal oblique

Piriformis

Gemellus superior

Gemellus inferior

Quadratus femoris

External oblique

Obturator externus

Obturator internus

FUNCTIONAL TRUNK TWIST

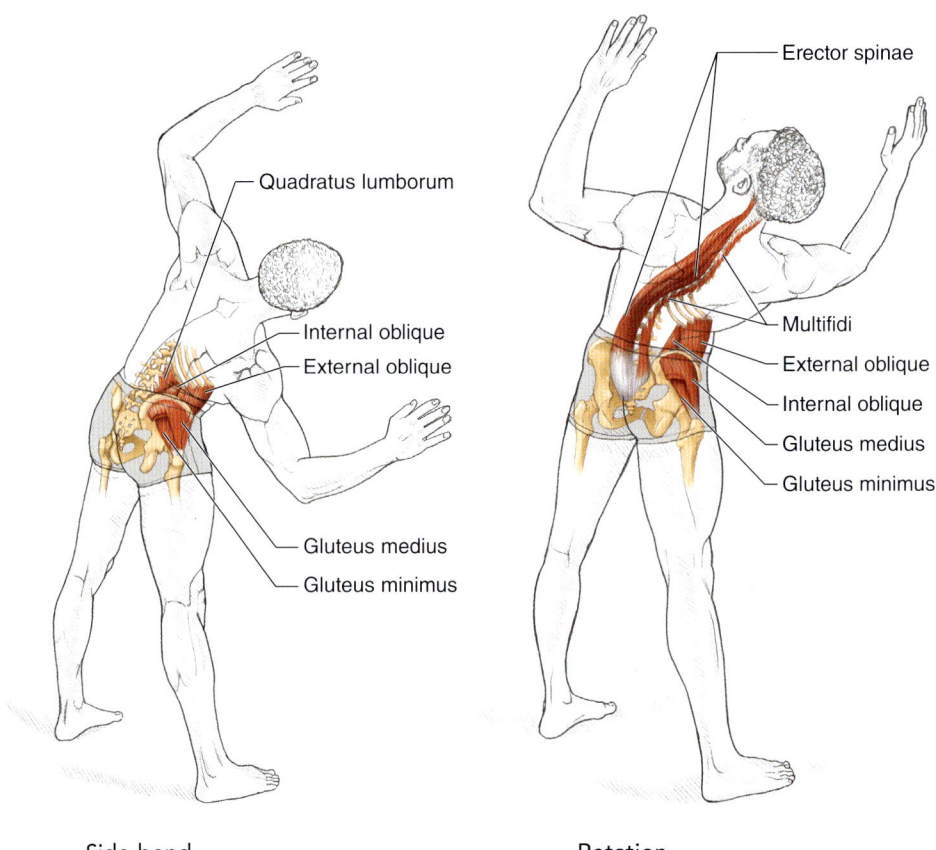

Side-bend.

Rotation.

EXECUTION

1. Stand with your legs parallel and your arms overhead in a high fifth position. Locate your neutral position, creating axial elongation. Move into a squat position with your trunk slightly flexed forward. Inhale to prepare.

2. On exhalation, move to an upright neutral spine position. Lift through the center and continue into a right side-bend along your frontal plane. Your arms and shoulders will now move into a 90/90 position. Visualize movement throughout the thoracic spine.

3. The movement continues into left rotation. Maintain control in your trunk and pelvis. Emphasize abdominal control while your spine side-bends and rotates.

4. Let your waist move along your transverse plane, opening the left shoulder. Allow your head and neck to follow. Maintain width through the chest and shoulders.

5. With control, inhale to reverse the movement and return to the squat starting position with the arms in high fifth. Repeat on the other side. Perform a total of 4 to 6 repetitions on each side. To increase the challenge, hold small weights in each hand.

> **SAFETY TIP:** Reemphasize lower-back support throughout the exercise to protect the lower segments of your spine. Maintain a firm stance and strong balance. This exercise involves full weight-bearing multijoint movement to emphasize abdominal control with segmental spinal movement.

MUSCLES INVOLVED

Side-bend: External obliques, internal obliques, quadratus lumborum

Rotation: Multifidi, erector spinae (iliocostalis, longissimus, spinalis), external obliques, internal obliques

Pelvic stability: Gluteus medius, gluteus minimus

DANCE FOCUS

Let this movement broaden your awareness. Don't just focus on how far you can bend to the side; instead, focus on articulation through every vertebra in your spine. While moving your trunk sideways to the right, you must also stabilize your pelvis so that your hip doesn't hike on the left side. This exercise promotes a skill used in jazz warm-ups requiring you to create isolations by separating certain body sections from others. In ballet, the grande cambré en rond requires effective movement through your upper back and stabilization through the pelvis.

Remember your planes of motion; stay within your frontal plane as you move to the side. Many dancers tend to let the lower back arch, open the ribs, and move forward off the frontal plane. To avoid this pitfall, visualize the four layers of your abdomen and the various directions of the muscle fibers creating your brace. The movement becomes more challenging when you add rotation with extension; now you are moving along your transverse plane. Notice how clean and organized the movement is when you visualize your planes and elongate through the spine and chest. Notice the beautiful curve that your spine creates.

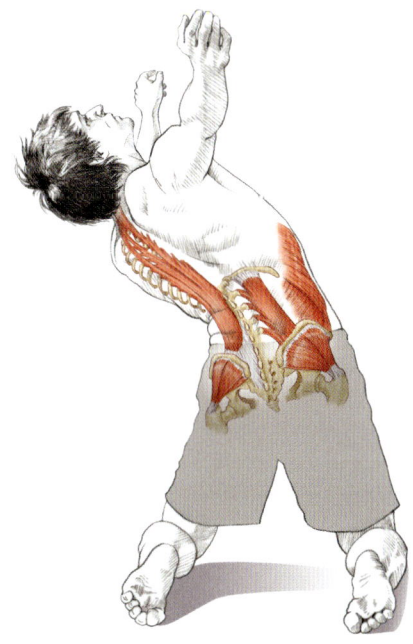

ABDOMINAL STRETCH

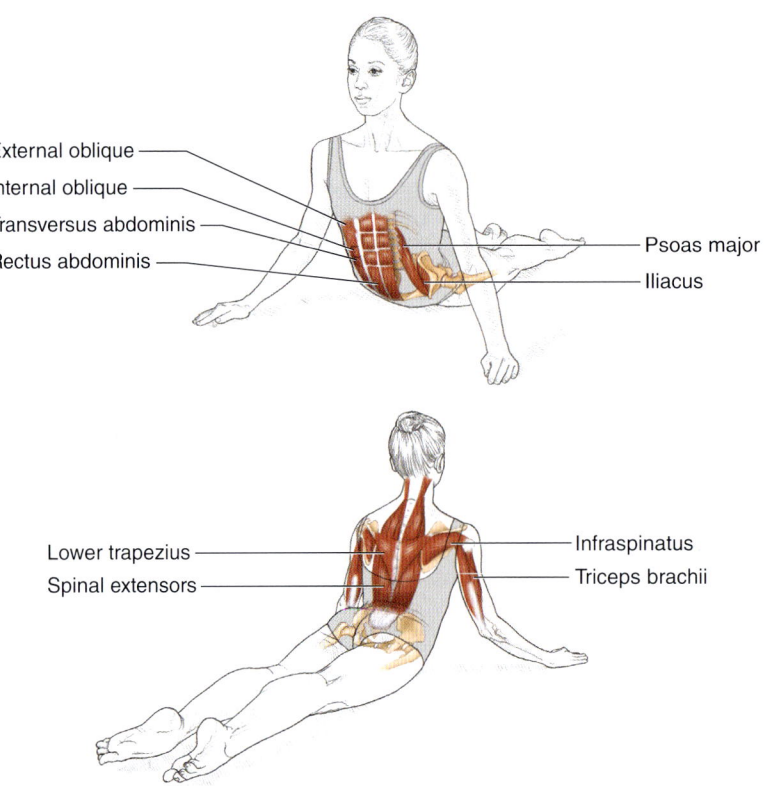

External oblique
Internal oblique
Transversus abdominis
Rectus abdominis
Psoas major
Iliacus

Lower trapezius
Spinal extensors
Infraspinatus
Triceps brachii

EXECUTION

1. Lie prone with your palms on the ground and your elbows bent and close to your sides. Glide your shoulder blades down toward your hips.

2. As you inhale, lengthen through your spine and press your hands into the ground to lift your chest off the ground. Keep the abdominals engaged toward the spine.

3. Move along the longest possible arc through your thoracic spine, avoiding overuse in the lower spine.

4. Continue, lifting your hips off the floor as long as you are supporting your entire spine and not just moving in the lower segments of the spine. Feel a nice stretch along the rectus abdominis.

5. On exhalation, reverse and bring yourself back down to the start position with control. Perform 4 to 6 times.

MUSCLES INVOLVED

Shoulders: Lower trapezius, triceps brachii, infraspinatus

Trunk: Transversus abdominis, internal obliques, external obliques, rectus abdominis, iliacus, psoas major, spinal extensors

DANCE FOCUS

As you have seen, the abdominals need to be strengthened for spine stability; they also need to be stretched. Beautiful extension of the spine is required, for example, in arabesque, cambré derrière, and tour jeté. To help you acquire that strong yet beautiful arch along your spine, this exercise emphasizes spinal extension through the entire spine with abdominal lengthening. You are using your arms to assist in the stretch and using your abdominals to support your spine. This exercise can also help you maintain posture as you get older.

CAMBRÉ SIDE

Cambré is a classical ballet movement that involves arching the back or bending at the waist to the front, side, or back. Let's break down an efficient cambré-side by emphasizing spinal stability, axial elongation, and effective breathing.

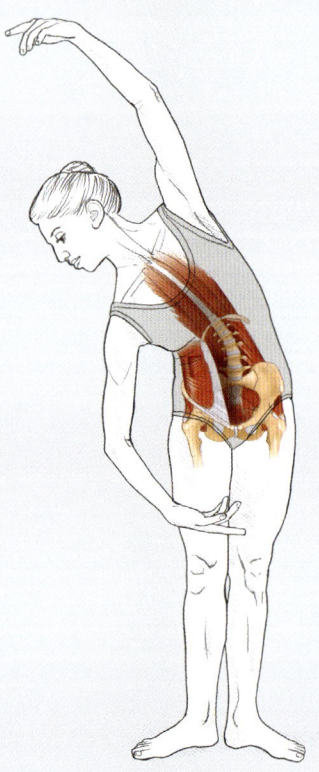

1. Stand in first position with your arms in second position. Locate your neutral spine position with your legs turned out while maintaining a firm stance. Distribute your weight equally between the first metatarsal, fifth metatarsal, and heel of each foot.

2. Inhale as you lengthen through your spine, as if to grow a bit taller. Begin to bring your left arm into a high fifth position and your right arm into an en bas position. Maintain a navel-to-spine feeling in your low abdominals for lumbar stability as you begin to side-bend to your right. Allow your head and neck to follow the line of your spine.

3. Maintain a neutral position in your pelvis and hips without allowing any lateral shift or anterior tilt. Keep lengthening through your spine as you cambré-side directly along your frontal plane.

4. Control your rib cage without allowing your chest or ribs to lift. You may rotate your head slightly toward your right shoulder or leave it centered.

5. Maintain an abdominal contraction to support your lower spine and feel yourself moving in the longest possible arc. Your left arm should also maintain a nice long arc over your head while the shoulder blades continue to glide down toward your hips.

6. Upon exhalation, begin to return to your starting position, reversing the cambré while continuing the axial elongation along your spine. Maintain the abdominal contraction to support your spine throughout the movement.

Muscles Involved

Side-bend (flexors contracting concentrically during right side-bend and eccentrically during left side-bend once body is off center): Internal obliques, external obliques, quadratus lumborum, erector spinae (iliocostalis, longissimus, spinalis)

Pelvis: Gluteus medius, gluteus minimus, deep external rotators (obturator externus, obturator internus, piriformis, quadratus femoris, gemellus superior, gemellus inferior, posterior fibers of gluteus medius), pelvic floor muscles (levator ani, coccygeus)

CHAPTER 7

Shoulders and Arms

All forms of dance require arm work for power, aesthetics, balance, and momentum; your arms are also vital for turns and changes in direction. Teachers and choreographers may tell you, "Isolate your arms from your shoulders" or "Keep your shoulders down," but do you really understand these cues? This chapter focuses on efficiency of movement within the shoulder complex through scapular stability. Once you understand the coordination of arm movement with the upper body, your shoulders will be more secure, thus enabling your arms, elbows, and wrists to move freely with style and grace.

The shoulder joint is an intricate and very mobile joint, and the associated muscles that create stability are just as intricate. Even more detailed movement is allowed by the elbow and wrist, which enables you to create fluidity when you move your arms from one position to the next. When you strengthen the muscles that control the shoulder joint, you can feel more secure in partnering movements as well as skills performed on your hands or elbows. Even though many dance injuries involve the lower extremities, the shoulder should not be forgotten; it deserves its own share of the attention.

Bony Anatomy

The bones that make up the shoulder complex are the clavicle (collarbone), scapula (shoulder blade), and humerus (upper arm; see figure 7.1). The humerus continues down to the elbow joint, where it meets the radius and ulna. These two bones continue down to meet the carpus (wrist), metacarpus (hand), and phalanges (fingers).

The clavicle bone of the chest forms a joint at the medial portion where it meets the sternum. The lateral end of the clavicle meets the acromion process, which is a small, bony protrusion of the scapula. Together, the clavicle bones create a beautiful line across the front of the sternum and can be seen clearly through the skin. This area is typically where instructors guide you to open the front of the chest for the remarkable sensation of presenting yourself to the audience.

The scapula is the triangular bone that glides along the back of your ribs. It includes a shallow socket, the glenoid cavity, where the humerus bone inserts. The scapula has both an anterior surface (which lies against the ribs) and a posterior surface (which has a slightly elevated portion referred to as its spine). The lateral end of the spine of the scapula becomes the acromion process. The scapula also includes another bony protrusion known as the coracoid process, which is important for its numerous muscle attachments. The scapula itself is an amazing bone that has numerous muscle attachments and serves as an anchor for your shoulder. Typically, inefficient movement of the scapula can result in shoulder injury.

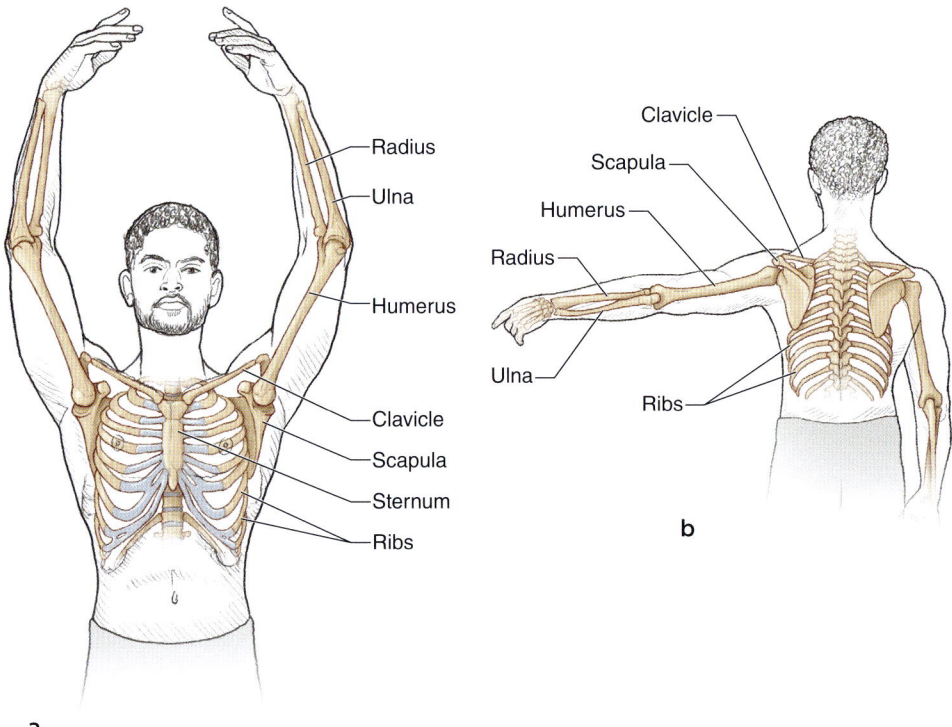

Figure 7.1   Bones of the shoulder complex: (a) front; (b) back.

Key Joint Motion

Several joints related to the shoulder complex can create movement. We will focus on two in particular: the scapulothoracic joint (where the scapula meets the thoracic spine) and the glenohumeral joint (where the humerus bone meets the glenoid cavity). In the scapula's position against the ribs, it can elevate, depress (travel downward), abduct/protract (move away from the center), and adduct/retract (move toward the center). It can also curve upward or move in a downward rotational pattern. You may have seen "winging" of the scapula, which occurs when the medial or lateral border of the scapula protrudes outward, thus giving the upper back a look of having small wings. Winging is displayed by some young, lean dancers with minor muscular imbalances that prevent the scapula from lying completely in contact with the rib cage. Scapula winging disrupts efficient movement patterns and can be a source of injury.

The glenoid cavity is a ball-and-socket joint that is held together by strong muscles. It is a relatively strong joint but has a shallow cavity; indeed, only one-fourth to one-third of the humeral head fits snugly into the glenoid cavity. The glenohumeral joint is capable of flexion and extension in the sagittal plane, abduction and adduction in the frontal plane, and internal and external rotation in the transverse plane. It can also move in horizontal abduction and adduction. Since the glenohumeral joint is not very deep, stability is important for reducing your risk of injury.

Take a moment to move your shoulders up and down. Visualize the movement occurring at each scapula and at the ribs. Move your arms to your sides and down again along the frontal plane. Visualize the movement of the scapula as it lies on the ribs. Rotate your humerus bone within the glenoid cavity; note the range of motion at this joint. The muscles that create movement at the glenohumeral joint connect between the humerus and the scapula. The muscles that allow movement to occur around the scapula connect between the scapula and the humerus, sternum, clavicle, spine, and ribs.

Strengthening the muscles that attach around the scapula will improve your upper-body placement and shoulder alignment. It will also allow the forces of energy and extreme range of motion to be distributed more efficiently through the glenohumeral joint. These improvements will give you better control and help you move more from your center. The basic dance warm-up may not provide you with enough shoulder stability needed for more demanding choreography. Therefore, many shoulder exercises are included in this chapter; use them both for warm-ups and for strengthening.

The joints between the humerus and the ulna and between the humerus and the radius work together as a hinge joint. Where the lower ends of the radius and ulna meet the carpal bones, a hinge-plus-rotary movement occurs, thus allowing for pronation (moving the palm down) and supination (moving the palm up). For some dancers, hyperextension (excessive movement past extension) at the elbow occurs when the arm and forearm are in a straight line. Hyperextension can stress the ligaments, especially when falling on an extended elbow. It's important for you to balance the strength between the elbow flexors and extensors to help control motion at the elbow joint.

This principle also comes into play with the numerous bones in the wrist. In particular, the scaphoid bone is at risk for injury during a fall and is difficult to see on an X-ray. Balancing flexibility and muscle tone along the forearm provides the beautiful fluidity needed for an elegant port de bras, creative contemporary arm work, strong partnering skills, and gesturing movements.

Muscle Mechanics

The beauty and style of your port de bras come from balanced and powerful shoulder musculature. You know how inspiring it is to create unique designs with your arms, but do you really know *how* to create the designs? Again, knowing which muscles to activate gives you a better understanding of each movement, which in turn enables more quality of movement.

Rotator Cuff

To understand the mechanics, let's break down the two primary joints that create movement within the shoulder. The glenohumeral joint is stabilized by four deep muscles known as the rotator cuff muscles (figure 7.2): the supraspinatus, infraspinatus, teres minor, and subscapularis. Their attachments connect the humeral head with the scapula and allow for stability, some rotational movement, and abduction. The supraspinatus, infraspinatus, and teres minor work together to create an amazing force that secures the shoulder joint and thus keeps the humerus bone from pinching against the acromion when you lift your arms. If the rotator cuff muscles are weak, however, this force is insufficient to create security for your shoulder joint. The resulting chronic pinching produces pain and swelling and can lead to impingement syndrome.

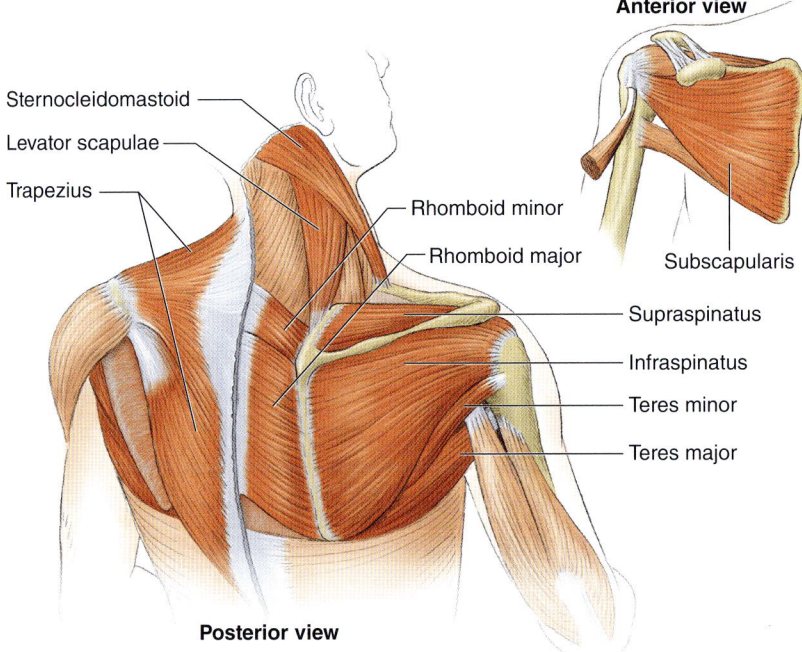

Figure 7.2   Muscles of the scapula and rotator cuff.

Scapula

You have learned that the scapula moves in many planes at the scapulothoracic joint. When the humerus begins to move, it rises first, followed by the scapula. For example, when lifting your arm into forward flexion, you have approximately 45 to 60 degrees of glenohumeral movement before the scapula begins to move. When lifting your arm to the side, you have about 30 degrees of glenohumeral movement before the scapula moves. The ratio of glenohumeral movement to scapula movement is two to one. Your shoulder blade and upper arm must work together within this ratio to keep the humerus bone from pinching against the acromion. If the muscles that connect to the scapula are weak, the scapula will be ineffective in its job of creating control for your shoulder joint. If you work on strengthening the muscles specified in the following discussion, the scapula will have a better chance of anchoring your arm movements.

Certain muscles play an essential role in upper-body placement by anchoring the scapula and creating efficient movement. The trapezius muscle originates at the base of the skull. The upper portion of the trapezius has attachments along the spinous processes of the first six cervical vertebrae. The lower portion of the trapezius has attachments along the spinous processes of the seventh cervical vertebrae and all 12 thoracic vertebrae. It has insertions on the lateral clavicle, upper acromion, and upper scapular spine. The trapezius is divided into upper, middle, and lower segments. If the upper trapezius is stronger than the other two segments, the shoulders elevate, which creates tension, imbalance, and fatigue. The tension can throw off jumps, turns, and balancing combinations. The lower and middle segments of this muscle are responsible for creating balance by bringing the shoulder blades down and inward. When you need to pull your shoulders down, think about gliding the scapulae down. You can also visualize this image when you are turning, lifting a partner, holding a prop, or raising your arms.

The levator scapulae and rhomboid muscles are located deep under the trapezius. They originate along various cervical and thoracic vertebrae and insert into the superior medial edge of the scapula. Because of the attachment location, these muscles can elevate the scapula, create downward rotation, and provide scapula adduction. The serratus anterior muscle connects the first to 8 and 9 ribs to the scapula, and the pectoralis minor connects ribs 3 through 5 to the scapula (figure 7.3). A winging scapula is related to weakness of the serratus anterior and lower trapezius muscles.

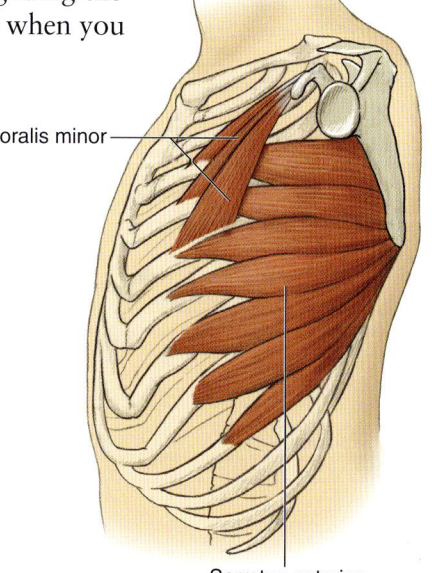

Pectoralis minor

Serratus anterior

Figure 7.3   Muscles with rib attachments.

Glenohumeral Muscles

The muscles that connect the humerus bone to the trunk are responsible for the larger dynamic movements of your arms. The pectoralis major is the large, superficial muscle in the front of your chest that connects the humerus bone to the sternum, clavicle, and various ribs (figure 7.4*a*). This muscle can pull your arms forward and together; for instance, in almost all turning combinations, it pulls the arms inward, thus generating some of the coordinated power for the turn.

The deltoid muscle divides into three sections: anterior, middle, and posterior. Each respective section creates movement to the front, side, or back. Hiding under the anterior deltoid and pectoralis major is the coracobrachialis, which is small yet capable of producing shoulder flexion and adduction.

The latissimus dorsi is the large muscle of the back that connects the humerus to the last six vertebrae of the thoracic spine, the five lumbar vertebrae, the ilium, the sacrum, and the lower three ribs (figure 7.4*b*). This muscle creates adduction, internal rotation, extension, and humerus depression.

Now that you know the significance of each muscle of the shoulder complex, you can understand how important it is to balance strength and flexibility in these muscles. That balance enables you to create the detailed and elaborate movements that dance choreography puts you through.

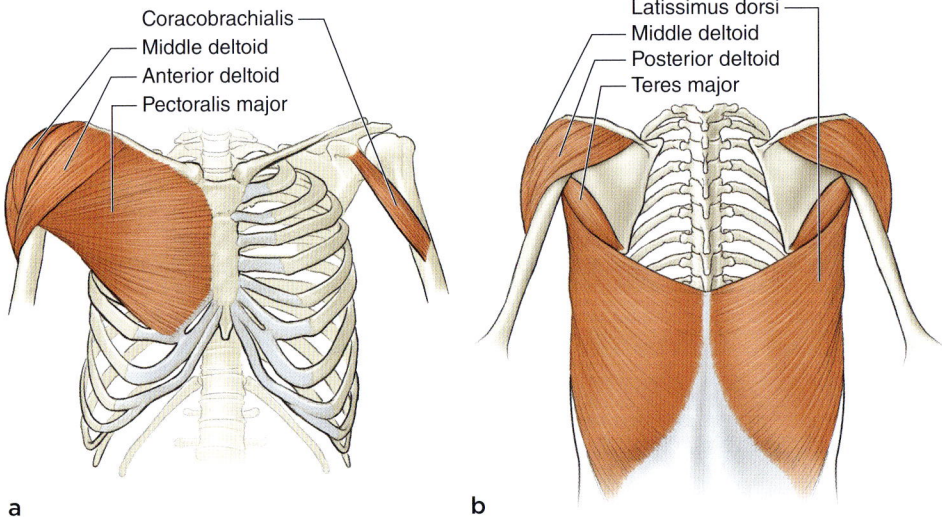

a

b

Figure 7.4   Glenohumeral muscles: *(a)* front; *(b)* back.

Arm Muscles

The elbow joint can flex and extend, and it is controlled by specific muscles that create those movements. The biceps brachii (figure 7.5a) flexes the elbow and connects the scapula with the radius. The triceps brachii (figure 7.5b) extends the elbow and the shoulder; it connects the scapula and upper humerus with the ulna. The biceps muscle has two originating attachments, or heads, and the triceps muscle has three. Hiding under the biceps is the brachialis, which connects the lower humerus with the ulna.

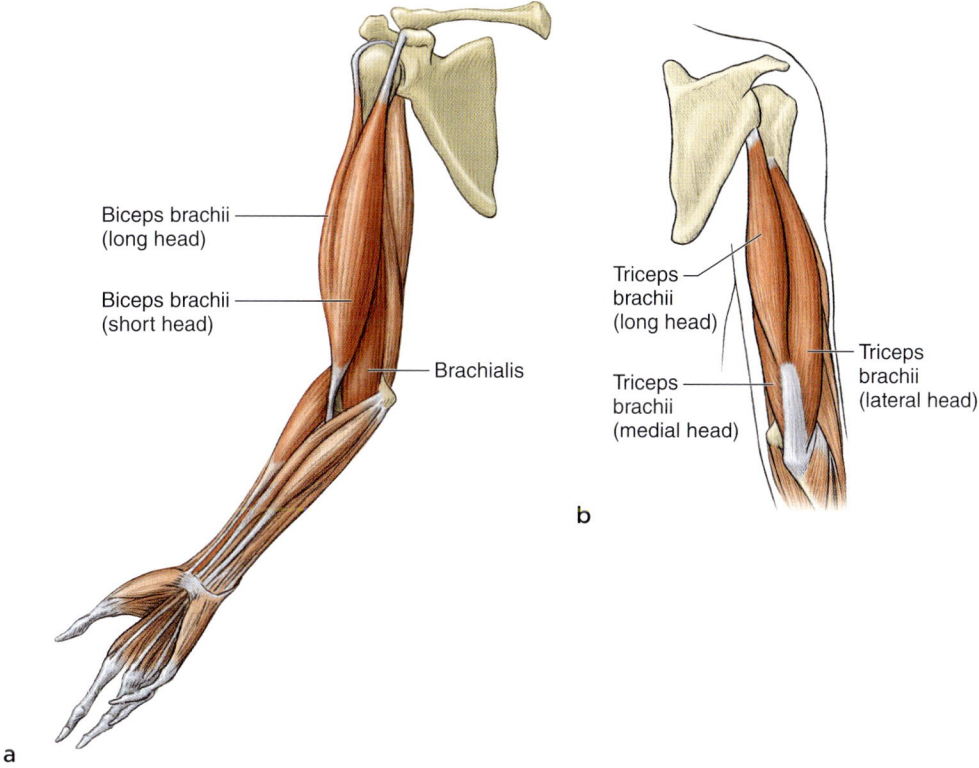

Figure 7.5   Muscles of the upper arm: (a) biceps; (b) triceps.

The forearm musculature (figure 7.6) allows for pronation and supination as well as flexion and extension of the wrist. Strengthening these various small muscles is important for some extreme choreography—for example, standing on your hands, lifting another dancer, or falling on your hands. Forearm strength is important for holding props and for partnering skills, and many of the various styles of couples dancing require coordinated movements of the hands and forearms. The exercises presented in this chapter help you stabilize the shoulder, elbow, and wrist.

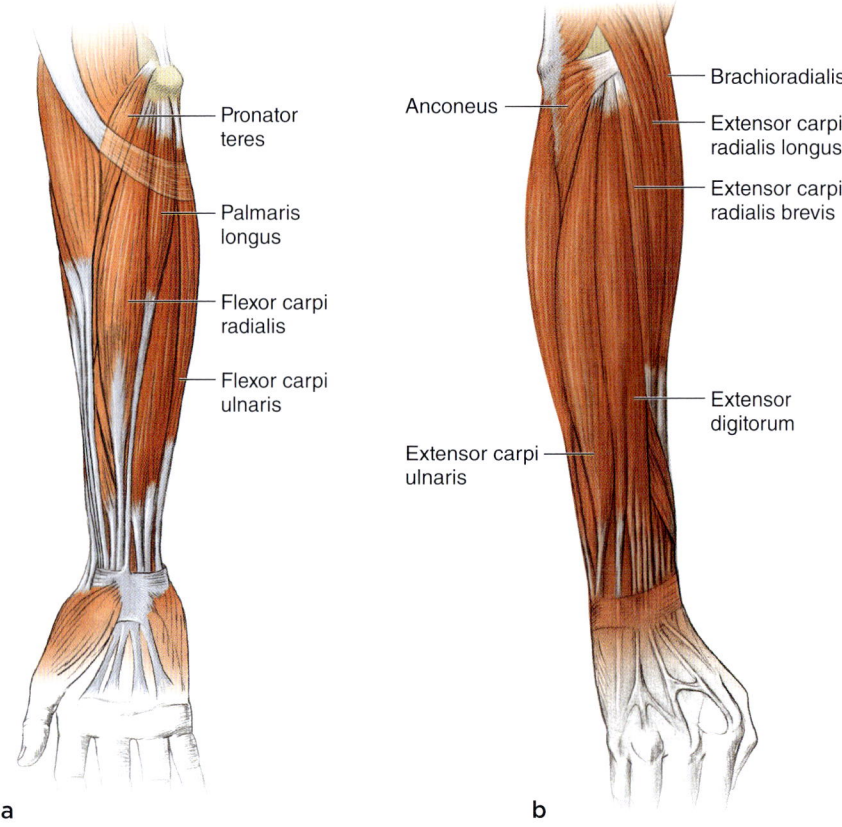

a   b

Figure 7.6   Muscles of the forearm: *(a)* flexors; *(b)* extensors.

Carriage of Arms

Carriage of the arms, termed *port de bras* in classical ballet, completes the movement in every style of dance. All port de bras movement should be fluid while incorporating scapular stability. When the arms move up into a high fifth position, the primary movers are the anterior deltoid and pectoralis major; the scapula must stabilize and move in an upward rotational pattern, not elevate. Balanced movement of the scapula and humerus is allowed by activation of the serratus anterior and lower trapezius. We tend to lift the arms with only limited control, thus allowing the humerus and scapula to elevate and leading to overuse of the upper trapezius muscles. Remember your two-to-one ratio of glenohumeral movement to scapula movement, stabilize the scapula, engage the lower trapezius and serratus anterior, and allow the humerus to move freely.

This strategy applies to all dance techniques and training; for instance, the same two-to-one principle is required for contemporary jazz movements in hip-hop choreography. Irish dancers, who dance primarily with the arms planted neatly at their sides, must also secure the upper body; that is, their scapulae must be anchored to their posterior ribs. In addition, since the elbows are fully extended, the triceps must be strong. To keep the arms securely at the sides, the pectoralis major must hold firm in an isometric contraction. All the scapular muscles are contracted to stabilize the shoulder blades.

Traditional modern dance takes the arms past their normal range of motion. Here, the arms are expected to perform in flexion, extension, internal and external rotation, and variations of all these positions. To consider one example, when you move your arm and shoulder into extension, the posterior deltoid and latissimus dorsi contract and the scapula needs to rotate downward and adduct slightly; therefore, the rhomboid and lower trapezius need to contract. Now you can see how important it is to strengthen the muscles throughout the upper body.

According to a 2016 study from the University of Birmingham by Abichandani and Hule, break dancers suffer knee, lower back, and shoulder injuries. A questionnaire was administered to break dancers between the ages of 17 and 30. Over 16% reported suffering from shoulder injuries. B-boy shoulder injuries typically result from the frequent falling and balancing of body weight on the hand, elbow, shoulder, or neck. The increased load on the muscles of the shoulder and forearm increases the risk of injury. Shoulder exercises for scapula stability, strengthening, and movement control such as those featured in this chapter can help to decrease injury risks.

Dance-Focused Exercise

In some dancers, the serratus anterior, rhomboid, and lower trapezius muscles tend to be weak. Therefore, many exercises presented in this chapter include additional repetitions to improve your strength; however, do not increase the number of repetitions if you are unable to maintain excellent form. Table 7.1 summarizes the muscles worked by the exercises. Focus on the alignment of the shoulder joint and ease in the neck and upper shoulders. To incorporate your core in the exercises, use the breathing patterns covered in chapter 5. When breathing, remind yourself to move the ribs in a three-dimensional pattern. Once you begin to feel stronger, you will find yourself working more efficiently from your center. Your instructors will also see improvement in how you incorporate corrections from their cueing.

When you receive a cue such as "Isolate your arms from your shoulders," remember that your scapula has numerous muscular attachments that allow for control so that the humerus, elbow, and wrist can move freely. When you hear "Get your shoulders down," focus less on the upper trapezius and more on the lower trapezius, serratus anterior, and rhomboids. If you are struggling with winging scapulae, focus on exercising the lower trapezius and serratus anterior.

After the following exercises, we will examine the beautiful en bas through first position into fifth position and see how the muscles addressed here are engaged during dance performances.

Table 7.1   Shoulder Muscles

Muscle	Origination	Insertion	Action
Biceps	Scapula (long and short heads)	Radius	Elbow flexion
Triceps	Scapula (long, medial, and lateral heads)	Ulna	Elbow and shoulder extension
Trapezius	Occipital, vertebrae C1 to C6, vertebrae C7 to T12	Clavicle, acromion, scapula	Upper fibers: elevate the scapula Middle fibers: retract the scapula Lower fibers: depress the scapula
Subscapularis (anterior)	Scapula	Humerus	Internal rotation
Infraspinatus (posterior)	Scapula	Humerus	External rotation
Teres minor (posterior)	Scapula	Humerus	External rotation, arm adduction
Supraspinatus (superior)	Scapula	Humerus	Abduction and shoulder joint stabilization
Deltoid	Clavicle (anterior), scapula (middle and posterior)	Humerus	Anterior: flexion and internal rotation Middle: abduction Posterior: extension
Serratus anterior	Rib 1 to ribs 8 and 9	Scapula	Protraction
Rhomboid	Vertebrae C7 to T1 and T2 to T5	Scapula	Retraction
Latissimus dorsi	Vertebrae T6 to T12 and L1 to L5, sacrum, iliac crest	Humerus	Shoulder adduction, shoulder extension, internal rotation
Pectoralis major	Clavicle, sternum	Humerus	Shoulder adduction
Pectoralis minor	Ribs 3 to 5	Scapula	Stabilizes scapula

EXTERNAL AND INTERNAL ROTATION

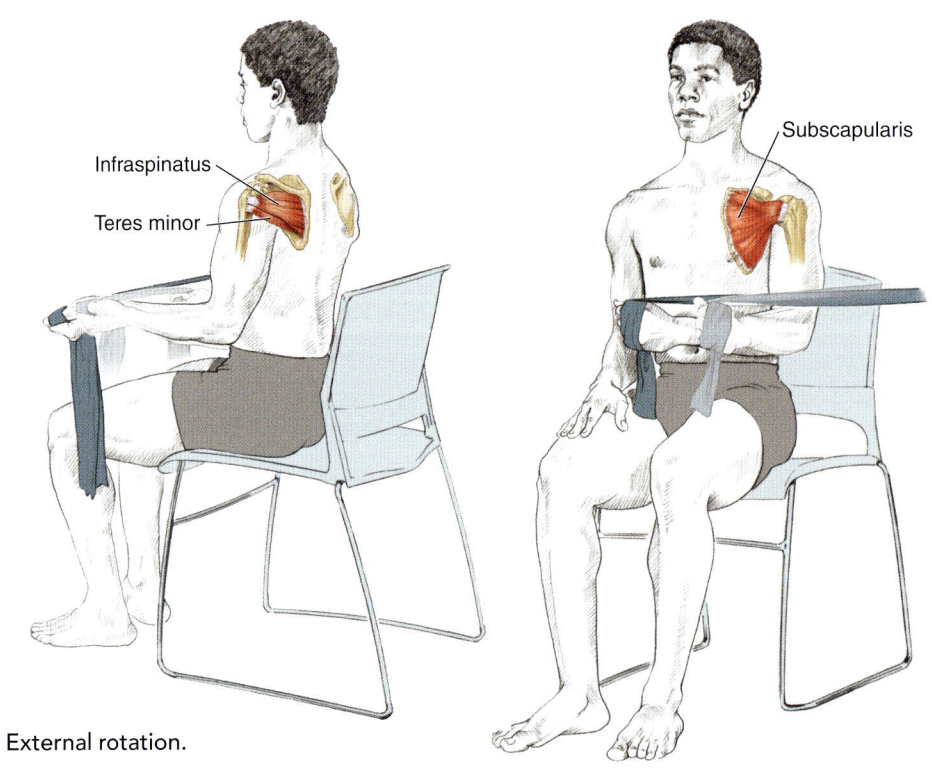

Infraspinatus

Teres minor

Subscapularis

External rotation.

Internal rotation.

EXECUTION FOR EXTERNAL ROTATION

1. Sit in a chair with your elbows flexed at 90 degrees and by your sides. Your forearms are forward with your palms facing inward. Hold an elastic band taut in both hands. Inhale to prepare.

2. On exhalation, glide the shoulder down and begin to externally rotate your arms against the resistance of the band, keeping your elbows snug against your waist. Hold for 2 to 4 counts and feel the strength within your shoulder joint. Open the front of the chest.

3. As you inhale, slowly return with control, keeping the shoulder blades down. Perform 12 times; work up to 3 sets of 12.

EXECUTION FOR INTERNAL ROTATION

1. Sit in a chair with the elbow of your working arm flexed at 90 degrees and by your side. In the hand of your working arm, hold a resistance band with resistance coming from the outside. Inhale to prepare; keep the shoulder blades down.

2. As you exhale, pull inward against the resistance of the band. Hold for 2 to 4 counts while keeping your elbow at the waist.

3. Inhale to return with control. Perform 12 times, then switch sides. Work up to 3 sets of 12.

MUSCLES INVOLVED

External rotation: Teres minor, infraspinatus

Internal rotation: Subscapularis

DANCE FOCUS

Dance classes alone may not provide enough strength for the rotator cuff, but extra conditioning can improve the workings of this joint. Shoulder injuries are not the most common in dance, but when they do occur, they require treatment, rest, rehabilitation, and improvement in technique; therefore, they put your career on hold. The glenohumeral joint is naturally weak because of its shallow cavity. If you are flexible in this joint, as some dancers are, then it is even more important for you to improve joint stability. Intense loads are placed on the shoulder in various styles of dance. Partnering, lifting, contemporary work, and break-dancing skills on your hands require strength in all ranges of shoulder motion. Falling onto your hands puts full body weight on your arms and shoulders. When executing any dance movement that places stress through the shoulder, visualize the deep rotator cuff muscles creating a firm brace for protection that allows stability in the shoulder joint.

VARIATION

External Rotation at 90/90

This is an excellent exercise if your instructors repeatedly ask you to pull your shoulders back and down. Stand with your arms elevated to the side at 90 degrees and your elbows bent at 90 degrees. Your forearms and palms are facing down. In both hands, hold a band that is resisted from the front. Maintain 90-degree angles at your shoulders and elbows. Inhale to prepare. As you exhale and engage your abdominals, externally rotate your arms up against the resistance of the band without allowing your humerus bones to drop. Hold for 2 to 4 counts. As you inhale, slowly return to start position. Maintain the lift in the humerus bones and don't allow your chest to drop. Keep lengthening through the upper back and neck and open through the front of the chest.

OVERHEAD LIFT WITH RESISTANCE

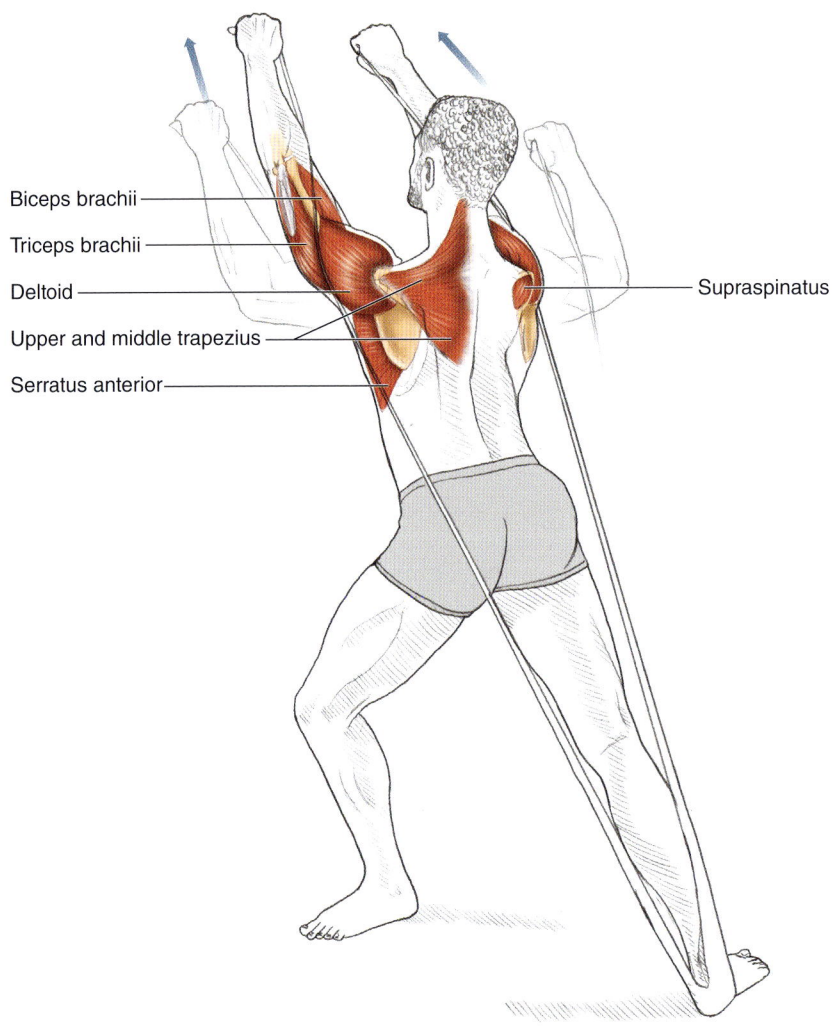

Biceps brachii

Triceps brachii

Deltoid

Upper and middle trapezius

Serratus anterior

Supraspinatus

EXECUTION

1. Stand with your right leg behind your left leg. Hold the ends of a long resistance band in your hands behind you with the center of the band secured under your right foot. Incorporate your strong neutral spine posture as you bring your arms to the 90/90 starting position used for the external and internal rotation.

2. Inhale to prepare. As you exhale, engage your abdominals to stabilize your spine. Lift both arms overhead by extending your elbows along your frontal plane against the resistance of the band.

3. Hold your arms straight overhead for 2 to 4 counts. Inhale to prepare. As you exhale, slowly lower your arms to the starting position. Perform 6 to 8 times, maintaining high-quality alignment. Perform 2 more sets.

MUSCLES INVOLVED

Deltoid, supraspinatus, biceps brachii, upper and middle trapezius, serratus anterior, triceps brachii

DANCE FOCUS

The shoulder joint is made up of a shallow glenoid socket and the humeral head supported by the rotator cuff muscles. Due to the overhead lifting performed in classical ballet, contemporary dance, and cirque, these dancers need to maintain strong shoulder joints to reduce the risk of injury involving shoulder impingement, such as irritation or pinching of the rotator cuff against the acromion as the arm is lifted. If impingement becomes chronic, it can result in rotator cuff tearing. Strengthening the rotator cuff and maintaining good, clean movement of the humeral head in the socket can help you feel strong and stable when lifting a dancer over your head. Weakness in the rotator cuff can cause the humeral head to move upward, thus exacerbating impingement. Strengthening the rotator cuff muscles helps keep the humeral head stable and functioning properly to avoid impingement.

Start working the rotator cuff with the external and internal rotation exercises, then advance to the overhead lift with resistance. Just remember, try not to elevate your scapula as soon as you begin to lift your arms. Your scapula does move in an upward rotation, but your arm needs to move first for at least 30 degrees. This balanced glenohumeral motion will help keep overhead lifting safe and allow your rotator cuff muscles to function most efficiently.

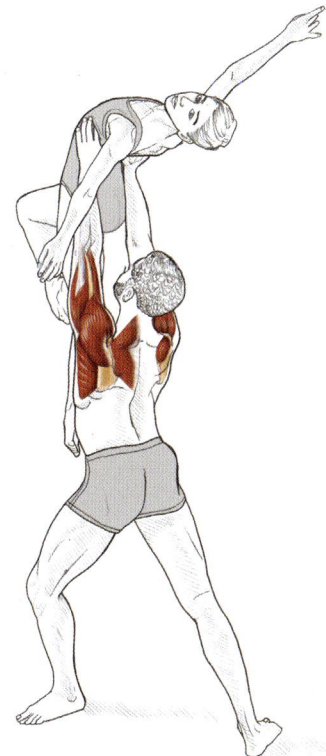

WALL PRESS

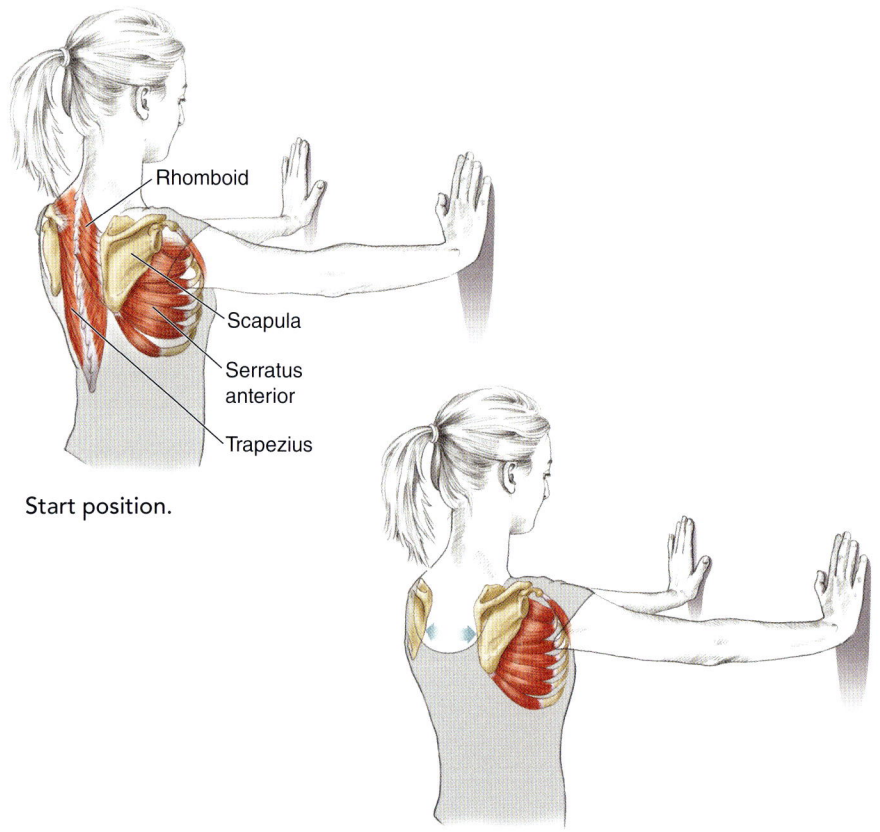

Start position.

Finish position.

EXECUTION

1. Stand facing a wall and lean into the wall with your hands wide at shoulder height; your elbows remain straight. Emphasize core control and inhale to prepare.

2. On exhalation, press against the wall while maintaining straight elbows. Allow both scapulae to move around the rib cage as if the outside edges are trying to pull to the front of your body (protraction); your upper back may round slightly.

3. As you inhale, allow the shoulder blades to move back and squeeze together (retraction). Movement occurs within the scapular region. Perform 10 to 12 times; work up to performing 3 sets.

MUSCLES INVOLVED

Protraction: Serratus anterior

Retraction: Rhomboid, middle and lower trapezius

DANCE FOCUS

Weakness in the serratus anterior can cause scapular winging, and weakness in the rhomboid and lower trapezius muscles can cause rounded shoulders. Both misalignments can occur frequently in all dancers. If you are an instructor, this information can help you provide important feedback. By imagining how the scapula works as it moves along the rib cage, you can help your students perform exercises to reduce winging and rounded shoulders.

It can be hard to understand cues or corrections about pulling the shoulders down when you are unsure of what muscles to use. Focus on sliding the scapulae down and inward, as if you want to drop them into opposite back pockets. Once you are comfortable with that movement, widen through the chest and visualize the scapulae lying against the ribs. Think about moving only the scapulae, not the spine, forward and back, like jazz isolations during warm-ups. You are separating the scapulae movement from spinal movement.

ADVANCED VARIATION

Plank Plus

Begin in a basic straight-arm plank position. Engage the core musculature to create stability along your spine. Your wrists should be aligned directly under your shoulders. Glide or depress your scapulae downward toward your hips. Inhale to prepare, maintaining trunk stability. On exhalation, feel as though you are pushing the floor away, engaging the serratus anterior muscles and pulling the scapulae into protraction around your rib cage. Keep the elbows softly locked. As you inhale, let the scapulae move back and try to pinch them together, emphasizing shoulder retraction. Maintain trunk stability and perform 10 to 12 times. Don't allow your lumbar spine to lose stability. If you can't maintain a stable neutral spine position, then stop, reorganize, and start over.

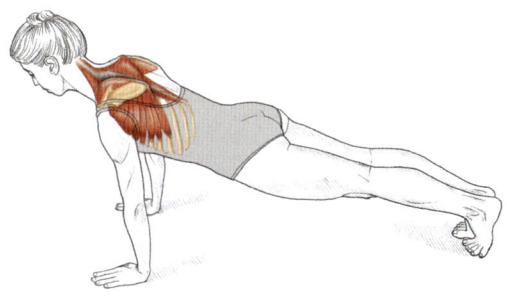

PORT DE BRAS

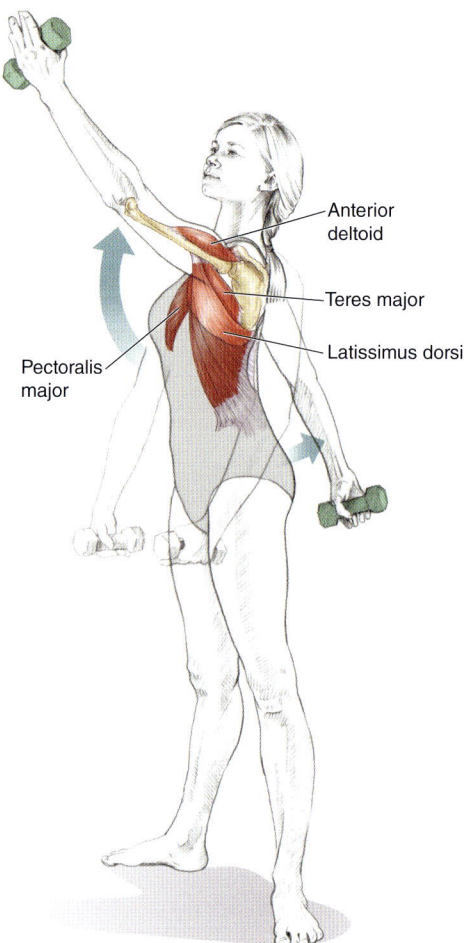

Anterior deltoid

Teres major

Latissimus dorsi

Pectoralis major

EXECUTION

1. Stand firm with legs hip-width apart and feet either parallel or turned out. Hold a small hand weight in each hand or wear wrist weights. Locate neutral position of the spine and pelvis. Inhale to prepare.

2. On exhalation move the left arm toward a high fifth position while moving the right arm into shoulder extension. Emphasize scapular stability. Your head and gaze can follow the top arm.

3. Hold for 2 to 4 counts. Feel width through the upper chest. Inhale and return with control to the start position before repeating on the other side. Perform at least 12 repetitions on each side.

SAFETY TIP: Organize your placement to maintain a stable spine for safety. While executing arm movements, avoid lifting the chest and extending in the lower back.

MUSCLES INVOLVED

Shoulder flexion: Anterior deltoid, pectoralis major

Shoulder extension: Pectoralis major, latissimus dorsi, teres major

DANCE FOCUS

Basic ballet emphasizes stylized arm positions isolated from the shoulders. The upper back is secure with a light, lifted effect. The scapula separates from the shoulder joint, emphasizing the stable body placement. As the shoulder moves forward, notice the activation of the anterior deltoid and pectoralis major—not the upper trapezius, which will cause your shoulder to lift. As the arm moves down from high fifth, gravity provides much of the assistance; however, as your arm moves behind your body, the shoulder extensors contract. Épaulement provides additional awareness by slightly twisting the trunk to give the carriage of the arms even more dimension. Regardless of the changing movement through the trunk, the arms maintain their elegance by emphasizing scapular stability. As the arm moves down and to the back, slight internal rotation will occur in the joint. Allow this motion to occur gently; feel smooth and easy movement in the joint.

ADVANCED VARIATION

Port De Bras on Ball

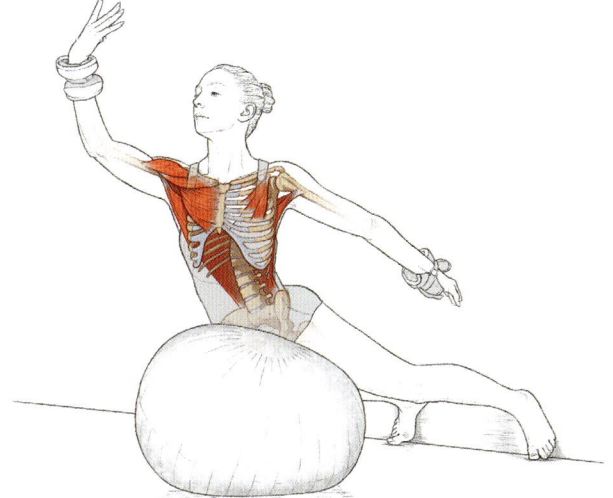

Rest your abdomen and hips on a stability ball. Place your feet against the floor or wall for balance, if needed. Gently pull your navel to your spine to help support the lower segments of the spine. While holding a small weight in each hand or wearing wrist weights, perform the port de bras exercise, focusing on spine and scapula stability while working the muscles of shoulder flexion and extension. The instability of the ball will challenge your balance and spine and scapula stability while increasing the muscle activity in the shoulders.

BICEPS AND TRICEPS BALANCE

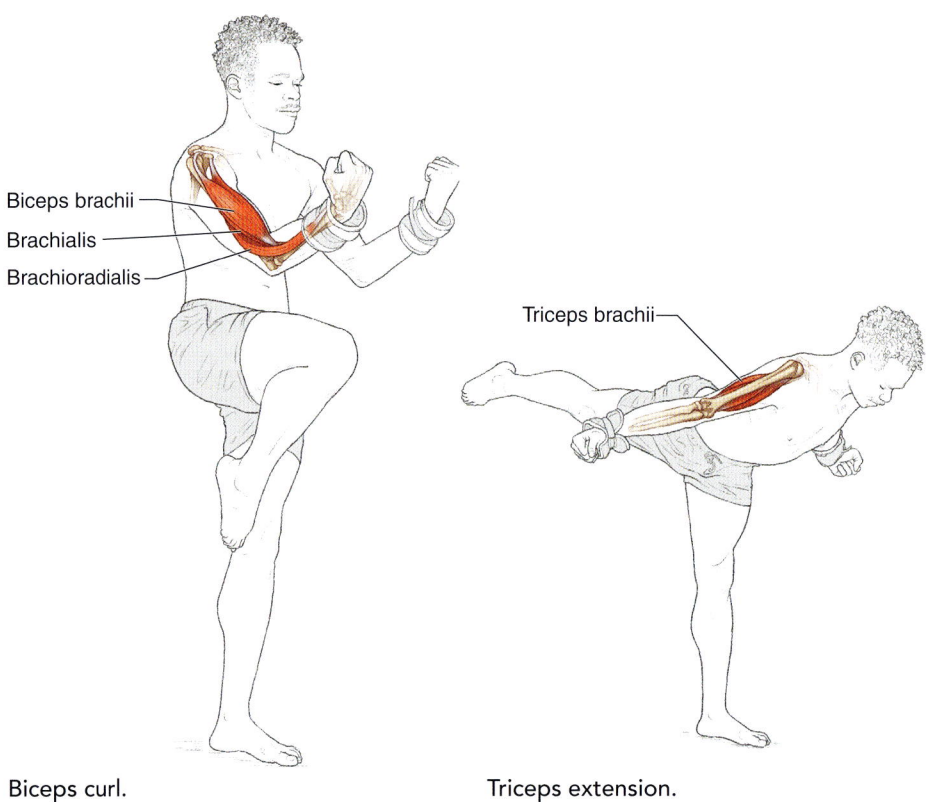

Biceps brachii
Brachialis
Brachioradialis

Triceps brachii

Biceps curl.

Triceps extension.

EXECUTION

1. While standing on your left leg, bring your right foot into parallel passé position. Hold a small hand weight in each hand or wear a set of wrist weights. Maintain your balance. Inhale to prepare.

2. As you exhale, flex the elbows, holding steady in the upper arms. Reemphasize scapular and spine stabilization while you balance.

3. Repeat for 2 to 4 biceps curls, focusing on your balance and the fibers of the biceps shortening with each curl. Return slowly with control to the start position. Inhale to prepare.

4. On an exhalation, with your arms by your sides, move into a flat back position, tilting forward and elevating your right leg to the back as if moving into a parallel arabesque position. Focus on your balance and spine stabilization, being mindful that your back leg should be aligned with your spine. Inhale to prepare.

5. On an exhalation, keeping your upper arms by your sides, flex and extend the elbows to work the triceps. Repeat the elbow flexion and extension for the triceps 2 to 4 times. Inhale to prepare.

6. On an exhalation, keeping your upper arms by your sides, focus on your balance and bring your body back to the upright start position.

7. Repeat this biceps and triceps balance series 5 times before repeating on the other leg.

> **SAFETY TIP:** Do not hyperextend the elbows, which places added stress on the small ligaments within the joint. Maintain firmness in the wrists; avoid hyperextending the wrists and causing strain in the hands and forearms. Keep your upper arms by your sides throughout the exercise.

MUSCLES INVOLVED

Biceps curls: Biceps brachii, brachialis, brachioradialis

Elbow extension: Triceps brachii

DANCE FOCUS

Elbow flexion is used in various dance movements, such as partnering, lifts, falls to the floor, resistance work with another dancer, pantomime movements, and when using props. Strength in the biceps protects the elbow from hyperextension injuries and aids with various shoulder flexion movements. Holding another dancer is challenging, especially when their body weight is completely supported by the anterior muscles of the shoulder and forearm. It is extremely important for the partner carrying the weight to use the biceps muscles in coordination with shoulder stabilization to reduce the risk of injury. Weakness in the biceps can cause faulty alignment and overuse of other structures.

The biceps and triceps balance exercise adds more dynamic movement and challenges balance and coordination skills while exercising the biceps and triceps. The triceps muscle plays an important role in elbow support and is also involved in shoulder extension and adduction. It helps you in the upward phase of push-ups by guiding the elbow into safe extension; in dance, numerous contemporary and break-dancing combinations require the elbow extensors to help raise your body from the floor. In addition, traditional Irish dance posture incorporates firm elbow extension with scapula stability to maintain security of the elbows with the arms by the sides. In contrast, weakness in this area allows the elbows to bend and move during the challenging and quick footwork that characterizes this style of dance. To create stability for the upper arm, visualize the three attachments: upper humerus, scapula, and elbow.

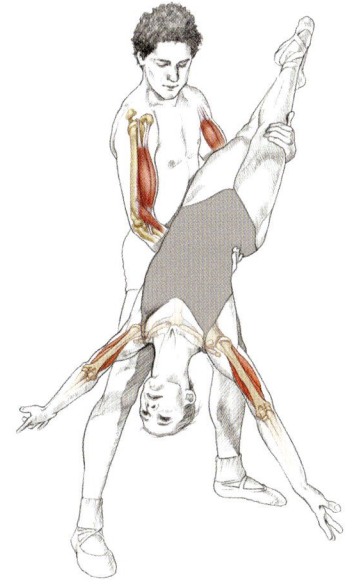

V'S

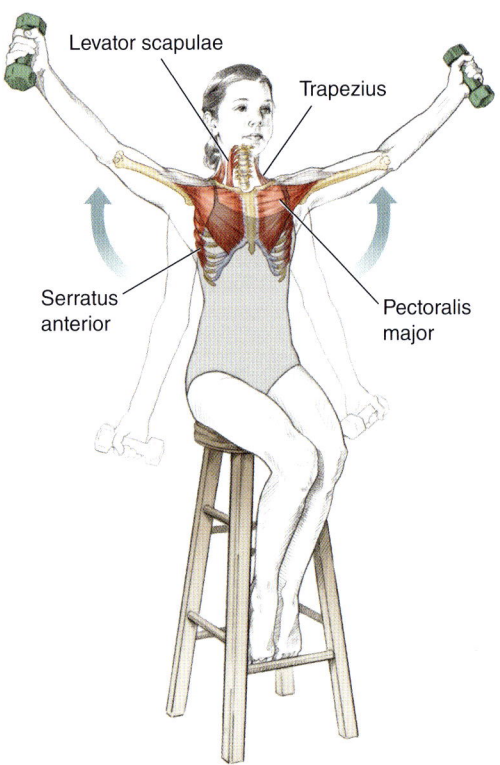

Levator scapulae

Trapezius

Serratus anterior

Pectoralis major

EXECUTION

1. Sit in a chair with erect posture in a neutral position. Keep your arms by your sides; your palms should face the front while holding hand weights or wearing wrist weights. The movement will occur along the frontal plane.

2. As you inhale, begin to lift your arms to the sides into a high V position. Emphasize scapular stabilization, widening through the chest. Feel axial elongation through the movement. Remain stable in the pelvis.

3. Hold at the top of the movement for 2 to 4 counts. Reemphasize the scapulae gliding down and inward toward your hips. Return slowly with control on the exhalation. Perform 10 to 12 times; work up to 3 sets.

SAFETY TIP: Maintain your neutral erect posture. Resist extending or arching through your spine, which means that you have lost core control. Practice elevating your arms without lifting through the chest and ribs. As your arms go up, maintain a strong connection with the oblique muscles and the rim of your pelvis. If lifting the arms without spinal extension is too difficult, try it without weights and exhale as the arms go up. Since inhalation can elevate your chest and facilitate spinal extension, try exhaling as the arms go up.

MUSCLES INVOLVED

Upward phase: Middle deltoid, supraspinatus, serratus anterior, trapezius
Downward phase: Pectoralis major, rhomboid, levator scapulae

DANCE FOCUS

This is such a beautiful movement, and it is seen in all styles of dance. You can perform this movement with jumps, on relevé, or with a partner—it's always invigorating. Freedom in the shoulder joint enables the grace of this arm movement. To achieve that freedom, focus your energy on the scapulae stabilizing with coordinated upward rotation so that the shoulder joints can move with less effort. Maintain placement through your center to show off the ability to isolate the shoulders from the trunk. On the upward phase, feel width through the shoulders without tensing the neck or overusing the upper trapezius. As you begin to bring the arms down, resist gravity and feel the strength through the upper back. Reemphasize deep inhalation on the upward phase and exhalation on the downward phase. Practice without hand weights while jumping as the arms move up—this is where you must control your placement and avoid arching your spine. Let your arms glide upward, keeping you in the air as if you could float over the stage.

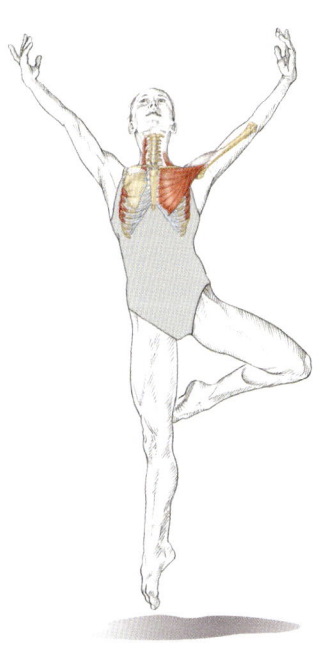

ADVANCED VARIATION

V's With Reverse Lunge

Stand with your feet parallel and hip-width apart. Breathe comfortably throughout the entire exercise. While wearing wrist weights or holding your hand weights, step your right foot back into a lunge position. Your right knee hovers over the floor with your body weight on your left leg. Your left hip, knee, and ankle should be at 90-degree angles. Hold this position as you repeat 5 V's with your arms, focusing on scapula and spine stability. Push through your front leg as you return to the start position before repeating with the opposite leg. This exercise adds more dynamics, challenges spine stability, and works the hamstrings, quadriceps, and glutes as well as challenging upper-body muscle activity. Repeat 10 to 12 times on each leg while maintaining your form.

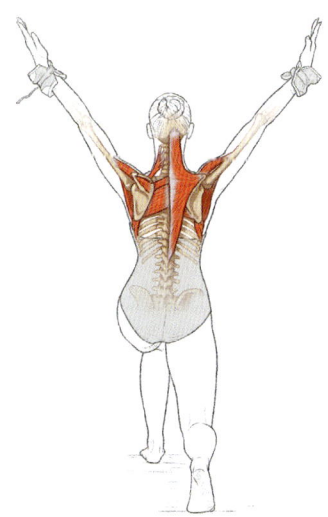

ROWING

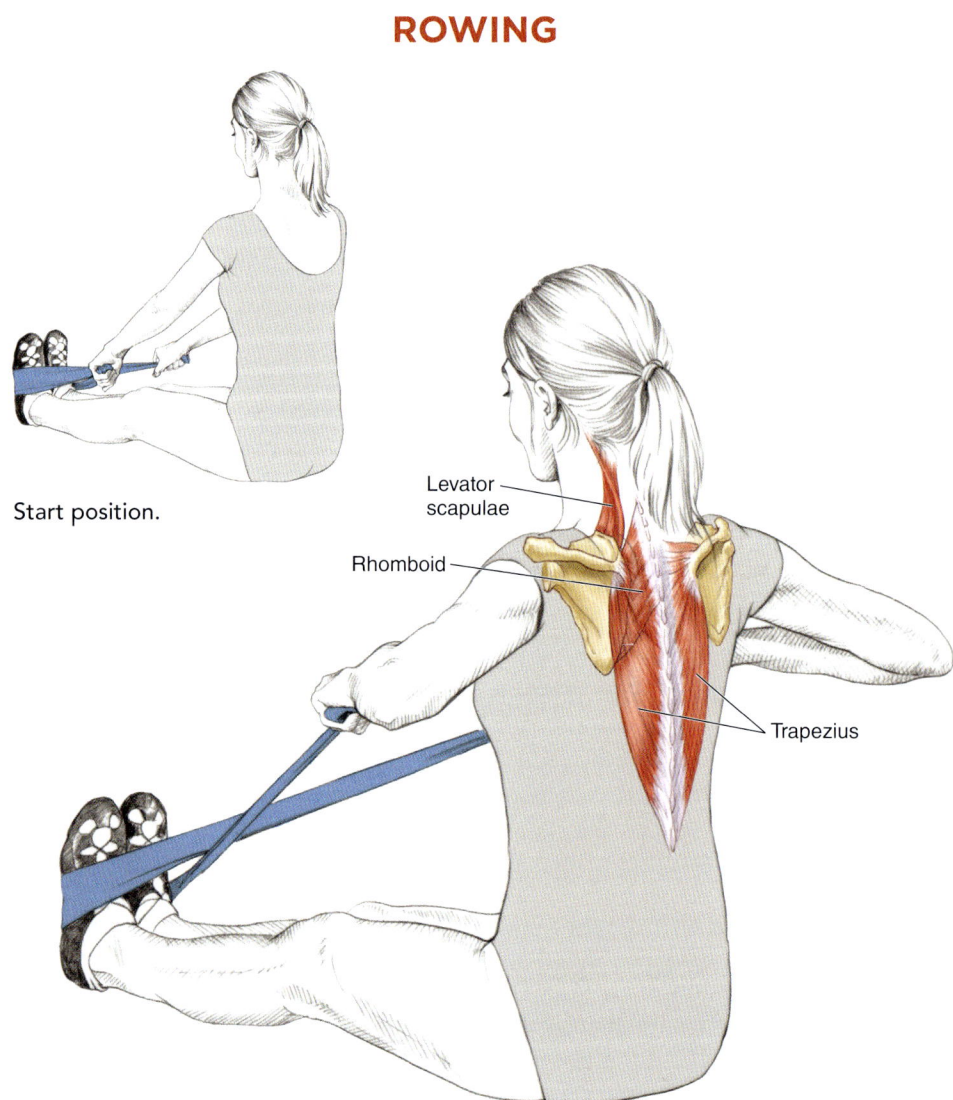

Start position.

Levator scapulae

Rhomboid

Trapezius

EXECUTION

1. While seated on the floor with erect neutral posture, secure a long resistance band around both feet with your legs extended to the front. Cross the band and hold it in your hands; your elbows are extended with your arms in front of you.

2. On inhalation, pull against the resistance of the band with the elbows bending at shoulder height and reaching to the back. Feel the scapulae pulling together. Widen through the chest and maintain a firm center.

3. Hold for 2 to 4 counts. Reemphasize scapular adduction; then, with exhalation, slowly return to the starting position. Perform 10 to 12 times; work up to 3 sets.

SAFETY TIP: Resist spinal extension. As the arms row back, reemphasize core control to maintain a stable spine. Isolate the middle and lower trapezius, not the upper trapezius.

MUSCLES INVOLVED

Retraction: Trapezius, rhomboid, levator scapulae

DANCE FOCUS

Moving the arms behind the body is common in dance, and again, maintenance of scapular control is crucial for resisting injury. Freedom in the shoulder and stability in the upper body allow for fluidity in all the styles of dance, especially jazz. As the shoulder blades move into retraction, let this movement open the front of the chest; resist compensation from the trunk. Remember that you are isolating the muscles that create this action; therefore, hold your core firm. Vary the speed of the rowing to simulate varying tempos; doing so creates more challenge for efficient scapular movement and body placement. Your arms function more effectively when you maintain a firm and balanced upper body. When you have a strong sense of the ability to perform rowing without compensation, increase the resistance of the band to give you more of a challenge.

VARIATION

Rowing Plank

Begin in a plank position while holding a small weight in each hand or wearing wrist weights. Maintain your plank alignment of the shoulders aligned over the wrists and the abdominals engaged for lumbar spine stability. While holding your hand weight, bring your right arm into a row position, engaging the rhomboids, trapezius, and levator scapulae. Complete 2 to 4 rows before repeating with the left arm. If you are looking to advance your strengthening program, rowing planks are a great variation to challenge your core and shoulders as well as the muscles used in shoulder retraction used for rowing.

PLANK TO STAR

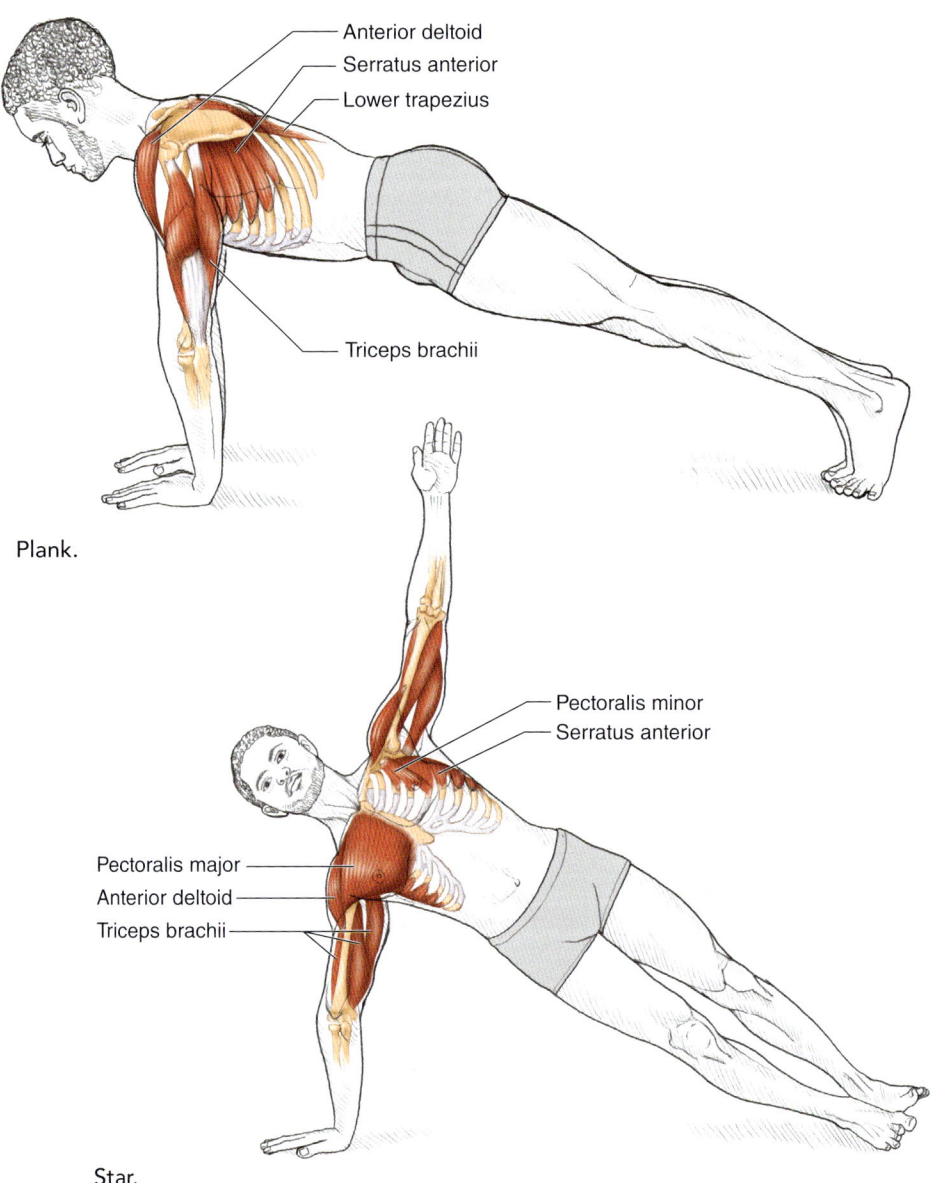

Anterior deltoid
Serratus anterior
Lower trapezius

Triceps brachii

Plank.

Pectoralis minor
Serratus anterior

Pectoralis major
Anterior deltoid
Triceps brachii

Star.

EXECUTION

1. Begin on your hands and knees. Walk your arms out slowly, maintaining control through your center, until your knees extend fully and your shoulders align directly over your wrists in a plank position. Your toes remain in a high relevé position on the floor.

2. Feel the scapulae gliding down toward your hips. Lengthen through the spine and keep your head in alignment with the spine.

3. Hold this position, breathing comfortably, for a count of 5. Feel the security in your shoulder joints and in the muscles surrounding the scapulae.

4. On exhalation, reemphasize deep abdominal contraction. Slide the right shoulder away from the ear as you begin to turn into a straight-arm side plank on the right side. Your feet will rotate, and your left foot will stack on your right. Your left arm will move to your side. Hold for 2 to 4 counts before slowly returning to your starting plank position.

5. Once you have reestablished your secure plank position, repeat the exercise on the other side. Perform 6 to 8 times on each side.

> **SAFETY TIP:** This exercise is advanced and requires firm control in your center. Gravity will try to pull your lower back toward the floor, causing extension in your spine, which can be harmful. Avoid arching throughout the spine; reemphasize deep abdominal contraction. When moving into the star position, maintain strong contraction of the scapula, pulling it down toward your hips. If you are unable to maintain safe, secure placement, stop to rest and reorganize.

MUSCLES INVOLVED

Shoulder flexion: Anterior deltoid, pectoralis major

Elbow extension: Triceps brachii

Scapular depression: Lower trapezius, pectoralis minor, serratus anterior

DANCE FOCUS

This is a very challenging movement that requires strength throughout the shoulder complex and the core. As you continue to gain strength and flexibility, you might be required to perform more challenging choreography. Some choreographers might want you to execute movement supporting your body weight with your arms. Break dancing is exciting to watch, and many of the moves require upper-body strength and scapular stability. To reduce your risk of injury, reemphasize scapula stabilization while practicing planks and this star exercise. Feel the deep stabilizing muscles along your back hugging your spine for support; remember the bracing effect provided by the abdominals for stability. Defy the gravity pulling you to the floor; push the floor away with your hands to feel strength through your forearms.

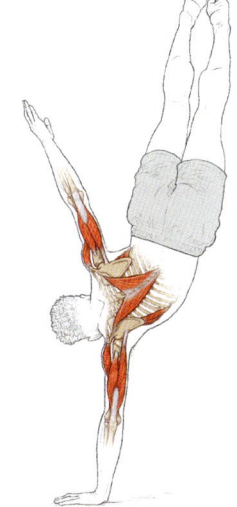

The front fall used in some modern techniques requires firm upper-body strength and control, as well as core strength. The fall should include a moment where the body is almost suspended in air before the hands and arms meet the floor. Without strength in the shoulder region, the front fall can result in an unfortunate accident! Remind yourself that technique class may not give you the needed strength for the shoulders; therefore, make time to condition your upper body. The pike freeze in break dancing requires the body's weight to be on one hand while the hips flex in the air and you are off balance. It takes core and shoulder strength to execute exciting and unique choreography.

REVERSE PLANK

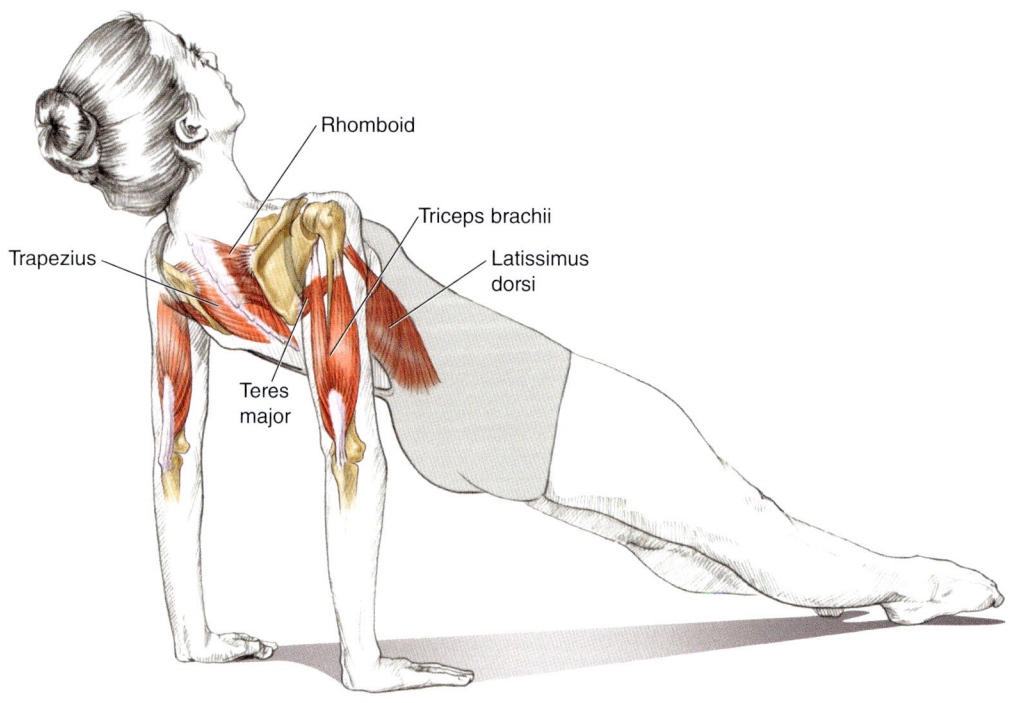

Rhomboid

Triceps brachii

Trapezius

Latissimus dorsi

Teres major

EXECUTION

1. Sit with your legs extended out front. Slightly lean back on your hands with your fingers facing forward. Your elbows should be in a soft but secure position aligned over your wrists. Inhale to prepare.

2. On exhalation, actively pull the scapulae downward and engage your abdominals. As you lift your hips to align with the legs, feel the hip extensors engaging to provide support. Continue to feel axial elongation and shoulder and scapular stability. Hold for a count of 5.

3. As you inhale, slowly return to the floor, resisting gravity. Maintain control and placement. Perform 6 to 8 times.

SAFETY TIP: Do not allow elbow hyperextension or knee hyperextension. To avoid overuse of the small elbow ligaments, maintain a strong isometric contraction throughout your biceps and triceps. To avoid overuse of the ligaments in the knee, maintain strong isometric contraction of the hamstrings and quadriceps.

MUSCLES INVOLVED

Elbow extension: Triceps brachii

Shoulder extension: Teres major, latissimus dorsi

Scapular adduction: Middle and lower trapezius, rhomboid

DANCE FOCUS

A creative pose such as this one is exciting and stimulating for audiences because it's not your typical dance move! It can be difficult to execute challenging skills in which body weight bears down into the wrists and hands if you lack good strength in the upper body to share the load. Think about distributing the forces throughout the entire hand and forearm to avoid straining the wrist. Push the floor away with the hands to feel more power in your forearms. As your body begins to elevate, allow the scapulae to move down to provide more upper-body security; this area is weak in many dancers. You may feel a wonderful stretch across the anterior aspect of your shoulder joint; this stretch derives from the eccentric pull of the biceps, pectoralis major, and anterior rotator cuff. Remember to breathe; you may need to focus your breath through the upper rib cage because of the downward pull of the scapulae and the eccentric lengthening of the abdominals.

ADVANCED VARIATION

Advanced Reverse Plank

Once you have elevated your hips, maintain stability of the scapula and a strong contraction in your triceps. Walk your feet back, bending your knees until you are in a lifted bridge position. Your knees will be bent at 90 degrees. Maintain a strong contraction in your abdominals, hip extensors, and scapula stabilizers. Hold an isometric contraction for 10 to 12 seconds, then slowly walk your feet back to your original reverse plank position. Slowly lower your hips to your original start position.

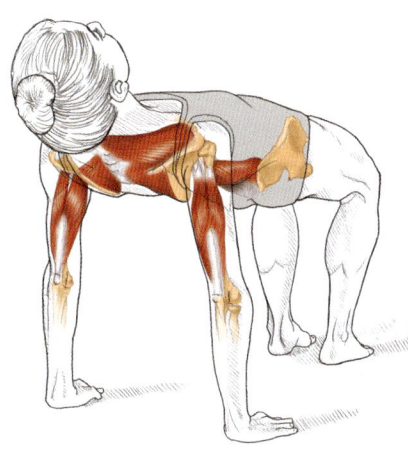

EN BAS THROUGH FIRST AND INTO FIFTH

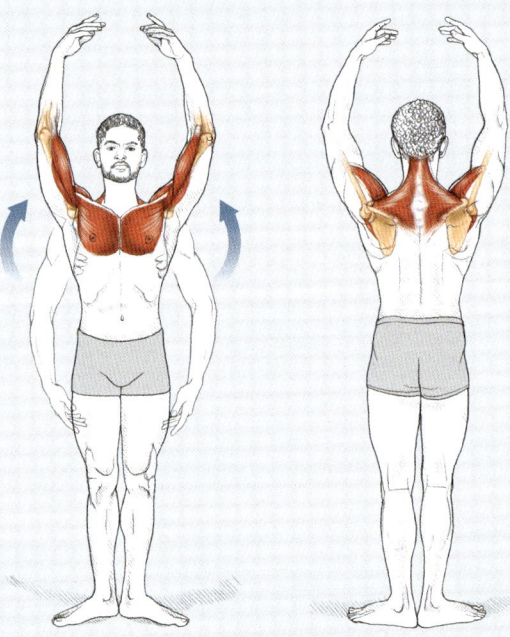

Let's look at the basics of moving your arms from en bas through first and into fifth position overhead.

1. Begin in the neutral spine posture with your legs in first position. Imagine lengthening through your spine as if to grow a bit taller. Your arms begin en bas, positioned slightly away from the front of your hips. The humerus bones are slightly internally rotated, the elbows are slightly flexed, and the wrists and fingers are soft to complete the line of the forearm.

2. As you begin to move into first position, the shoulders begin to flex, engaging the anterior deltoid and pectoralis major. The rotator cuff muscles begin to contract to secure the head of the humerus in the glenoid cavity. The trapezius and serratus anterior also engage to allow scapula abduction and upward rotation.

3. As your arms continue into fifth position overhead, the upper trapezius continues to contract; try to engage the lower trapezius as well to keep the scapula from elevating too high. Feel your neck lengthening, without overrecruitment of the upper trapezius muscles. Release unwanted tension. Allow your arms to provide a gentle frame around your face, and you have mastered a lovely port de bras!

4. Maintain a lengthening feeling along your spine and breathe comfortably. Allow the humeral head to move freely in the glenoid cavity. Try to keep your shoulders away from your ears but maintain tone in your deep scapula stabilizers.

Muscles Involved

Anterior deltoid, pectoralis major, biceps brachii, coracobrachialis, subscapularis, upper and middle trapezius

CHAPTER 8
Pelvis and Hips

Dance requires unusual repetitive movement around the hip joint, and that movement demands extreme control. For example, fast and fancy hip movement is the signature for spicy Latin dance. Modern dancers have the strength and agility to work their hips in all planes while shifting weight and still maintaining balance. Tap dancers can move their feet and legs with impeccable speed while the pelvis holds steady. Ballet dancers show off the height of the développé by maintaining strength and flexibility in their hips. Break dancers perform intricate footwork as well as power moves with their hips in unique positions. All dancers need to understand how the forces of leg movement are distributed through the hip joints and pelvis. At various times, each dance style requires the thigh to work in parallel or internally and externally rotated positions. You can enhance your technique by understanding how your pelvis works in coordination with your legs. Your goal is to achieve the desired movement of your legs without losing control of your pelvis.

This chapter focuses on understanding pelvic alignment and femur (thigh) movement. Your pelvis is powerful when organized and balanced. All core musculature inserts into the pelvic region, and most muscles of the thigh originate from this region; it is quite a powerful intersection! Think about it: Your body's core musculature inserts into the pelvic region, and your leg muscles begin at the pelvis. In other words, your pelvis provides the link between your trunk and your legs.

You must learn to move from your center, and your pelvis is the base of your center. It is made up of the ilium, ischium, and pubic bones on each side (figure 8.1). The sacrum is also discussed as part of this group because it connects the spine with the pelvis; it is wedged between the two pelvic bones at the base of the spine. Just in front of your sacrum lies your center of gravity. To maintain your balance on one foot, you must maintain your center of gravity in a vertical line that passes through your foot to the floor. Visualize your pelvis and sacrum located over your standing leg.

Located along the side of the pelvis is the acetabulum, or deep hip socket. This is the cuplike socket where the head of the femur (thigh bone) inserts. Your femur is the strongest and longest bone in your body, and the deep hip socket or acetabulum allows the femur to lift forward or extend back into arabesque. The acetabulum also allows your thigh to perform battement to the side as well as turn in or out.

The head of the femur angles downward, forming the neck, then creates two bony prominences: the lesser trochanter, which is located medially, and the greater trochanter, which is located laterally. Both prominences are important because of the muscles located at these points. These muscles help create pelvic stability for your standing leg as well as dance movement for your gesture leg.

Before we continue with alignment and muscles, let's get familiar with the term *hip dissociation*, which refers to isolating movement at the hip, separate from the pelvis or spine. Try to tighten or contract the gluteus maximus muscle and maintain that tightness while you kick your leg to the front. What happens? It's next to impossible to get any height out of the thigh as long as the muscles of your buttocks remain tight! If your hamstrings are short and tight, it will also be difficult to kick your leg up without compensating in the lower back or hiking up your hip. Try it again, but this time release

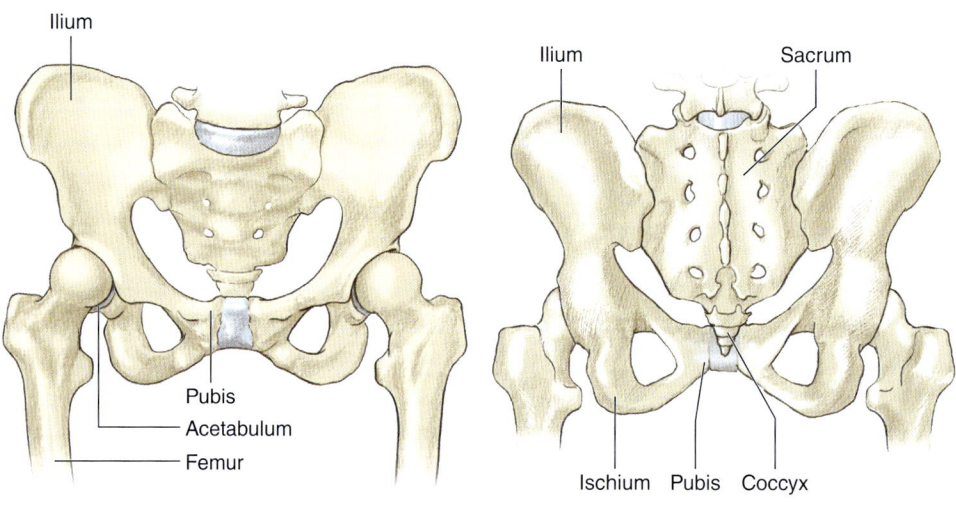

a   b

Figure 8.1   Bones of the pelvis: *(a)* front; *(b)* back.

and lengthen the muscles of your buttocks and hamstrings as the leg goes up. If you use your core muscles to stabilize your pelvis and learn to release your buttocks and hamstrings, your leg can move independently of the pelvis or spine. Think about executing a large fan kick. A stable pelvis allows the working leg to move freely in the socket to produce fluidity and greater range of motion; it also allows the hip joint to better absorb forces that might be harmful to the lower spine.

When you kick your leg to the front (battement), the anterior muscles contract and the posterior muscles release and lengthen. Think back to the discussion of concentric and eccentric muscle work presented in chapter 1. Concentric contraction involves shortening the muscle with contraction, whereas eccentric contraction involves lengthening the muscle fibers but maintaining strength and muscle tone. When you kick your leg to the front, the gluteus maximus and the erector spinae in your lower back can also be trained to lengthen eccentrically while you engage your core to maintain lower-back and pelvic stability. Hip dissociation, then, is the ability to isolate movement at the hip joint, independent of your pelvis and spine.

Pelvic Link

You already know that many of your injuries occur in the lower extremities. If these injuries are not acute (i.e., they do not occur suddenly), then they result from overuse. Overuse injuries can relate to faulty technique, which usually results from poor alignment in the lower spine and pelvis. The iliopsoas muscle (figure 8.2a) serves as the magic link that connects the lower spine and pelvis with the femur; specifically, the psoas major connects the lower spine to the femur at the lesser trochanter, and the iliacus connects the pelvis to the femur at the lesser trochanter. The psoas major plays an important role in stabilizing the lower spine. Weakness and tightness can result in misalignments of the lower back and pelvis, which then trickle down to the legs.

For example, the iliopsoas crosses over the hip joint and can cause snapping as the leg comes down from développé or grand battement. The snapping usually occurs when the iliopsoas tendon moves over the head of the femur or the lesser trochanter; it can produce pain and possibly develop into an injury that needs to be assessed by a health care provider. In some cases, the snapping occurs due to the iliopsoas being tight and weak. The iliopsoas can, however, function in a position that reduces snapping if you maintain stability of the lumbar spine and strength throughout the entire range of motion. Snapping of the iliopsoas tendon can also be prevented by maintaining flexibility of the iliopsoas.

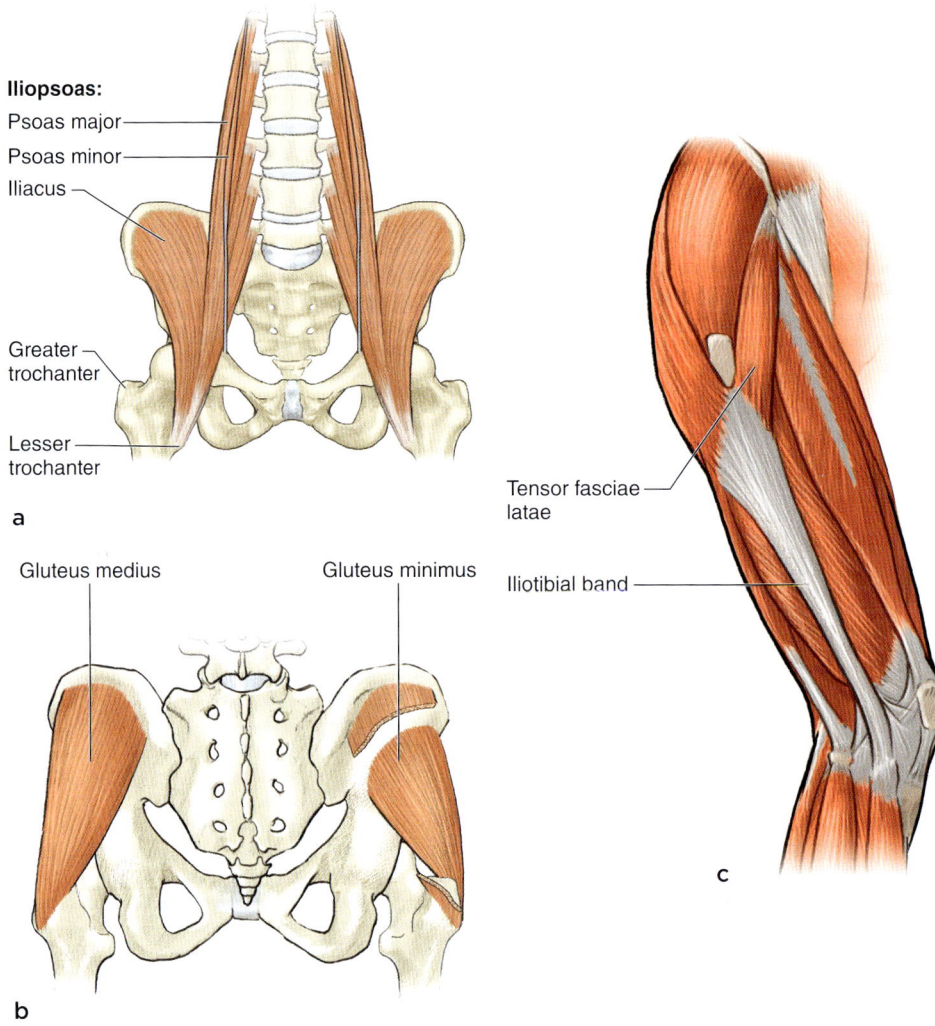

Iliopsoas:
Psoas major
Psoas minor
Iliacus

Greater trochanter

Lesser trochanter

a

Gluteus medius

Gluteus minimus

b

Tensor fasciae latae

Iliotibial band

c

Figure 8.2 Muscles of the pelvis: *(a)* front; *(b)* back; *(c)* side.

The iliopsoas muscle is the major hip flexor; it flexes your hip and helps to stabilize your lower back so that you can lift your leg above 90 degrees. Developing strength in the deep lower-back muscles (multifidus) will allow the psoas major to work more efficiently as a hip flexor. Visualize the location of the iliopsoas as it travels from the lower spine to the inside of your upper femur. Imagine the muscle fibers shortening as you flex at the hip, bringing the femur closer to your trunk. To compete, audition, or advance your technique as a dancer, it's important to get your legs up in the air! Gaining strength and stabilization in the lower segments of your spine will allow for the iliopsoas to work at getting your legs up. Nothing is more frustrating than fighting with your thighs to get your legs above 90 degrees (this problem is addressed further in chapter 9).

Since the iliopsoas originates on the anterior aspect of the lower-spine vertebrae, tightness in this muscle can pull your lower spine into extension, which tilts the front of the pelvis forward. Therefore, even if you understand the concept of trying to hold your pelvis in a neutral position, doing so is next to impossible when your iliopsoas is tight. Dancing with this anterior pelvic tilt and lower-back arch can also result from inactivity of the abdominals as well as the adductors (inner thigh muscles). The anterior tilt also causes tightness in the lower-back musculature and creates shearing force against the vertebrae.

This book focuses on dance-specific exercises, but the hip flexor stretch in this chapter is an important addition. Try it after your warm-up to encourage effective movement through your hips before you start your center work. Remember your plumb line (chapter 4) and your core work (chapter 6). Reemphasize engaging your core to locate your neutral pelvis position.

When you get a cue or correction from an instructor such as "Don't arch your lower back," you may tend to overcompensate and tuck your pelvis under to limit the arch. The problem is that tucking the pelvis can create muscle imbalances. It overworks your gluteus maximus and causes tension in the hamstrings and unusual pressure on the discs of the lower spine. How can you advance your technique when you are constantly fighting to find your pelvic placement? Remember to create length through your spine; locate your neutral pelvis position while engaging the deep core to support the lower back. Abdominal strengthening, combined with iliopsoas and lower-back stretching, can help you overcome arching your back. This reorganization of your placement allows you to move on and advance your skills.

Lateral Hip Power

The gluteus minimus and gluteus medius (figure 8.2*b*) connect the outer surface of the ilium with the lateral area of the greater trochanter. These two muscles help with hip abduction and hip stabilization; for example, they work during performance of parallel side lunges, chassés to the side, and the wings that tap dancers execute to the side. Typically, these two muscles are very strong in modern dancers because of the numerous side leg lifts and parallel leg work.

Another small muscle, the tensor fasciae latae (figure 8.2*c*), connects the outer ilium with the iliotibial band. This band runs from the ilium down the side of the thigh to the lateral femur, patella, and tibia. It is a very strong band of fascia that in some respects may work as an external rotator along with the tensor fasciae latae. Nevertheless, a large portion of pelvic stability, which you need for your supporting leg strength, comes from the gluteus medius and minimus. Thus, when you execute the side-lying passé press exercise presented in this chapter, visualize the location of the hip abductors while focusing on maintaining spinal and pelvic stability.

Control of Pelvic Floor Muscles

The pelvic floor muscles form the bottom of the core and are critical in supporting the pelvis, yet they are overlooked in dance technique. Many instructors are unfamiliar with the function of these muscles, and dancers are often uncomfortable discussing this area of the body. You never hear an instructor give cues about the pelvic floor in technique classes!

The pelvic floor is a series of muscles that line the base of the pelvis. Remember the pelvic diamond? Visualize the two sit bones, the pubic bone, and the coccyx; visualize the muscles that connect the bones of the diamond and form a basin. In a basic modern contraction, the pelvis rocks posteriorly and the sit bones of the pelvic diamond move together very slightly with the contraction of the pelvic floor muscles. On arching the lower back and tilting the pelvis forward, the sit bones move apart, eccentrically lengthening the muscles. There is also a very slight movement of the sacrum, which creates the connection of the diamond from the coccyx to the pubic bone.

For example, a demi-plié in second position should start with the pelvis in neutral. On the downward phase, the hips dissociate, the sit bones move away from each other, and the pubic bone and coccyx move away from each other. The opposite occurs on the upward phase. In other words, on the downward phase, the pelvis stays neutral and the diamond widens; on the upward phase, the pelvis continues to stay neutral and the diamond shrinks.

Rotation of the Femur

The femur must turn in and out to accommodate all styles of dance; therefore, you need an excellent balance of strength and flexibility between the internal and external rotators. A large role is played in both turnout and stabilization of the hip joint by six small muscles located deep under the gluteus maximus. One of these, the piriformis muscle, connects the sacrum and the posterior ilium with the greater trochanter. Two others, the obturator internus and obturator externus, connect the ischium and pubic bone with the greater trochanter. The fourth and fifth muscles, the gemellus inferior and gemellus superior, connect the lower ischium and sit bones with the greater trochanter. The sixth muscle, the quadratus femoris, connects the sit bones with the greater trochanter. We can refer to these hip external rotators muscles as the "deep six."

Internal rotation of the femur is shared by several muscles. Some of these are discussed in the next chapter, but let's introduce them now. Two of the hamstring muscles (the semitendinosus and semimembranosus) can rotate internally. In addition, internal rotation can be assisted by the anterior fibers of the gluteus medius and gluteus minimus, as well as the tensor fasciae latae. Remember that the femur moves in various directions without tucking or tilting the pelvis. Excellent hip dissociation skills allow for more effective hip movement and more core stability.

Turnout

Turnout of the legs is used in movements performed by ballet dancers. There are a few anatomical factors that determine your turnout: strength of the external rotators, flexibility of the internal rotators, ligament laxity, and the bony alignment of the femoral head and neck. Much of the turnout must come from movement in the hip socket. The International Association for Dance Medicine & Science states that on average, 60 percent of turnout comes from the hip, 20 to 30 percent from the ankle, and the remaining 10 to 20 percent from the knee and tibia (Krasnow and Wilmerding 2011). The strength in your deep hip external rotators can help you achieve quality turnout. Whenever you are required to lift your leg while it's turned out, initiate the movement by contracting the deep external hip rotators to fully turn out within the hip socket. Maintain the muscle contraction through the entire movement of the leg while other muscles assist.

For example, in arabesque, the deep rotators contract and the gluteus maximus assists as a turn-out muscle to help bring the hip into extension. Without the contraction of the deep six rotators, your leg would extend in a parallel position! When executing plié, allow the rotators to contract to keep the femurs open along the frontal plane and aligned over the toes. On the downward phase, the inner-thigh muscles assist by working eccentrically; on the upward phase, they work concentrically.

Visualize the location of the small external rotators as they connect the femur with the sacrum and lower pelvis. As the muscle fibers contract and shorten, the femur rotates laterally in the socket. The femur can turn out in the hip socket without unwanted movement in the lower back or pelvis, which supports the hip dissociation theory. Practice moving your femur inward and outward while sitting, lying down, and standing. Focus on movement only deep within the socket; notice that you don't need to twist your pelvis or tuck it under to actively rotate your femur in the joint. Just move your femur, not your pelvis or spine.

Understanding the passive range of motion in your hip can help you understand your natural hip rotation, but your functional turnout—what you can hold and work with—is more useful. As mentioned previously, 60 percent of your turnout comes from your hip and the rest from your knees, tibia, foot, and ankle. To attain ideal functional turnout, you must work within your means and with proper skeletal alignment. Ideal turnout of 180 degrees is physically challenging and can create compensation and potential injuries for some dancers. Not all hips are created the same. You must learn to work with and control the range you have. Focusing on turning your legs out from your hips can minimize the stresses on the knee and ankle. Consider these keys for better functional turnout:

- Always align your patella (kneecap) over your second toe, to avoid screwing or twisting through the knee joint.
- Keep your weight placed equally over your heel and first and fifth metatarsals to avoid overpronating the feet.
- Maintain your pelvis in a firm, neutral position, using your abdominals and deep hip external rotators to avoid anterior pelvic tilt.

Bone Alignment

If you struggle with turnout, you will benefit from learning about femoral anteversion, which, when applied to the angle of the femur, involves turning forward. This placement in the hip socket causes an internal rotation of the femur, or toeing in, which makes it anatomically challenging to execute turnout for ballet. This alignment issue can also create an anterior tilt of the pelvis as you try to force your turnout. Forcing more turnout will cause screwing of the knees and rolling in at the foot and ankle. Femoral anteversion may be a result of your personal bony anatomy. You can learn to work within your hip's natural range of motion. Make sure that your thigh and patella are always aligned over your toes while using your hip external rotators. Engaging the hip external rotators will help you to use less forced turnout at your foot and ankle.

The opposite of femoral anteversion is femoral retroversion, in which the angle of the femur allows for more external rotation or toeing out. This condition, of course, is more suitable for ballet. Another anatomical factor associated with turnout is the placement of the hip socket, or acetabulum. The socket is generally located slightly forward; however, if your socket is located directly to the side, you may have more turnout.

Regardless of anatomical factors, compensating to achieve a look of perfect turnout will not help you. Focus instead on good pelvic alignment and work to strengthen your core and hip-rotation muscles.

Dance-Focused Exercise

While executing the following exercises, think about maintaining stability in your pelvis and lower spine, and allow the femur to move freely in the hip socket. The legs can be directed into many amazing moves and angles, and you can learn to work the relevant muscles effectively. As one group of muscles works to create the movement, the opposing side must lengthen; meanwhile, the core must secure the movement. While working through the exercises, visualize each muscle's location (use table 8.1 as a reference). Focus on the muscle action and how it makes your femur move. To challenge your balance skills, close your eyes for some of the repetitions. Repeat some of the repetitions at a faster pace and notice how changes in tempo challenge your stability. Each exercise relates directly to your technique; use the illustrations to learn which muscles work together. The chapter concludes with a detailed look at passé.

Table 8.1 Deep Hip Rotation Muscles

Muscle	Origination	Insertion	Action
Piriformis	Sacrum, ilium	Greater trochanter	External rotation, hip abduction
Gemellus inferior	Ischial tuberosity	Greater trochanter	External rotation
Gemellus superior	Ischial spine	Greater trochanter	External rotation
Obturator internus	Pubic bone	Greater trochanter	External rotation, hip abduction
Obturator externus	Outer surface of pubic and ischium bones	Greater trochanter	External rotation, hip stabilization
Quadratus femoris	Ischial tuberosity	In between greater and lesser trochanter	Hip external rotation, hip adduction

PRONE PLIÉ HEEL SQUEEZE

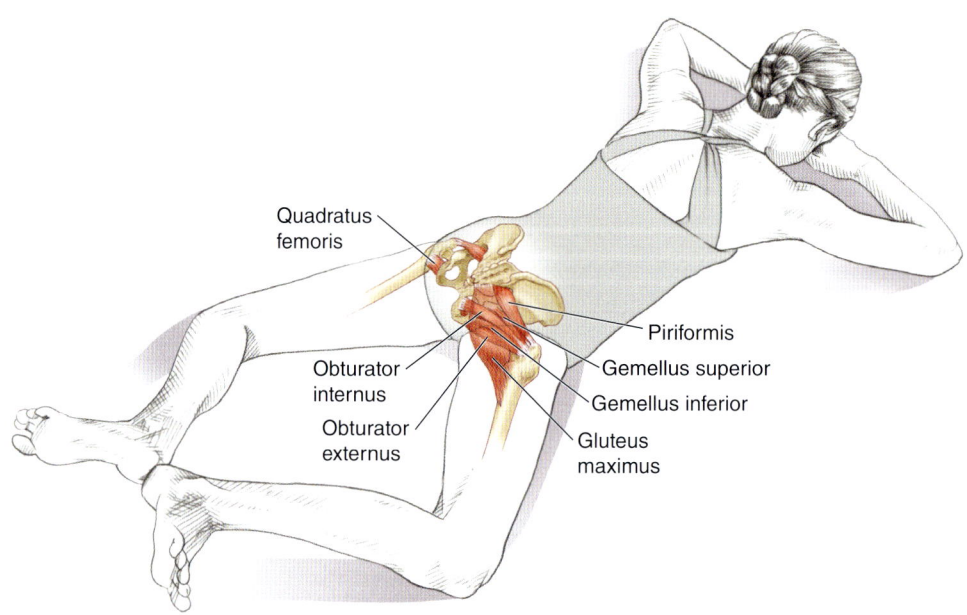

Quadratus femoris

Obturator internus

Obturator externus

Piriformis

Gemellus superior

Gemellus inferior

Gluteus maximus

EXECUTION

1. Lie facedown in a slight demi-plié position with your forehead resting on your hands. Your pelvis must be neutral, not tipped forward with an arch in your lower back. Your heels are touching each other. Inhale to prepare.

2. On exhalation, coordinate contraction of the deep abdominals and press the heels together to create an isometric contraction for the deep rotators and the lower fibers of the gluteus maximus. Hold this position for 6 counts.

3. Relax the contraction as you inhale and prepare to repeat. Push and relax 10 to 12 times.

SAFETY TIP: Avoid arching the lower back, which shortens the deep hip flexors and tightens the lower back. Remain in your natural, neutral pelvic position with your abdominals engaged.

MUSCLES INVOLVED

Obturator internus, obturator externus, piriformis, quadratus femoris, gemellus inferior, gemellus superior, posterior gluteus medius, lower fibers of the gluteus maximus

DANCE FOCUS

One of your goals is to understand the principle of hip dissociation and how it can improve your performance as a dancer in any style of movement. Let this exercise help you focus on the deep six muscles that externally rotate your legs while resisting the need to tip your pelvis forward or backward. Visualize the femurs working separately from the pelvis. The strength of the contraction and short-ening of the deep six should give you the effect of the femurs almost hovering slightly over the floor without strain in the upper thighs or hip flexors. Imagine a grande plié, in which the thighs are moving along the frontal plane, directly to the side. Also imagine a pas de chat, in which you are completely turned out along your frontal plane and have a per-fectly neutral pelvis.

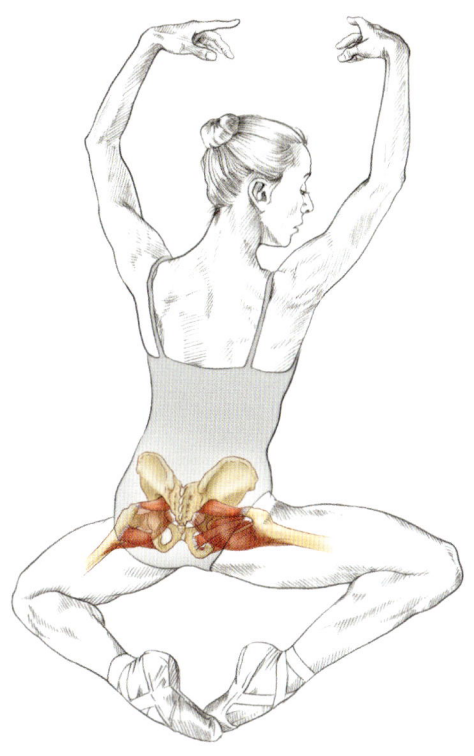

ADVANCED VARIATION

Prone Passé

Once you have mastered isolating the deep rotators with the prone plié heel squeeze, advance the exercise by lying prone on a table and allowing your right leg to be off the side of the table. Position yourself to maintain good balance, locate your neutral spine position, and bring your right leg into a parallel passé position off the edge of the table. Inhale to prepare. On exhalation, move your right leg into passé turned out, hold, and feel the turn-out muscles lifting your thigh against gravity into passé. Hold for 6 counts, then slowly return to parallel. Perform 10 to 12 times on each side. To advance even further, add a weight around your thigh to provide resistance.

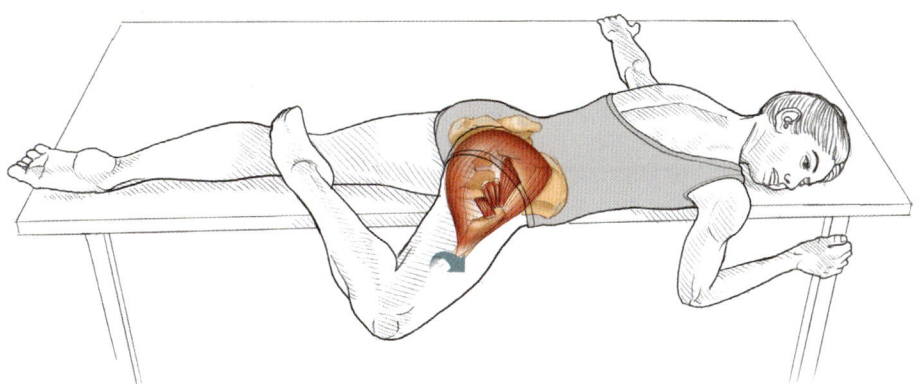

WEIGHTED COUPÉ TURN-IN

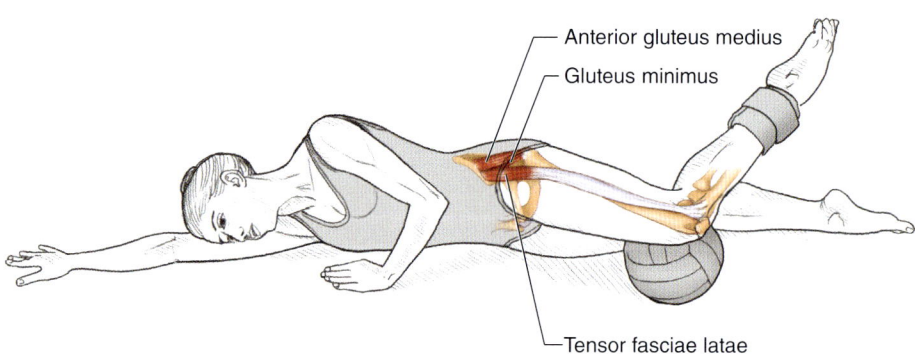

Anterior gluteus medius

Gluteus minimus

Tensor fasciae latae

EXECUTION

1. Begin while lying on your right side. Your head rests on your bottom arm, which is extended overhead on the floor. Your top arm is on the floor in front of you. Locate your neutral position. Your top (left) leg is placed in a parallel coupé position; the foot is just above the opposite ankle, and the knee is placed on a ball. Place a 2- to 5-pound (about 1- to 2.5-kg) weight around your left ankle. Organize your trunk and inhale to prepare.

2. On exhalation, reemphasize stability through the core and pelvis. Maintain a strong lift along the waistline on the floor. Gently press the left knee into the ball, contracting the internal rotators. Allow the lower leg to move away from the bottom leg, encouraging more turn-in. Hold for 6 counts.

3. As you inhale, slowly return with control and pelvic stability. Perform 10 to 12 times; work up to 3 sets. Focus on hip dissociation.

SAFETY TIP: Anchor your pelvis by reemphasizing core control; avoid any movement in the lower back. This firm base allows for more fluidity and range of motion in the hip joint and reduces the risk of injury in the lower back. Avoid pelvic tilt; maintain a natural, neutral position with the hip flexed.

MUSCLES INVOLVED

Tensor fasciae latae, anterior gluteus medius, gluteus minimus

DANCE FOCUS

Strengthening the turn-in muscles is important for maintaining pelvic postural balance. As you're working in a turned-in position, visualize the front of the thigh turning toward the mid-sagittal plane and the head of the femur gliding in a slightly posterior direction. You don't have to compensate by moving your lower back. Because turn-in exercises work the gluteus medius and the gluteus minimus, which also provide stabilization for the standing leg, they will give you multiple positive results. Internal rotation of the hips is used in hip-hop styles of dance and numerous modern movements.

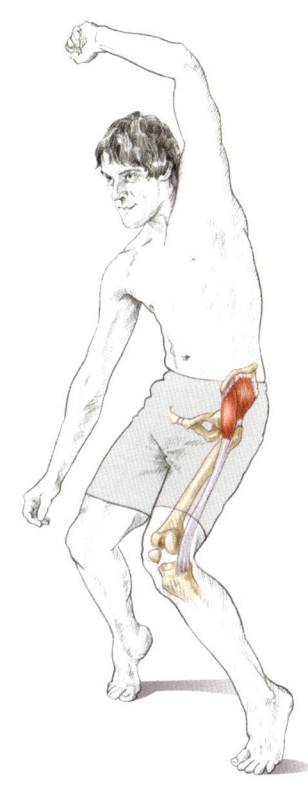

ADVANCED VARIATION

Advanced Weighted Coupé Turn-In

Begin in the same starting position but place your right arm closer to your side with your forearm straight ahead. Inhale to prepare. As you exhale, engage your deep abdominals, glide your right shoulder blade down away from your ear, and lift your body into a side plank. Once you have established good balance, press your left knee into the ball and turn your thigh in, encouraging more internal rotation. Maintain deep abdominal contraction and scapula stability. Perform 10 to 12 times before controlling and returning to the start position.

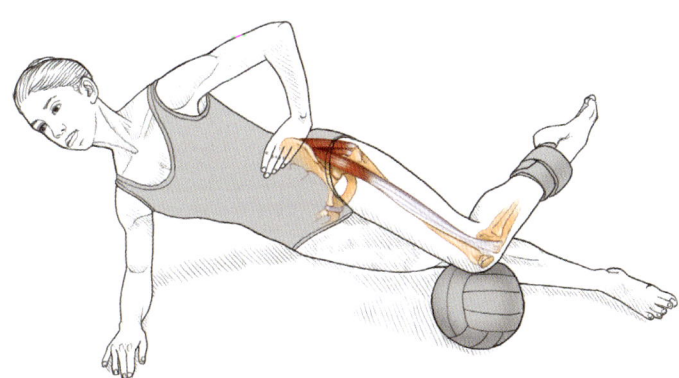

SIDE-LYING PASSÉ PRESS

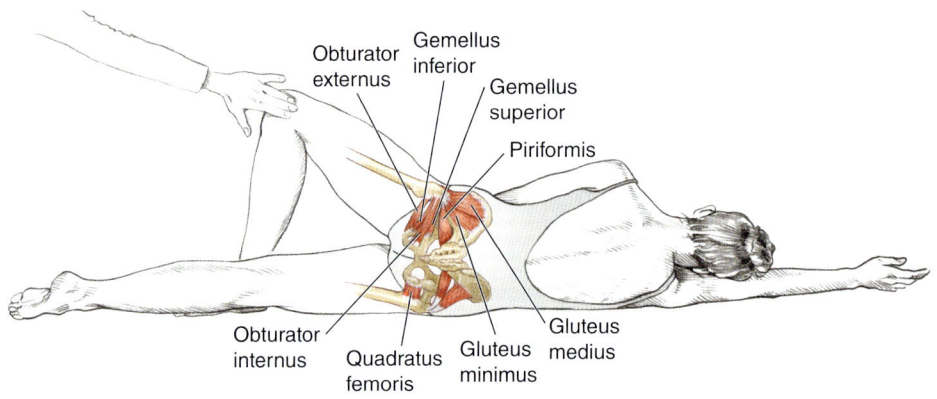

Obturator externus
Gemellus inferior
Gemellus superior
Piriformis
Obturator internus
Quadratus femoris
Gluteus minimus
Gluteus medius

EXECUTION

1. Begin on your right side with the bottom arm overhead and your head resting on it. Your top arm is on the floor in front of you. Place the top (left) leg into passé position and the left foot on the floor in front of the bottom leg, which must remain turned out. Feel the outside edge of the left foot against the bottom leg. Reorganize your trunk by engaging your core to feel an added lift along your right side. Inhale to prepare.

2. On exhalation, engage the deep abdominals and begin to contract the deep six rotators, opening the thigh along the frontal plane. Continue the contraction, pressing your leg into the resistance of a partner's hand. Hold for 6 counts, then slowly return. Perform 10 to 12 times.

3. As the deep contraction occurs, feel the separation of your thigh from your pelvis and supporting leg. Keep the turnout working with the bottom leg as well. Resist twisting of the pelvis—you are moving your thigh, not your pelvis.

4. To advance this exercise, repeat in a standing position, as in the variation.

SAFETY TIP: Maintain trunk stability to support the lower back. Keep the pelvis level to emphasize the deep rotators and hip abductors.

MUSCLES INVOLVED

Obturator internus, obturator externus, piriformis, quadratus femoris, gemellus inferior, gemellus superior, posterior fibers of the gluteus minimus and gluteus medius

DANCE FOCUS

As you perform this exercise, visualize the strength of the passé leg giving you the power to sail in multiple en dehors pirouettes. Turning requires coordination of force, balance, timing, and strength. Even while performing en dedans pirouettes, you must make an excellent coordinated effort of the working leg turning out in passé and the standing leg turning out. If you lose turnout in either hip, the pirouette comes to an unattractive end. This exercise reinforces the oppositional work between the passé leg turning out and the supporting leg turning out and stabilizing the body.

VARIATION

Standing Passé Press

From a standing turned-out position while facing the barre, bring the left leg into a passé. The right leg remains secure and turned out. Reemphasize the deep external rotators and deep abdominal muscles for excellent posture. With assistance from a friend, press your passé leg into the resistance of her hand while firmly maintaining turnout and stability on the standing leg. Hold for 4 counts. Slowly relax and prepare to repeat. Your goal is to execute stability all the way down the chain of the supporting hip and leg and to isolate the deep six rotators of the passé leg. Perform 6 times. Avoid any twisting in the knee of the supporting leg by reemphasizing the stability of the standing leg and the turn-out muscles of the standing leg.

SIDE-LYING QUADRATUS FEMORIS ROTATION

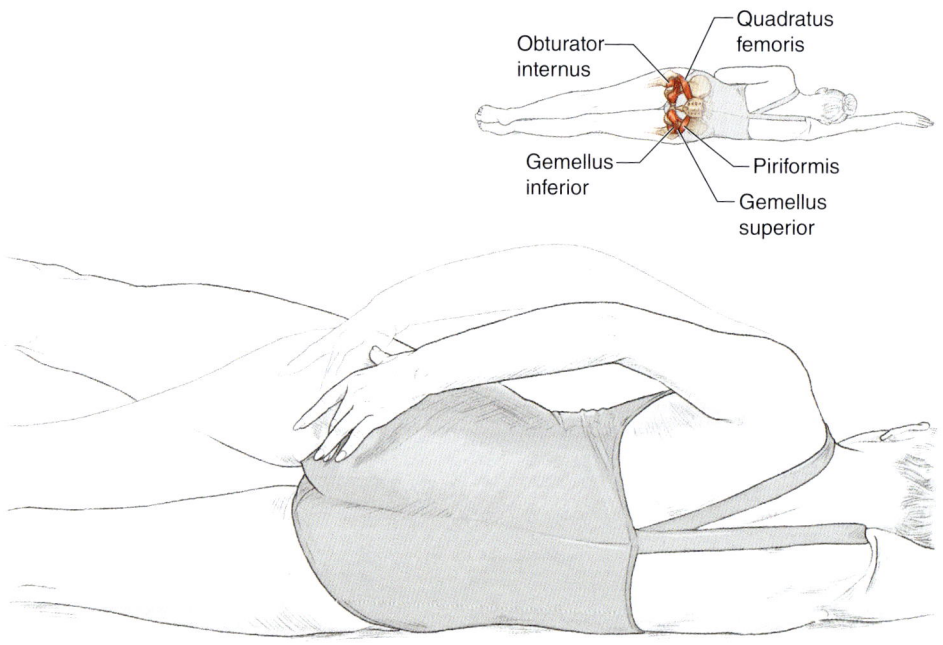

EXECUTION

1. Lie on your right side with the bottom arm overhead and your head resting on it. Both legs extend in parallel alignment, with the top leg resting on the bottom leg.

2. To locate the quadratus femoris muscle, place your thumb on the greater trochanter and reach your fingers toward the sit bone of your top leg. You will need to engage your core, remain in a neutral pelvic alignment, and hold your balance.

3. Slightly lift the top leg while holding your hand in the same place. Slowly begin to turn out (externally rotate) the top leg; the line between your greater trochanter and sit bone represents the quadratus femoris.

4. Alternate turning the top leg out and returning to parallel as you visualize the quadratus femoris and try to feel it contract on external rotation. Try not to recruit the gluteus medius or minimus, and focus on the location of the quadratus femoris. Repeat 8 to 10 times before moving to the other side.

MUSCLES INVOLVED

Quadratus femoris, obturator internus, obturator externus, piriformis, gemellus inferior, gemellus superior

DANCE FOCUS

While this exercise engages the deep six external rotators, our focus is on the quadratus femoris. If you can visualize the appropriate location of this muscle, it can help you maintain a stable turned-out hip, particularly on the standing leg. You may need to gently move your legs into hip flexion while on your side to better locate this muscle, but finding its location will help you improve the strength of your turnout. Due to its anatomical location, connecting the lateral margin of the ischial tuberosity with the greater trochanter, this small but powerful muscle helps you hold your turnout on the standing leg. Try to visualize the muscle contracting and rotating your femur into a turned-out position while performing any ballet type exercise.

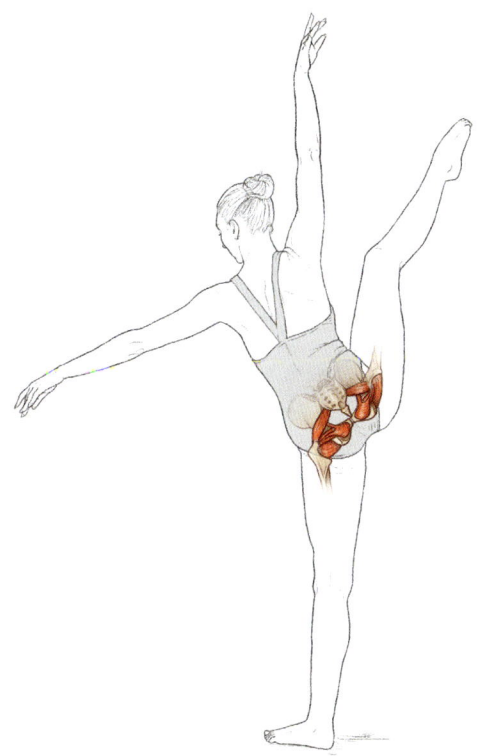

STANDING INNER-THIGH PRESS

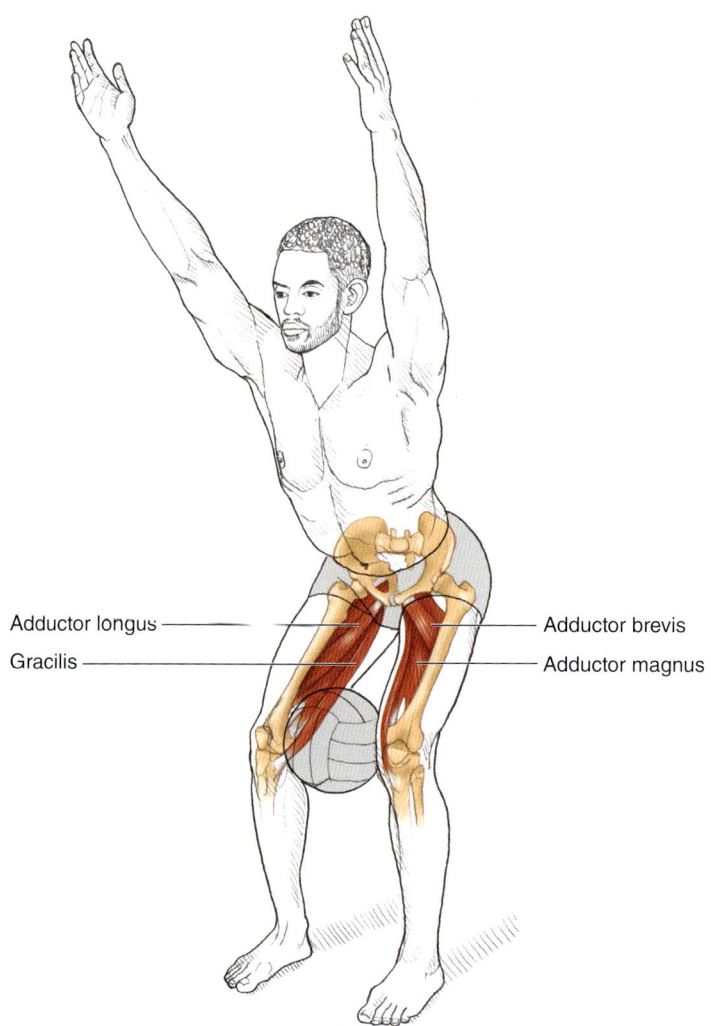

Adductor longus
Gracilis
Adductor brevis
Adductor magnus

EXECUTION

1. While standing with your arms overhead, lengthen through your spine and organize your trunk to locate your healthy neutral position.

2. Flex the hips to 90 degrees, going into a parallel squat position, and place a ball between the inner thighs. Inhale to prepare.

3. Breathing comfortably, engage the lower abdominals, squeeze the ball with the hip adductors, and begin a small squat; maintain the hip adductor squeeze. Execute 10 squats, then hold with knees at 90 degrees for a 10-second isometric contraction. Perform 5 more times.

SAFETY TIP: Avoid arching the lower back; work to stay in a natural, supported pelvic position by engaging the deep abdominals. Maintain weight into the heels and don't squat deeper than 90 degrees.

MUSCLES INVOLVED

Adductor longus, adductor brevis, adductor magnus, gracilis

DANCE FOCUS

Fast and firm adductors are required for various movements, such as those that bring the legs together, crossing positions of the legs, and jumps that involve leg beats in the air. The up phase of the plié requires the adductors to contract concentrically, and the downward phase requires them to contract eccentrically. In the lower ranges of leg height, the inner thighs also help with hip flexion and extension. Some of the muscle fibers lie in a position to produce flexion; others lie in a position to produce hip extension. Maintaining a balance between the hip adductors and hip abductors provides another mechanism for pelvic security. You may spend a lot of time stretching the inner thighs for more flexibility, and it is just as important for you to strengthen this area.

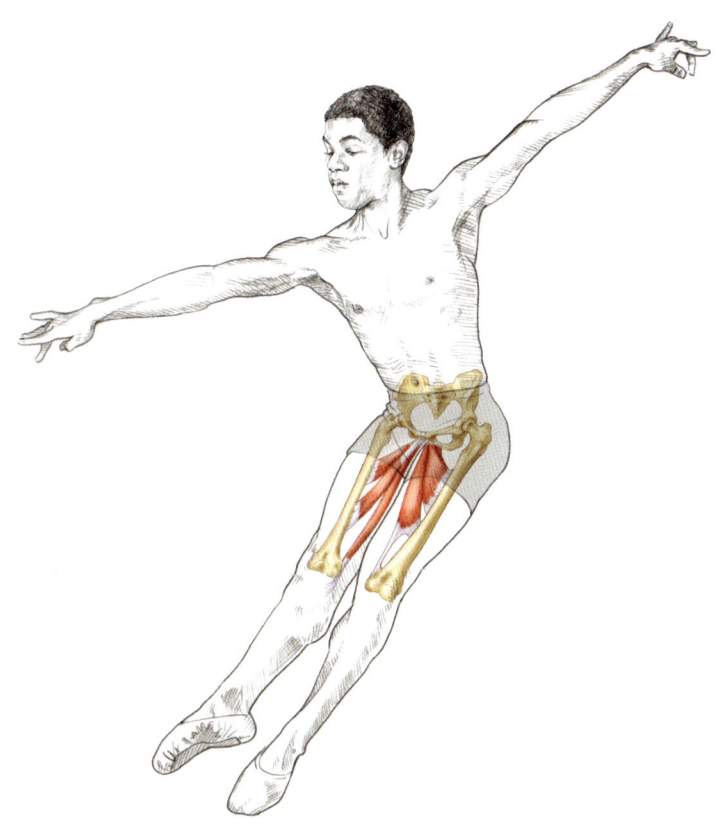

ARABESQUE PREP

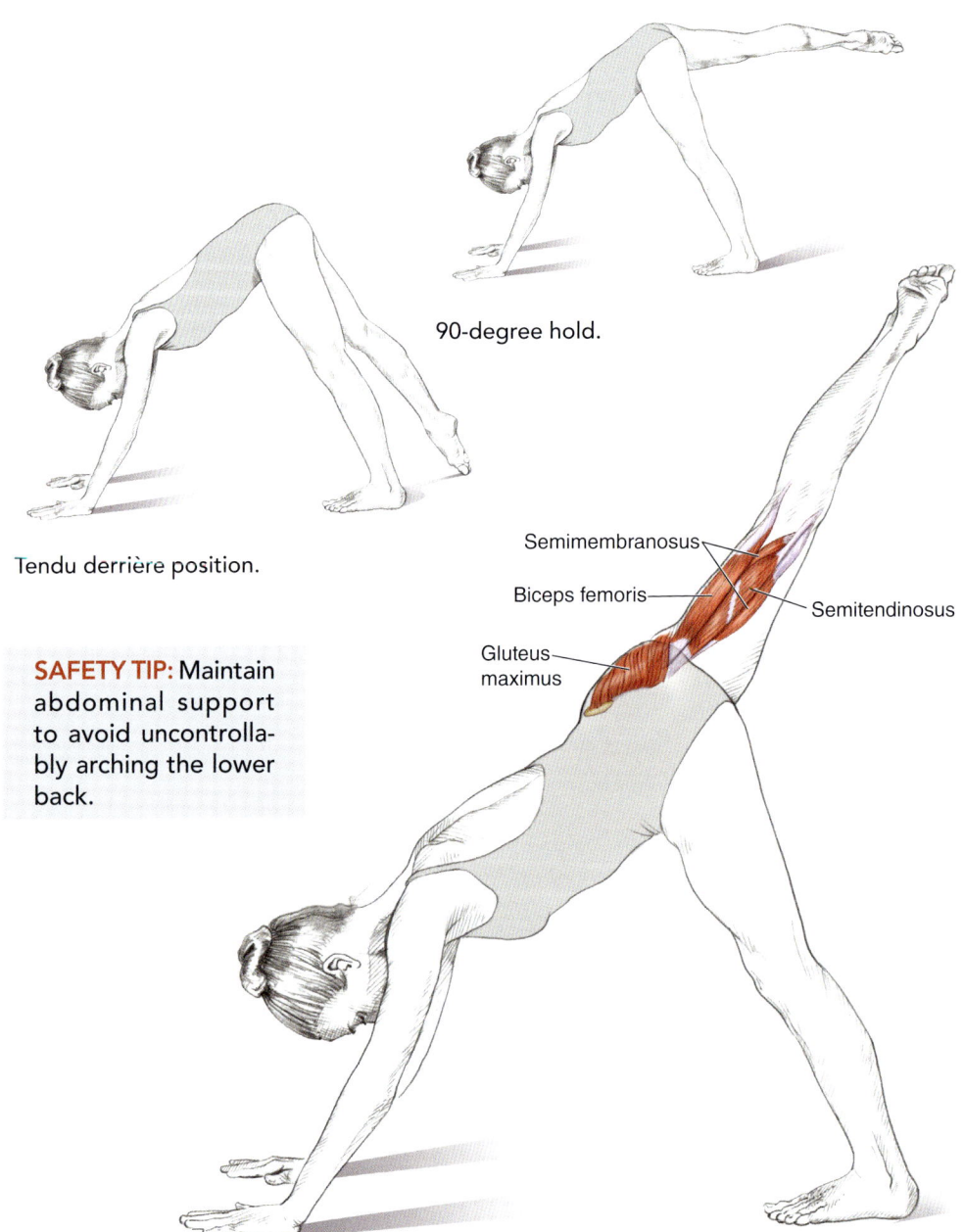

90-degree hold.

Tendu derrière position.

Semimembranosus

Biceps femoris

Semitendinosus

Gluteus
maximus

SAFETY TIP: Maintain abdominal support to avoid uncontrollably arching the lower back.

EXECUTION

1. From a standing position with your legs hip-width apart, slowly roll down until your hands are touching the floor (that is, into an inverted-V position). Reorganize your trunk for balance awareness. Move your right leg into tendu derrière position.

2. As you inhale, move from tendu to arabesque, stopping the movement at 90 degrees. Hold this position for 4 counts as you exhale. On inhalation, continue to lift the leg as high as you can, focusing on the hip extensors.

3. Hold this position for 4 counts as you exhale. Return with control to tendu as you inhale. Resist gravity on the downward phase and focus on eccentric lengthening of the hip extensors. Perform 3 times parallel and 3 times turned out on each side.

MUSCLES INVOLVED

Gluteus maximus, hamstrings (semitendinosus, semimembranosus, biceps femoris)

DANCE FOCUS

Arabesque can be an amazing movement to watch. Executing it requires detailed coordination of hip extension with spinal extension. In keeping with the principle of hip dissociation, remember to work the thigh while resisting uncontrolled lower-back arch and pelvic twisting. Once you achieve support from your core, hip extensors, and hip rotators, let that power

support any pelvic rotation or anterior tilt as the leg goes higher. Feel the arabesque movement being initiated by the hip extensors, as well as the eccentric lengthening of the abdominals to protect your spine. Your upper body must tilt forward slightly to correlate with the leg elevating. Here occurs a graceful tug-of-war, in which the gluteus maximus and hamstrings lift the thigh while the anterior structures of the core lengthen but maintain control of the lower back. This is a beautiful example of strength, flexibility, and coordination.

VARIATION

Resisted Arabesque

Repeat the main exercise but add a resistance band to the foot of the arabesque leg. The foot of the supporting leg stands on the other end of the band. The resistance band should tighten as the leg moves upward from 90 degrees. Reemphasize lumbar control, using the hamstrings and gluteus maximus for hip extension. Perform 3 or 4 times.

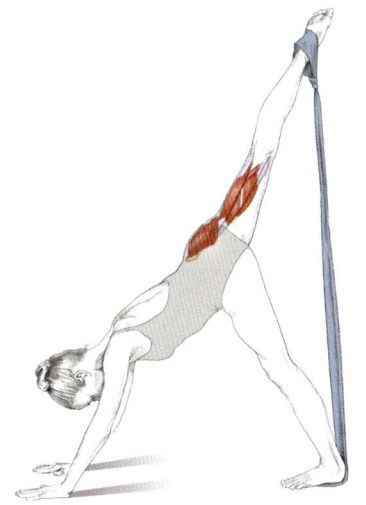

STANDING HIP FLEXOR LIFT

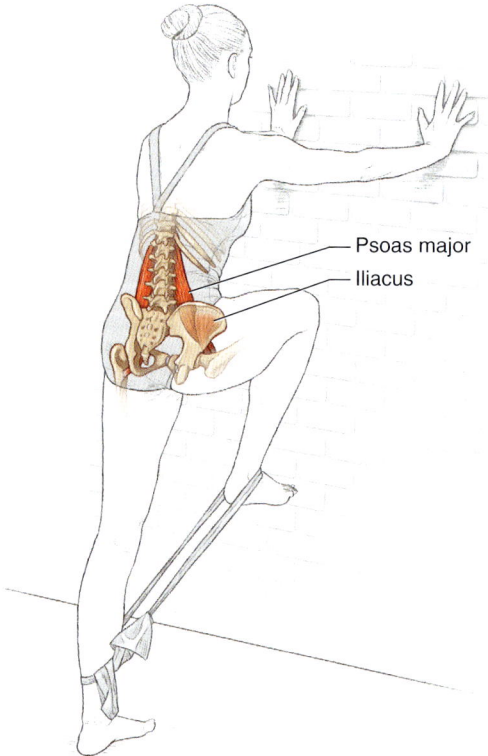

Psoas major

Iliacus

SAFETY TIP:
To emphasize pelvic stability and protect the lower spine, avoid lateral tilt (hip hike) of the working leg.

EXECUTION

1. Begin this exercise without a resistance band to fully understand the contraction of the iliopsoas, then add the band for resistance. Lean forward with both hands on a wall with a neutral spine position, legs hip-width apart, a light resistance band around both ankles (if appropriate). Inhale to prepare.

2. On exhalation, tighten your waist to activate your deep transversus abdominis and lift your right leg into hip flexion. As you lift, try to feel the right thigh folding at the hip and femoral head pulling deep into the hip socket. Maintain a strong and stable lower back and try not to over-activate the outside of the right hip and thigh.

3. Lift the thigh above 90 degrees without losing stability in the lower back. Hold for 2 to 4 counts. Focus on the deep iliopsoas contracting to lift the thigh.

4. Slowly return the right leg to the start position before repeating this exercise 8 to 10 times. Repeat with the other leg.

MUSCLES INVOLVED

Psoas major and iliacus

DANCE FOCUS

Isolating the iliopsoas will be your secret to getting those legs up in the air. If you have flexible hamstrings and lumbar and pelvic stability combined with strength and awareness of the iliopsoas, then you can be confident that your leg height will improve. The standing hip flexor lift exercise serves as preparation for a better développé if you can maintain a stable lumbar spine.

Feel the thigh lifting as high as it can from deep under the low abdominals while visualizing the thigh being pulled deep into the hip socket. Coordinate the lifting of the right thigh with the dropping of the right sit bone and maintaining softness of the lateral thigh muscles; doing so reduces the tendency of the hip to hike up a bit, which takes the work away from the iliopsoas and places it into the tensor fasciae latae and gluteal muscles.

The psoas major has attachments along the anterior aspect of the lumbar spine before moving down to insert at the lesser trochanter. The deep multifidus muscles need to help stabilize your lower back as you move into any développé or high kick movement so your deep iliopsoas can be activated to lift your leg. Lifting the thigh involves concentric contraction; holding the leg up involves isometric contraction.

The bridge with resistance variation (described next) is an excellent way to develop stabilization of the spine and pelvis of the supporting or standing leg as well as strengthening of the gluteus maximus and hamstrings. The resistance of the band against the leg that is lifting will focus on the iliopsoas if you do your best to pull the head of the femur into the socket as you elevate the thigh.

VARIATION

Bridge With Resistance

Begin supine, with your knees bent and your feet on the floor hip-width apart. Place a resistance band around your thighs, close to your knees. Relax your neck and shoulders and locate your neutral spine position. Inhale to prepare. On exhalation, engage your abdominals and elevate your hips into a bridge. The hips should lift high enough to be in alignment with the shoulders and knees. Once you have established a balanced position, inhale to prepare.

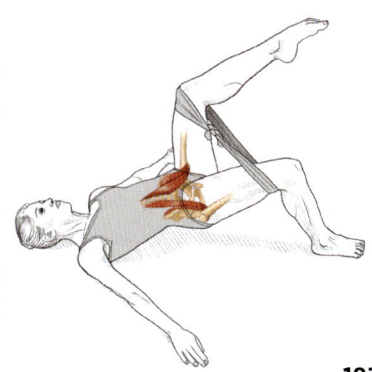

On exhalation, lift one knee toward your chest against the resistance of the band while maintaining pelvic stabilization. Feel the iliopsoas contracting to lift your leg above 90 degrees. Visualize the femur folding at the hip joint against the pelvis while you also feel the resistance of the band against your supporting leg. Hold for 6 counts before lowering the leg and returning to the start position. Perform 10 to 12 times on each side. This exercise simulates the functional work of lifting your leg devant while needing the stability of the supporting hip. You can also turn out the lifted leg.

HIP FLEXOR STRETCH

SAFETY TIP: For comfort, you may place a cushioning pad under the kneeling knee. Keep the exercising knee at a 90-degree angle to avoid compression forces in the knee joint.

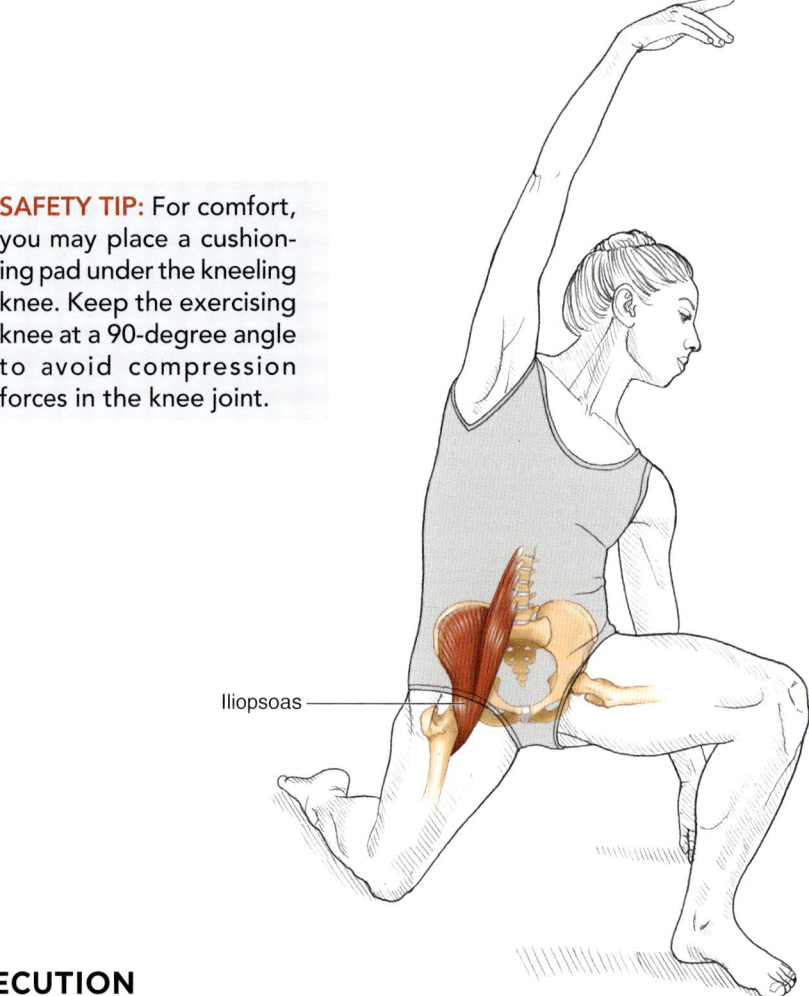

Iliopsoas

EXECUTION

1. Kneel on your right knee. Place your left foot forward on the floor with your left knee bent at 90 degrees. Organize your trunk and lengthen through your spine.

2. Create strong posterior pelvic tilt with the abdominals. While lifting through the waist, focus on balance skills. The right leg is in slight hip extension.

3. Begin a long side cambré to the left with the right arm overhead. Reemphasize the posterior tilt. Hold the right hip stretch for 45 seconds, taking three long, deep breaths. Feel lengthening through the anterior hip and along the right side of your waist. Slowly return. Perform 3 to 5 times on each side.

MUSCLES INVOLVED

Iliopsoas, tensor fasciae latae

DANCE FOCUS

Working intensely on the deep hip flexors may create unwanted tension. Your goal is to isolate the iliopsoas for lifting the legs above 90 degrees—not to create an overuse syndrome. You may need to repeatedly stretch the hip flexors while you work on strengthening the deep hip flexors. Just remember that you will receive more benefits from your stretches if your body is warmed up. Stretching the front of the hip is also beneficial for working the legs in hip extension. Remain in the posterior tilt for the entire stretch. If your pelvis begins to compensate and tilt forward, you lose the effectiveness of the stretch; in fact, you shorten the hip flexors!

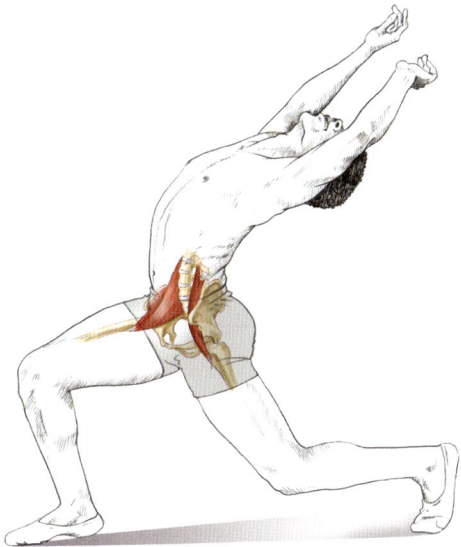

VARIATION

Tensor Fasciae Latae Stretch

Return to the start position. Once you have reestablished a posterior tilt with your hip flexor stretch, gently glide your hips to the right along your frontal plane to feel a gentle stretch in your tensor fasciae latae. You can continue into the cambré left as you hold the posterior tilt and the hip glide to the right. Hold for 30 to 45 seconds before returning to the start position. Perform 3 to 5 times on one side before the other side.

PASSÉ

This chapter focuses on stabilizing the pelvis and separating movement from the pelvis and the femur—that is, hip dissociation. Let's now examine passé, the name of which is the French word for "passed," which in this case refers to the foot of the gesturing leg passing by the inside of the knee of the supporting leg.

1. Begin with your legs and feet in fifth position and the right leg in front. (We will leave the arms and lower legs out of this discussion to focus on alignment of the trunk, pelvis, and thigh.) Organize your posture to incorporate your neutral spine and pelvis position. Feel the deep abdominals gently pulling in. Feel the hip adductors working, as well as the deep hip external rotators.

2. Before you begin, feel lengthening through your spine and think about the supporting leg. Feel the hip external rotators contracting deep in the back of your hip. Feel turn-out rotation all the way down the legs to the floor. Maintain a strong foot and arch, keeping your weight evenly placed on all five toes and the heel.

3. Move the front leg (right) into a cou-de-pied position, maintaining external rotation of both the gesturing leg and the supporting leg. Feel opening in the front of the hips and a deep, strong contraction of the hip external rotators of both legs. The pelvis should remain stable. The gluteus medius of the supporting hip should feel a deeper contraction to maintain your balance and pelvic stability. Continue to feel the anterior hip bones and the pubic bone aligned directly along the frontal plane.

4. Point your right foot so that its fifth toe can be placed lightly along the inside of your supporting leg and continue to slide the right leg up the inside of the supporting leg. While maintaining turnout, think about the femur moving directly along the frontal plane. Think about the outside of the right leg contracting to keep the knee directly out to the side.

5. Once your gesture leg has reached the inside of the knee of the supporting leg, continue the movement of opening the front of the hips while maintaining strong hip external rotation. Now, gently move the right foot behind the knee of the supporting leg to begin the controlled descent back to fifth position derrière. Maintain a strong contraction of the hip external rotators of both legs working through the entire range until you can close into fifth position with the right foot in the back.

Muscles Involved

Trunk placement: Deep transversus abdominis, internal oblique, external oblique, multifidus, pelvic floor muscles (coccygeus, levator ani)

Cou-de-pied to passé: Obturator internus, obturator externus, piriformis, quadratus femoris, gemellus inferior, gemellus superior, posterior gluteus medius, sartorius

Supporting leg: Obturator internus, obturator externus, piriformis, quadratus femoris, gemellus inferior, gemellus superior, posterior gluteus medius, lower fibers of the gluteus maximus, adductor longus, adductor brevis, adductor magnus, gracilis

CHAPTER 9

Legs

Though of course you dance with your entire body, the magic of dance reveals itself in the beauty of the legs and feet. All dance styles show off the capabilities of the legs, which defy gravity and challenge the limits of what is humanly possible. This aesthetic quality provides you with the means to communicate with your audience. This chapter discusses the anatomy of your legs and focuses on precision—that is, the degree of refinement in the movement of your legs. Precise movement requires accuracy and coordinates speed in your muscle contractions.

Let's explore the bones and muscles that contribute to the beauty of your legs. The femur, which is the longest and strongest bone in the body, angles down from the pelvis to form the top of the knee joint (figure 9.1). It has numerous muscle attachments that help create the precision of your dance movements and skills. The knee joint, the largest joint in the body, is a hinge joint supported by strong ligaments. Your knee joint may receive a load equal to three times your body weight, or more, during jump landings. The patella (kneecap) is a free-floating bone within the tendon of the thigh muscle group (quadriceps femoris) that inserts into the tibia. The patella functions as a pulley for your thigh muscles; good thigh strength is important for stability and alignment of the patella. During knee flexion and extension, the patella moves in a gliding pattern, meaning that as the knee flexes, the patella glides downward, and as the knee extends, it glides upward. If an imbalance of strength exists within the quadriceps muscles, the patella could glide abnormally, putting you at risk for injury. For example, the patella may exhibit abnormal lateral glide if the lateral thigh muscles are overused on landing from jumps, which could put you at risk for injury.

There are four main ligaments of the knee joint: the medial collateral ligament (MCL), which connects the femur and tibia; the lateral collateral ligament (LCL), which connects the femur and fibula; and the anterior cruciate ligament (ACL) and posterior cruciate ligament (PCL), which cross over each other and connect the femur and the tibia. These four

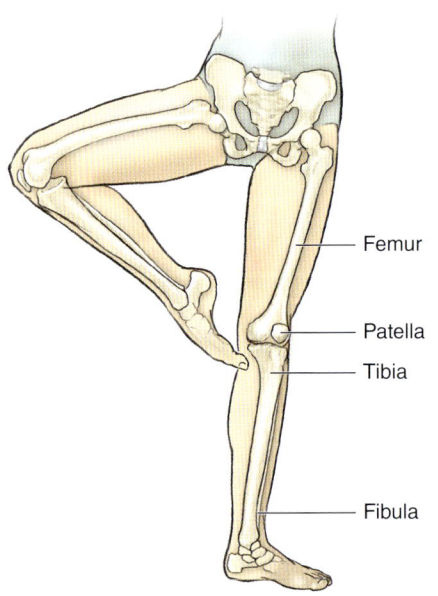

Femur

Patella

Tibia

Fibula

Figure 9.1 Bones of the leg.

ligaments provide support and can be severely injured if alignment is compromised, especially when landing from jumps. The MCL is very strong and helps provide stability for the medial (inner) side of your knee joint, whereas the LCL provides stability for the lateral (outer) side of the joint.

The knee ligaments and their connections relate to precision of movement. For one thing, the femur needs to stay aligned over the tibia, especially during landings; any deviation allows the femur and tibia to twist abnormally, which causes serious stress in the ligaments. The ACL helps keep your tibia secure from moving anteriorly, and the PCL keeps it from moving posteriorly. Serious forward displacement of the tibia from the femur can cause the ACL to tear, which can require a long recovery. To prevent injury, it is crucial to maintain healthy, strong knees that can take on the forces of jumping, twisting, and turning. The wall sit exercise presented in this chapter emphasizes aligning the knees directly over the toes while moving the legs along the sagittal plane.

The femur is held tightly in the hip socket, or acetabulum, by strong ligaments—iliofemoral, pubofemoral, and ischiofemoral—whose names correlate with the bones they connect. When you lift your leg to the front, all three ligaments relax a bit to give you a greater range of motion; they become tight, however, when you lift your leg to the back or tuck your pelvis under. The iliofemoral ligament, also called the Y ligament (due to its shape), is extremely strong and therefore contributes to hip stability and control of body placement. Tightness in this ligament can limit turnout in your hip. To loosen the Y ligament and enable more turnout, some dancers tilt the pelvis forward, which takes them out of a healthy neutral position. This unnecessary anterior tilt of the pelvis also creates tightness of the iliopsoas and erector spinae muscles of the lower spine.

Muscle Awareness

In chapter 8, we examined the lateral hip muscles, the deep external rotators, and the iliopsoas. Now let's look at the anterior (front), medial (side), and other posterior (back) leg muscles. The anterior muscles of the thigh make up the quadriceps group. The largest muscle in this group is the rectus femoris, which runs from the iliac spine to the tibia, crossing the hip joint (figure 9.2*a*). The other three quadriceps muscles are the vastus medialis, vastus intermedius, and vastus lateralis; here again, the names relate to the locations. These muscles originate along the medial, lateral, and anterior upper femur and insert into the patellar tendon.

These muscles flex the hip and extend the knee. The vastus medialis is especially important for maintaining healthy alignment of the patella; it also contracts during knee extension in the last 15 degrees of the upward phase of demi-plié. We can also include here the sartorius muscle, which begins at the upper iliac spine and runs down to the inside surface of the tibia. This muscle, which is the longest in your body, helps extend the knee and participates in turnout. All these muscles are very strong and help you maintain full extension in the knee of your supporting leg. They extend the knees on the upward phase of the plié and complete a développé movement.

The adductors, or inner thighs, begin along various aspects of the pubis bone and attach along various aspects of the medial femur. They include the adductor longus, adductor brevis, adductor magnus, pectineus, and gracilis (figure 9.2*b*). They adduct your thigh and can also bring your leg to the front and back at lower levels. Many seasoned ballet dancers believe that the adductors are important for holding their legs in external rotation, especially when both legs are on the floor. For example, in first position relevé, activating the adductors provides added pelvic stability and security of turnout.

The hamstring muscles line the back of the thigh. The biceps femoris originates along the femur and the ischial tuberosity, or sit bone, and inserts into the lateral tibia and fibula. The semitendinosus and semimembranosus originate at the ischial tuberosity and insert into the medial tibia. The hamstring muscles flex the knee and extend the hip; the biceps femoris activates strongly in arabesque movements. The hamstrings also play an important role in body placement. If you activate the hamstring muscles and the abdominals while standing, you can coordinate excellent alignment of the pelvis. This effect allows your standing leg to be more stable so that you don't have to overuse or grip the quadriceps muscles.

We must not forget the gluteus maximus. It originates from the posterior surface of the ilium, sacrum, and coccyx bones and inserts into the femur; it also has fibrous attachments along the iliotibial band. Together, the gluteus maximus and the hamstrings initiate every swing kick to the back, battement derrière, and arabesque. The hamstring curl exercise focuses on engaging the deep abdominals while activating the back of the thighs and buttocks. The gluteus maximus is the strongest hip extensor, and some of the lower fibers can play a role in external rotation. But note that if you are unable to locate and use the deep external rotators, you might overuse the gluteus maximus and tuck your pelvis under, thus limiting turnout.

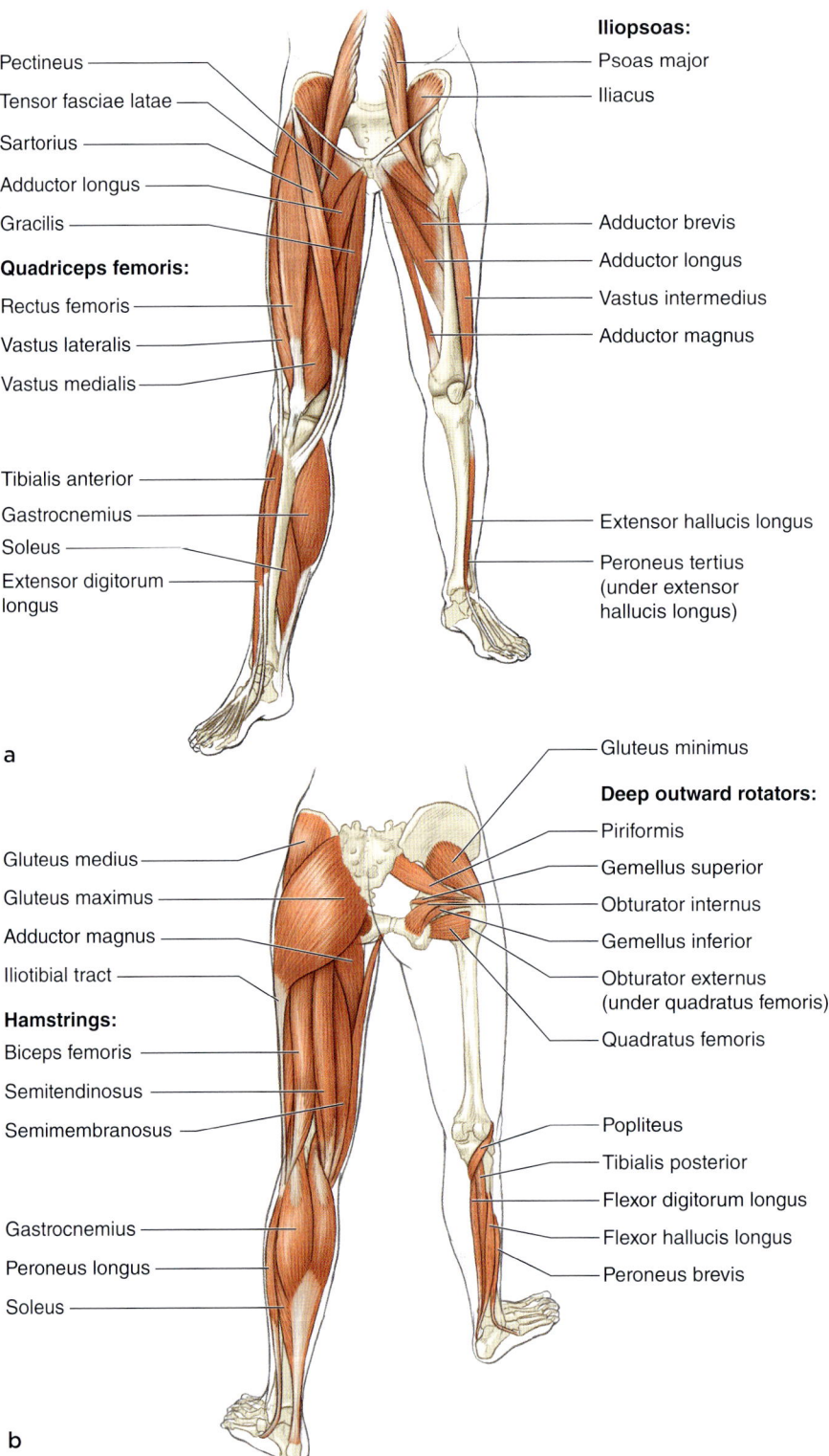

Pectineus

Tensor fasciae latae

Sartorius

Adductor longus

Gracilis

Quadriceps femoris:

Rectus femoris

Vastus lateralis

Vastus medialis

Tibialis anterior

Gastrocnemius

Soleus

Extensor digitorum
longus

Iliopsoas:

Psoas major

Iliacus

Adductor brevis

Adductor longus

Vastus intermedius

Adductor magnus

Extensor hallucis longus

Peroneus tertius
(under extensor
hallucis longus)

a

Gluteus medius

Gluteus maximus

Adductor magnus

Iliotibial tract

Hamstrings:

Biceps femoris

Semitendinosus

Semimembranosus

Gastrocnemius

Peroneus longus

Soleus

Gluteus minimus

Deep outward rotators:

Piriformis

Gemellus superior

Obturator internus

Gemellus inferior

Obturator externus
(under quadratus femoris)

Quadratus femoris

Popliteus

Tibialis posterior

Flexor digitorum longus

Flexor hallucis longus

Peroneus brevis

b

Figure 9.2   Muscles of the leg: (a) front; (b) back.

Precision of Leg Movement

In chapter 8, we examined the need to improve extensions, but so many dancers struggle with overuse of the quadriceps when attempting to lift the legs higher than 90 degrees. With any leg lifts to the front, especially when turned out, the head of the femur must drop down and in as the leg begins to elevate (figure 9.3).

Visualize the sit bone of the lifting leg reaching down and pulling inward. The iliopsoas engages to produce a concentric contraction while the gluteus maximus and global lower-back muscles lengthen. The deep local lower-spine muscles must contract to stabilize your lower back. The supporting leg must hold steady by engaging the hamstrings and hip abductors. Any time you begin a movement with a hip hike (i.e., elevate the hip), you engage the anterior fibers of the gluteus minimus, gluteus medius, and tensor fasciae latae, which will begin to turn your leg inward. The deep external rotators must work to keep the femur rotated externally throughout the entire range. Remember the principle of axial elongation from chapter 4: Lengthen through your spine and engage the core musculature.

If you are performing a développé type of movement, the head of the femur must, again, glide downward and inward and continue to turn out with help from the iliacus. The knee comes into the ribs as high as possible, with the focus on the psoas. Then you can begin the concentric contraction of the thighs to straighten the knee. Once you have tightened the quadriceps, they cannot assist you in creating more elevation—your développé is finished.

When your knee is bent, the supporting ligaments loosen, which means that the stability of your knee depends on the strength of the muscles. As the knee extends, a small amount of anatomical rotation occurs within the joint. With

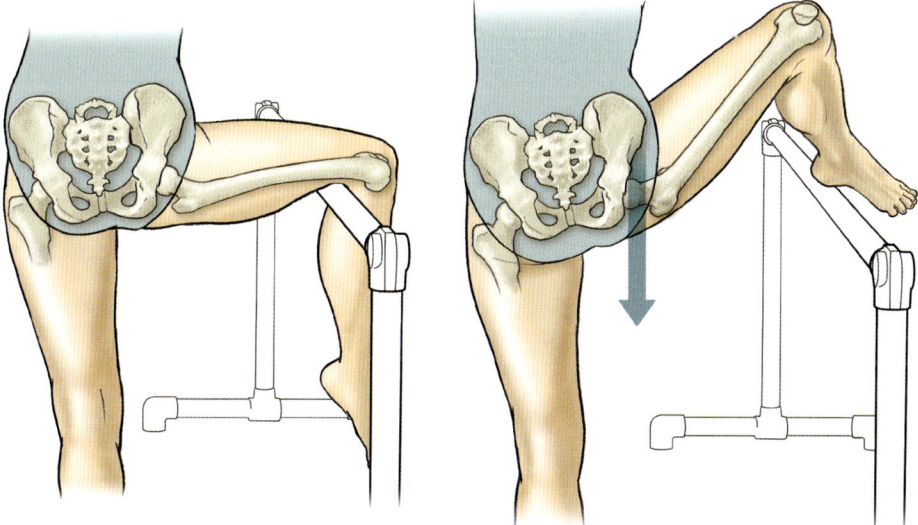

Figure 9.3　Movement of the femur in the hip socket.

this in mind, you can reduce the risk of knee injury by performing controlled landings in which the thighs and knees align over the toes. Whenever the legs are coming downward, whether from an aggressive kick or from a jump, think about precise control. In this situation, your muscles need to change course quickly and contract to resist gravity. For instance, returning from grand battement to the front requires both reorganization through the trunk and concentric contraction of the hip extensors. The descending battement exercise is a great way to think about control on the return of a movement.

Returning safely and effectively from a jump requires eccentric control through the quadriceps, hamstrings, and lower-leg plantar flexors (discussed in the next chapter). For now, remember (from chapter 1) that an eccentric contraction involves the muscles working but lengthening at the same time. Eccentric contractions occur on the downward phase of a movement for the purpose of control. On landing, for example, your knee is responsible for about one-third of the muscle work. The landing can be softened by rolling through the toes and forefoot and into the heels with eccentric control. The knee and hip can then bend with control to absorb the rest of the forces. Avoid putting all your effort into the takeoff phase and therefore lacking the capacity to exercise control during the landing phase. So many injuries occur while landing from jumps!

Dance-Focused Exercise

Each of the following exercises relates directly to your technique. As you perform them, think about moving along the most efficient path. In other words, engage your core musculature for supportive placement, and recruit only the muscles needed to accomplish the movement. Table 9.1 summarizes these muscles. Unwanted muscle activity wears you out, whereas energy conservation allows you to dance longer with precision. For example, you don't need to overwork your neck and shoulders just to lift your leg to the back. Overuse of the neck and shoulders is a hindrance, causes fatigue, and increases injury risk. Instead, use your new principles of dancing:

1. Plumb-line placement for spinal and postural awareness
2. Hip dissociation for thigh movement without spinal or pelvic movement
3. Trunk stabilization to increase controlled movement
4. Effective breathing for engaging core muscles

Of course, this is a lot to think about, but once you practice new movement strategies, using them becomes automatic. You will then be able to stabilize one body part, move freely with another, and enhance your performance. You will see this benefit in the développé, which is analyzed in detail after the last of the following exercises.

Table 9.1   Leg Muscles

Muscle	Origination	Insertion	Action
Rectus femoris	Anterior superior iliac spine	Patella to tibial tuberosity	Flexes thigh, extends knee
Vastus medialis	Intertrochanteric line of femur	Patella to tibial tuberosity	Extends knee, stabilizes patella
Vastus intermedius	Anterior and lateral shaft of femur	Patella to tibial tuberosity	Extends knee
Vastus lateralis	Greater trochanter	Patella to tibial tuberosity	Extends knee
Sartorius	Anterior superior iliac spine	Proximal tibia	Flexes thigh, flexes knee, externally rotates thigh, abducts thigh
Adductor longus	Pubic bone	Middle third of femur	Adducts thigh, flexes thigh, externally rotates thigh
Adductor brevis	Pubic bone	Proximal third of femur	Adducts thigh, flexes thigh
Adductor magnus	Pubic bone	Medial femur	Adducts thigh
Pectineus	Pubic bone	Femur: distal to lesser trochanter	Adducts thigh, flexes thigh, externally rotates thigh
Gracilis	Pubic bone	Medial tibia	Flexes knee, adducts thigh
Biceps femoris	Long head: ischial tuberosity Short head: posterior femur	Head of fibula	Flexes knee, extends hip, externally rotates at hip and knee
Semitendinosus	Posterior medial ischial tuberosity	Proximal medial tibia	Extends thigh, flexes knee, internally rotates knee when knee is flexed
Semimembranosus	Superolateral ischial tuberosity	Posterior medial tibia	Extends thigh, flexes knee, internally rotates hip and knee
Gluteus maximus	Posterior ilium, posterior superior iliac crest, posterior inferior sacrum, coccyx	Posterior femoral surface, iliotibial band	Extends hip, abducts thigh, externally rotates thigh
Gluteus medius	Ilium	Greater trochanter	Abducts thigh; anterior fibers internally rotate thigh, posterior fibers externally rotate thigh
Gluteus minimus	Ilium	Greater trochanter	Abducts thigh; anterior fibers internally rotate thigh, posterior fibers externally rotate thigh
Tensor fasciae latae	Anterior superior iliac spine, anterior iliac crest	Iliotibial band	Internal and external fibers rotate thigh, stabilize hip

ASSISTED ATTITUDE DEVANT

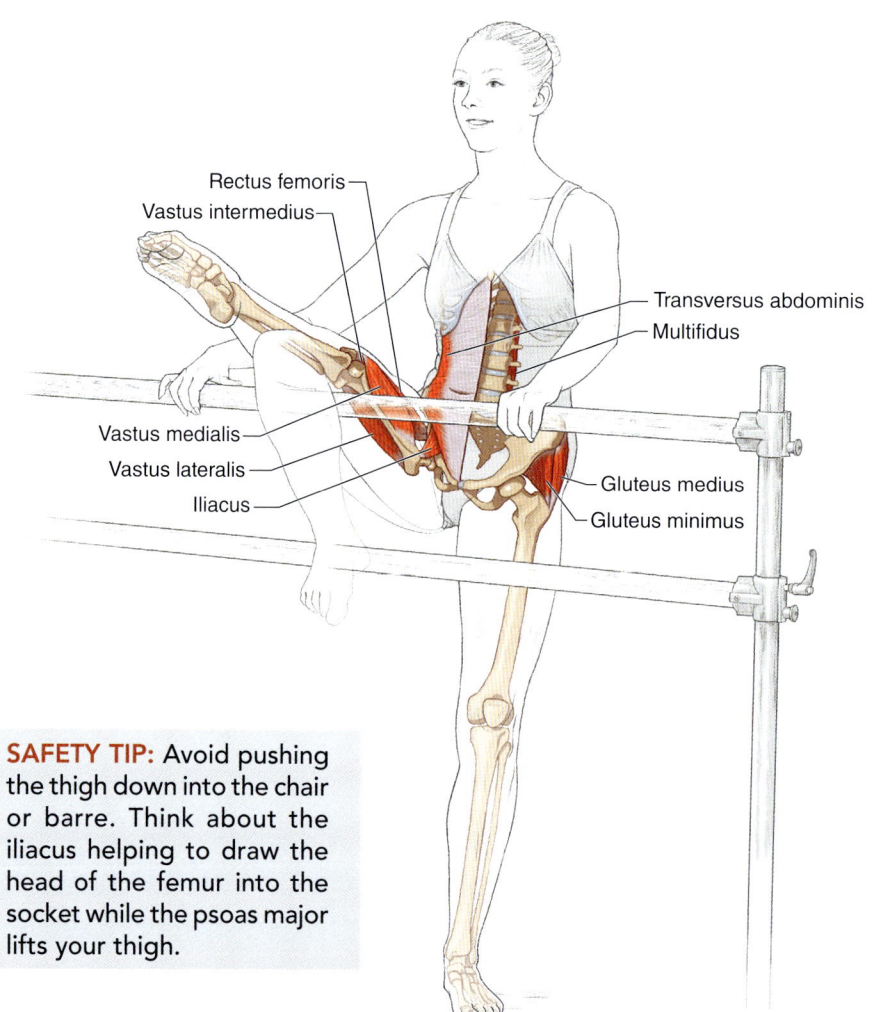

Rectus femoris
Vastus intermedius
Transversus abdominis
Multifidus
Vastus medialis
Vastus lateralis
Iliacus
Gluteus medius
Gluteus minimus

SAFETY TIP: Avoid pushing the thigh down into the chair or barre. Think about the iliacus helping to draw the head of the femur into the socket while the psoas major lifts your thigh.

EXECUTION

1. Begin with the right leg flexed at the hip and the knee bent, resting on the back of a chair or low barre, as if in an attitude devant position.

2. Maintain a strong stable spine and supporting leg; arms can be overhead or the hands can grasp the bar for support.

3. Inhale to prepare. On exhalation, engage your deep transversus abdominis and feel the action of the right head of the femur pulling in and down to avoid right lateral hip hike.

4. Engage the deep iliopsoas to begin to gently elevate the thigh off the back of the chair. Extend the knee using the quadriceps as if you are completing a développé devant.

5. Visualize the inside of the right femur at the lesser trochanter slightly floating up and away from the back of the chair or barre with the contraction of the iliopsoas. Visualize the lower leg floating up with the contraction of the quadriceps.

6. Feel the hip abductors of the left leg holding your pelvis stable. Hold for 2 to 4 seconds, then slowly lower the thigh and bend the knee to return to the start position.

7. Repeat this exercise 10 times before moving to the other side.

MUSCLES INVOLVED

Standing leg: Gluteus medius, gluteus minimus

Gesture leg: Rectus femoris, vastus medialis, vastus intermedius, vastus lateralis, iliacus

Spine: Psoas major, transversus abdominis, multifidus

DANCE FOCUS

We are starting the chapter's exercises with one of the most challenging and functional movements of all. While there are so many muscles involved in this exercise, the muscles for you to emphasize have been listed here. The transversus abdominis and multifidus work to stabilize your spine and pelvis with the goal of allowing the gesture leg to float up. Visualizing the head of the femur pulling down and in while floating the leg up can help you be mindful of not hiking the hip up. You should feel the thigh folding at the hip joint. As the knee extends, isolate the quadriceps to help straighten the knee but not to drive the thigh down.

Think back to segments of this book discussing the importance of the small, local stabilizing muscles that need to work to hold the lower back stable so the iliopsoas can lift your leg above 90 degrees. Coordinating all the muscles needed to stabilize the spine and supporting leg while the muscles can function to lift the thigh and extend the knee will help you improve your alignment, strength, and height of développé devant.

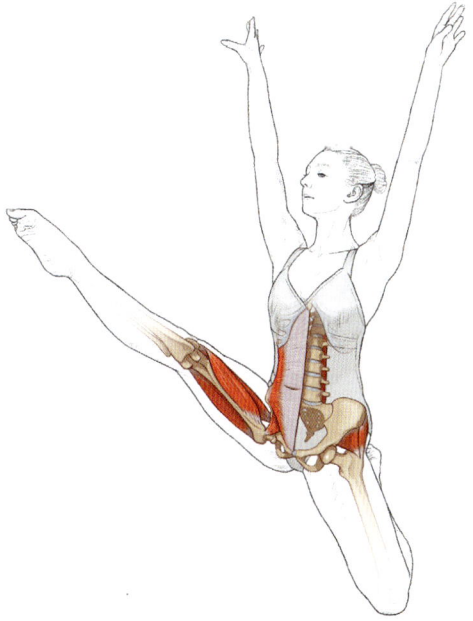

WALL SIT

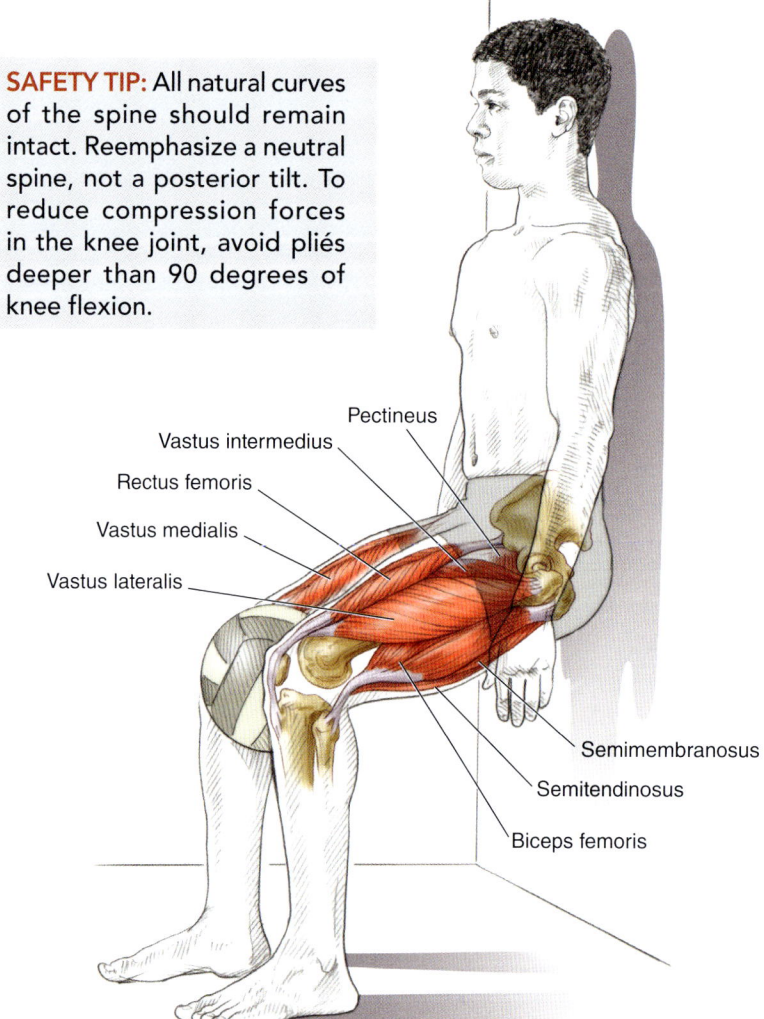

SAFETY TIP: All natural curves of the spine should remain intact. Reemphasize a neutral spine, not a posterior tilt. To reduce compression forces in the knee joint, avoid pliés deeper than 90 degrees of knee flexion.

Pectineus

Vastus intermedius

Rectus femoris

Vastus medialis

Vastus lateralis

Semimembranosus

Semitendinosus

Biceps femoris

EXECUTION

1. Stand with your back against a wall. Bring your heels away from the wall about 2 feet (0.6 m). Place a small ball between your knees and lean into the wall. Inhale to prepare.

2. Exhale and perform a parallel demi-plié by sliding down the wall. Feel your weight placed evenly throughout your feet. Reemphasize pressure through the heels if necessary. Contract the adductors into the ball.

3. Hold that position for 2 to 4 counts, thus creating an isometric contraction. Slide up the wall to return. Repeat with a deeper demi-plié, taking the thighs parallel with the floor. Hold for 2 to 4 counts, then slide up the wall. Perform the series 10 to 12 times.

MUSCLES INVOLVED

Hamstrings (semitendinosus, semimembranosus, biceps femoris), quadriceps (rectus femoris, vastus lateralis, vastus medialis, vastus intermedius), adductor longus, adductor brevis, adductor magnus, gracilis, pectineus

DANCE FOCUS

You will notice that the challenge of this exercise increases when you bend your knees a little deeper. Whether performed in parallel or turned out, a deeper squat will create compression under the patella. Grand plié can be a wonderful exercise as long as your quadriceps are strong enough. As your knee flexes and your plié deepens, the contraction of the quadriceps increases and the compressive forces of the patella against the femur increase as well. In fact, grande plié can put a load equivalent to about seven times your body weight directly into the knee joint. Imagine how much that is with every grande plié! Perhaps, then, grande plié should be used a little later in a ballet technique session; doing so would allow more time for the legs to get warmed up.

You need impeccable quadriceps strength to execute the hinge of the Horton modern technique, where your body weight is back and your knees and thighs maintain your weight and stability. Strong quadriceps contraction is also needed, along with deep knee flexion, for the grande cambré lunge in classical ballet. In addition, various contemporary styles of choreography may require you to bear all your weight on your knees while turning. If you are using this exercise to emphasize alignment of the knees over the toes, then 10 to 12 repetitions may be enough. But if you are looking to gain strength, then repeat it to fatigue.

KNEELING HAMSTRING CURL

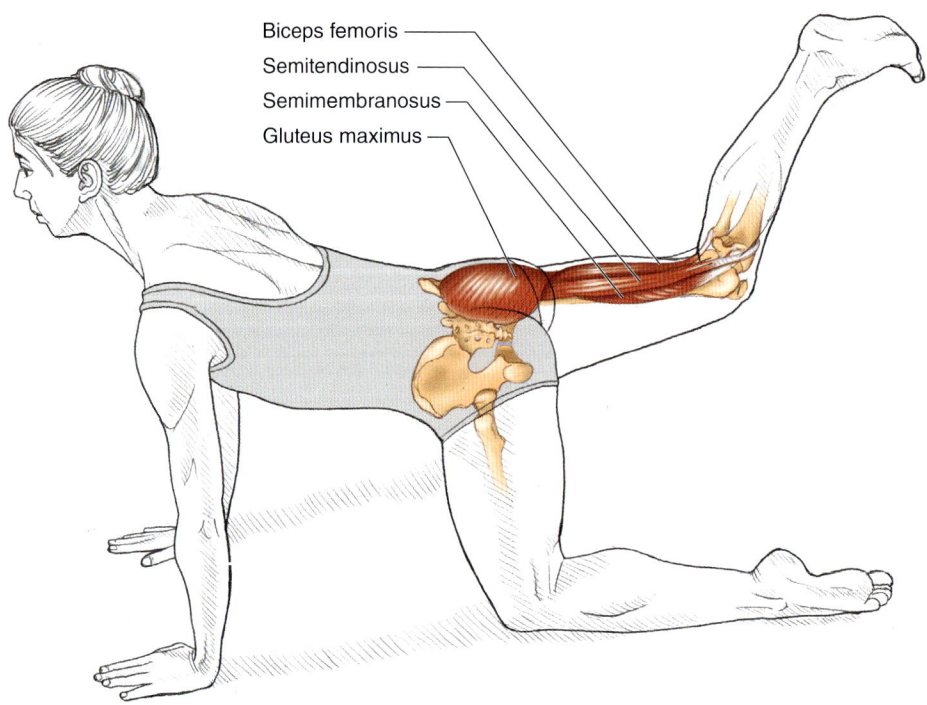

Biceps femoris
Semitendinosus
Semimembranosus
Gluteus maximus

EXECUTION

1. Get into a quadruped position (i.e., on your hands and knees) with your shoulders directly over your wrists and your hips directly over your knees. Extend your right leg to the back while maintaining a strong deep abdominal contraction to support your lower spine. Inhale to prepare.

2. On exhalation, engage the deep abdominals and lengthen through the spine. Engage the hamstrings and gluteus maximus as you bend the right knee without allowing the thigh to drop. Hold for a count of 4, then extend the knee. Perform 10 to 12 times, then switch legs.

3. Lengthen through the front of the right hip as the knee bends and feel a strong cocontraction of the hamstrings and gluteus maximus. To advance this exercise, add an ankle weight.

SAFETY TIP: Reinforce the deep abdominals to protect the lower spine. This exercise also engages stabilizing muscles along the spine. Resist arching the lower back—work to stay in your neutral pelvic position.

MUSCLES INVOLVED

Hamstrings (semitendinosus, semimembranosus, biceps femoris), gluteus maximus

DANCE FOCUS

In addition to providing support in perfect body placement, the hamstrings flex the knee and extend the hip. Thus, they execute two-joint action! Some dancers have hyperextended knees, meaning that the knees can continue past full extension because of laxity and posterior gravitational torque. To help control hyperextension, activate the hamstrings a little sooner. The hamstrings work for you each time you execute coupé, passé, and attitude positions in ballet, as well as barrel turns and stag leaps in jazz. The biceps femoris also helps with turnout; you should feel it contract with attitude derrière and turned-out arabesque. Try to think about hip dissociation: Move your thighs to the back as far as you can without any movement in your lower spine. Challenge yourself to move the thighs against the resistance of the pelvis and spine.

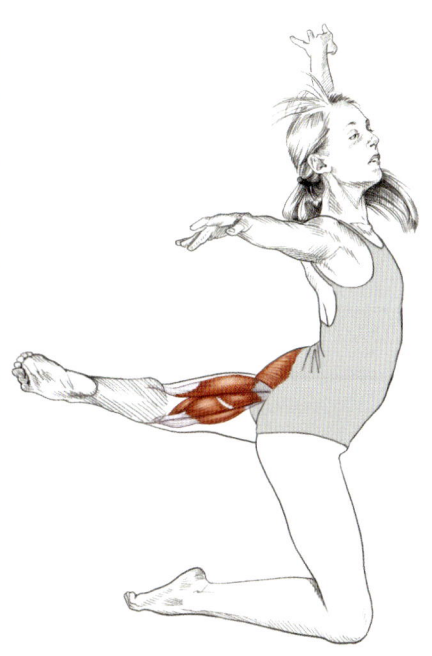

VARIATION

Hovering Plié

Lie facedown in a turned-out demi-plié position while maintaining a neutral position of the pelvis. Reemphasize deep abdominal support for the lower back. Inhale to prepare. On exhalation, contract the lower abdominals. Lift both legs slightly off the floor by engaging the deep rotators and the biceps femoris. Let your legs hover 1 to 2 inches (2.5 to 5 cm) over the floor. Emphasize the deep low external rotators. Hold this position for 2 to 4 counts, then slowly return the thighs to the floor with control. Perform 10 to 12 times.

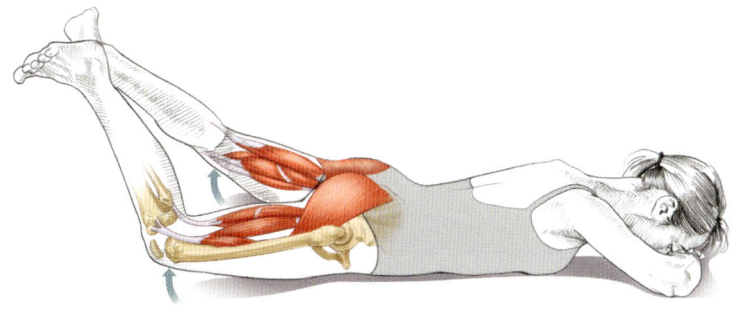

SUPPORTED HAMSTRING LIFT

Start position.

SAFETY TIP: Activate the abdominals, then lift with the hamstrings. Don't release tension in the lower back, which will let your momentum carry your leg up without control! Doing so will eventually wear down the lower segments of your spine and tighten the lower back, thus causing an overuse injury.

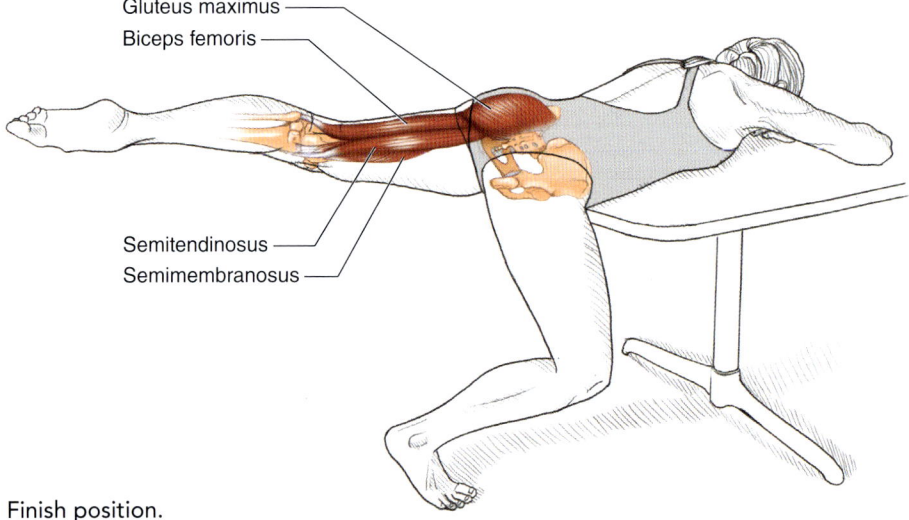

Gluteus maximus
Biceps femoris
Semitendinosus
Semimembranosus

Finish position.

EXECUTION

1. Lay your upper body over a table with your feet on the floor. The table's edge should be firm against the hip flexors, and your hands should be under your forehead. Inhale on the preparation.

2. On exhalation, engage the deep abdominals and lift one leg off the floor with a straight knee, contracting the hamstrings and gluteus maximus. Do not allow your pelvis or lower back to move. Hold for 4 counts, then return slowly. Perform 10 to 12 times, then switch sides.

3. Feel a strong cocontraction of the hamstrings and gluteus maximus and focus on lumbar stabilization. To advance this exercise, add an ankle weight.

MUSCLES INVOLVED

Hamstrings (semitendinosus, semimembranosus, biceps femoris), gluteus maximus

DANCE FOCUS

As you have learned, your hamstrings originate at the sit bones. Because of their connection with the pelvis, weakness in the hamstrings can cause ineffectiveness in pelvic alignment. Think about your plumb line. Weakness of the hamstring complex allows your pelvis to tilt forward, thus moving you out of optimal body placement. In contrast, a firm balance between strength in the hamstrings and the lift of the abdominals facilitates a balanced pelvis and lower back. Thus, while you need extreme flexibility in your hamstrings, it is also important to maintain strength.

Your hamstrings help you with arabesque and powerful jumps. To give you added support, practice leg lifts to the back with a new awareness of engaging the hamstrings with the abdominals. As the arabesque goes higher, maintain lower abdominal support, and shift your upper torso for-

ward, thus emphasizing spinal extension in the upper back and chest while maintaining that abdomen–hamstring connection. The exercise variation supports the spine and allows you to isolate the hamstrings and gluteus maximus without engaging the spinal extensor muscles. The back of your thigh contains fast-twitch muscles that move your knee and hip through all levels of rapidly changing dance movements. Sometimes, the top of your thigh will overpower your hamstrings. Continually work on strengthening your hamstrings.

VARIATION

Resisted Dégagé

While standing, firmly attach one end of a resistance band to a fixed point in front of you with the other end around your right ankle. Hold the supporting leg in a small demi-plié. Move your right leg into tendu derrière. Breathing comfortably, lift your right leg into dégagé while focusing on the hamstrings and gluteus maximus. Maintain abdominal control to support your lower spine. Lengthen through the front of the right hip and allow no movement in your pelvis and lower spine. Hold for 2 to 4 counts, then slowly return to tendu. Perform 10 to 12 times, then switch legs.

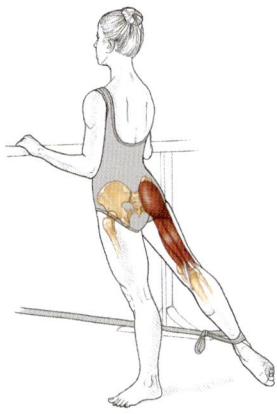

SIDE SCISSOR

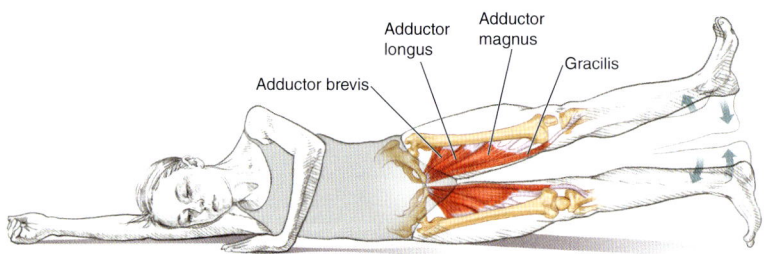

EXECUTION

1. Lie on your right side with your right arm extended overhead, your head on your right arm, and your left arm on the floor in front of you. Both legs are extended. Maintain a neutral spine and maintain lift in the waist on both sides of your body. Stack your knees one on top of the other. Inhale to prepare.

2. On exhalation, turn out and lift the top leg, then turn out and lift the bottom leg. Engage your deep abdominals to maintain a secure trunk. If balance is compromised, bring the legs forward slightly by flexing the hips. Keep your spine and pelvis neutral.

3. Execute small inner-thigh pulses. Feel contraction in the pelvic floor, deep transversus abdominis, and adductors. Perform the pulses for 10 to 12 counts before slowly returning with control. Perform the series 6 to 8 times, increasing the tempo with each set.

SAFETY TIP: The bottom leg must remain turned out to avoid compression of the greater trochanter against the floor. Maintain a deep abdominal contraction for spinal stability.

MUSCLES INVOLVED

Adductor longus, adductor brevis, adductor magnus, gracilis

DANCE FOCUS

Most dancers seem to spend more time stretching the adductors than strengthening them. The adductors and the gluteus medius work together to aid pelvic stability. Visualize the originations and insertions of the inner-thigh muscles; they line the medial portion of the femur to connect with the pelvis. Even though they lose effectiveness at leg heights above about 50 degrees, they are very active at lower levels in flexion, extension, and, of course, adduction. Irish dancers use the adductors frequently when the legs are crossed to give the audience an illusion of seeing only one knee from the front. The same principle applies when performing bourrées in ballet: The adductors remain contracted to cross the legs. Strong inner thighs are also required for jumping combinations with leg beats and for fourth and fifth positions in ballet, which call for contraction of the inner thighs for pelvic stability. Begin practicing the side scissor slowly and with control, then increase the speed of the leg beats to improve precision.

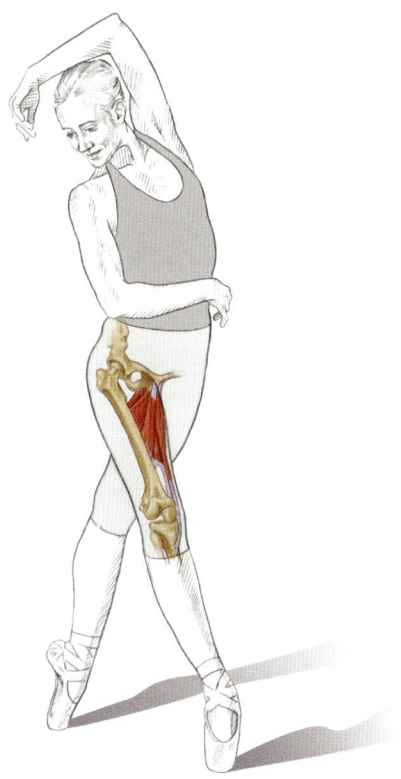

ASSISTED DÉVELOPPÉ

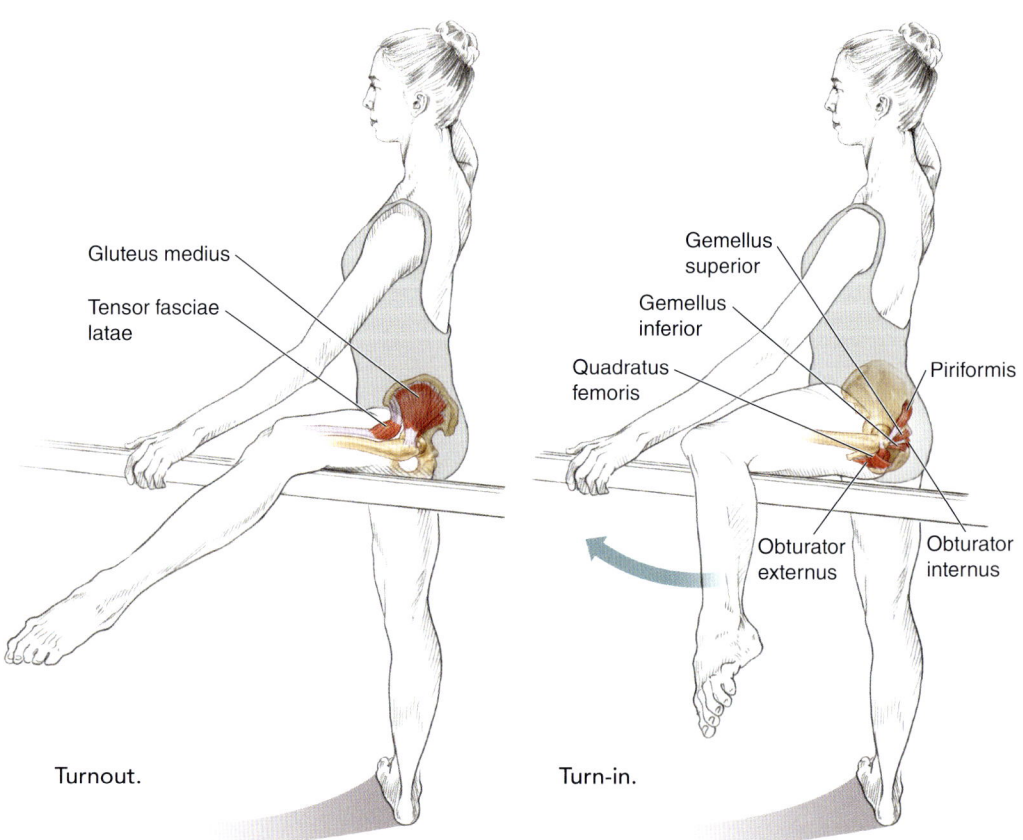

Turnout.   Turn-in.

EXECUTION

1. Stand with your left hand on the barre. The inside (left) leg lies over the barre beneath your knee in à la seconde. Organize your placement: Turn out your standing (right) leg and place your right hand on your shoulder. Your left thigh, which is on the barre, must be higher than 90 degrees.

2. Turn the left thigh inward and then outward, noting the hip hike with turn-in and the deep low rotators on the external rotation. Perform 4 times.

3. After completing the last turnout of the thigh, begin to extend or unfold the knee by lifting the lower leg, not by allowing your thigh to bear down into the barre. Keep your leg resting on the barre while engaging the deep hip external rotators and the iliopsoas.

SAFETY TIP: Avoid twisting in the knee of the supporting leg.

MUSCLES INVOLVED

Internal rotation: Anterior fibers of the gluteus medius and gluteus minimus, tensor fasciae latae

External rotation: Obturator internus, obturator externus, piriformis, quadratus femoris, gemellus inferior, gemellus superior

Knee extension: Quadriceps (rectus femoris, vastus lateralis, vastus medialis, vastus intermedius), sartorius

DANCE FOCUS

So, you have figured out how to get your thigh to your chest, but then you start to extend the knee and the femur starts to drop. You feel an intense overuse of the quadriceps. Remember, once you have contracted the quadriceps, they cannot help you elevate your leg any higher—your développé is done!

Visualize your femur glued to your ribs; increase the deep psoas major contraction to keep your thigh glued to the ribs and keep the deep low rotators contracting very strongly to maintain turnout of the thigh. Throughout the movement, there is a spiraling effect of the thigh in the hip socket. It may help to remember to aim your sit bone down toward the floor and let the outside of the thigh rotate downward as well. Feel the head of the femur wanting to pull in and down with the help of the iliacus to minimize a lateral hip hike.

Now, just lift the lower leg; visualize the tibia, foot, and ankle floating up; let the pull of the quadriceps contraction elevate the lower leg. It's important to maintain contraction of the iliopsoas and the deep rotators to provide support for the femur above 90 degrees. Remind yourself to keep turning out the back of your thigh. You will also notice the supporting-side gluteus medius helps to stabilize your pelvis. Let them all work together to give you an amazing développé!

DESCENDING BATTEMENT

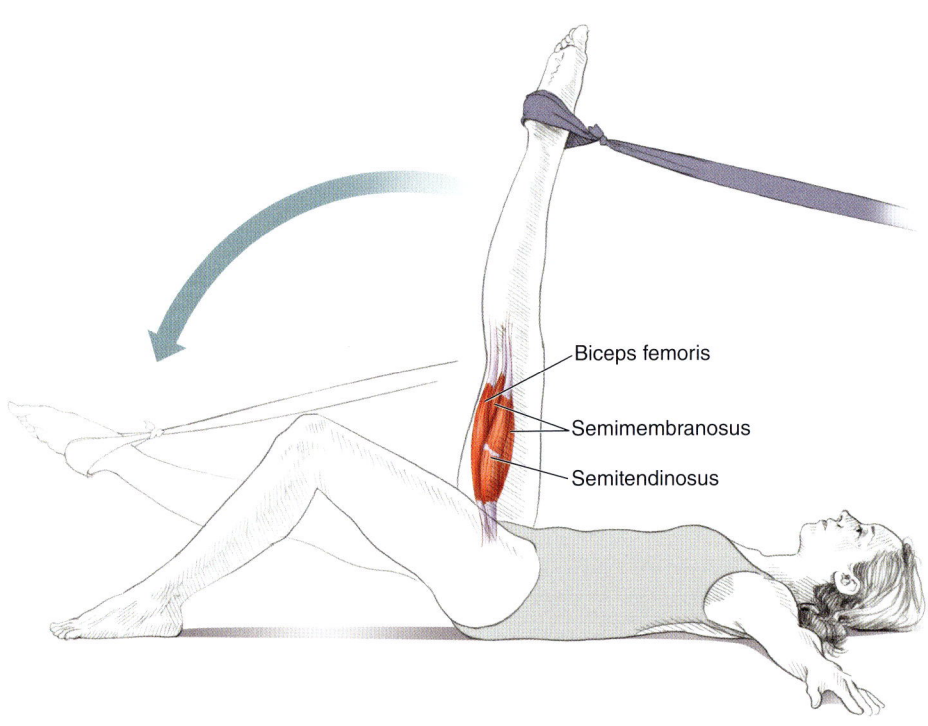

Biceps femoris

Semimembranosus

Semitendinosus

EXECUTION

1. Lie on your back with your left knee bent and your left foot on the floor. Your right leg begins at 90 degrees of hip flexion and turned out; the knee is fully extended. Secure one end of an elastic band around the right forefoot; the other end should be stabilized high and behind you. Inhale to prepare.

2. On exhalation, engage the deep abdominals to secure your lower back. Bring your leg down with control against the resistance of the band as if returning from grand battement.

3. Inhale as the leg goes up. Feel as though you are lifting the leg with the upper inner thigh. Increase speed on the upward phase but keep the movement slow and controlled against the resistance on the downward phase. Reemphasize trunk control with each battement. Perform 10 to 12 times.

SAFETY TIP: To maintain pelvic security, avoid both anterior pelvic tilt and lateral pelvic tilt. Move only the thigh, not the pelvis or spine.

MUSCLES INVOLVED

Hamstrings (semitendinosus, semimembranosus, biceps femoris)

DANCE FOCUS

When you use control in coming down from high kicks, grand jetés, or traveling leaps, your work looks as if you are defying gravity. In this exercise, use the band to focus on the concentric contraction of the hamstrings as the leg comes down. Allow the band to assist you as the leg goes back up while you maintain the eccentric lengthening through the hamstrings

and gluteus maximus. Fight to hold your turnout through the entire range; doing so will keep the hip from elevating. At the top of the battement, feel as though your leg could lengthen; hover and lift before it slowly begins to come down. Maintain an anchored pelvis and reaffirm the principle of hip dissociation. Remind yourself to turn out the back of the thigh.

This exercise can also be repeated while lying on your side for à la seconde. While executing these two exercises, close your eyes for a couple of repetitions to focus on the work of the deep transversus abdominis hugging your spine like a corset. This support enables you to move your legs freely.

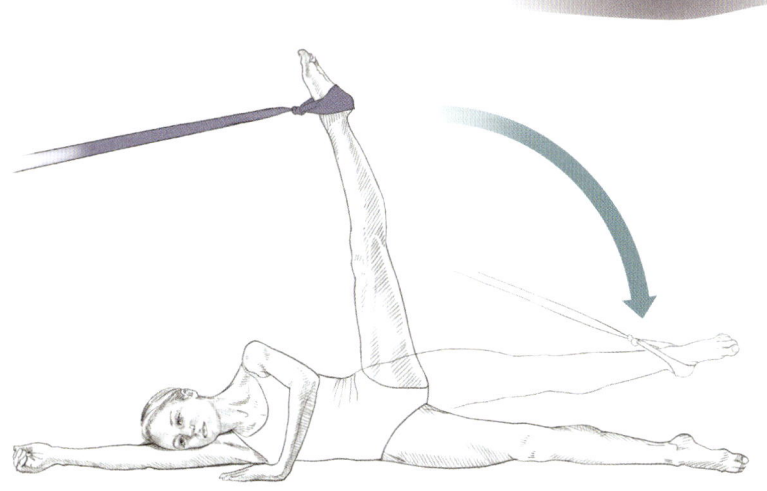

Variation: Side-lying battement.

DÉVELOPPÉ

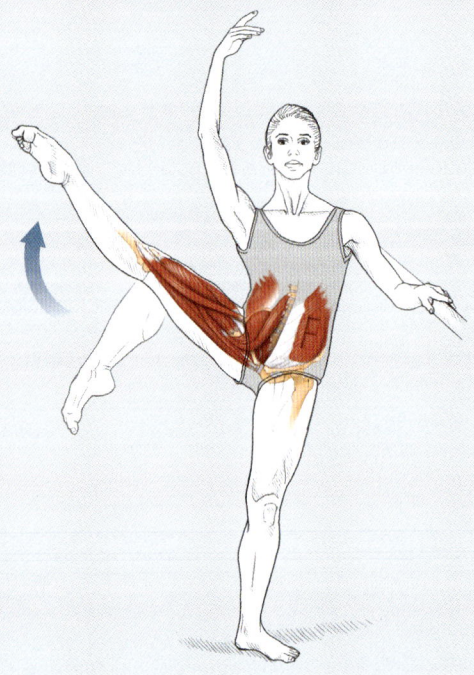

Développé is a classical ballet term meaning "to develop." If you have the necessary strength in your hip flexors and hip abductors, as well as flexibility in your hamstrings, then there should be no reason why you can't have gorgeous high développés. Let's create a développé à la seconde; from the passé movement created in the preceding chapter, you will continue into attitude à la seconde to développé.

1. Begin with your legs and feet in fifth position and the right leg in front. (We will leave the arms out of this description to focus on pelvis and thigh alignment.) Organize your posture to incorporate your neutral spine and pelvis position. Feel the deep abdominals gently pulling in. Feel the hip adductors and deep hip external rotators working.

2. Before you begin, feel lengthening through your spine, and think about the supporting leg. Feel the hip rotators deep in the back of your hip. Feel turnout rotation all the way down the legs to the floor. Maintain a strong foot and arch, keeping your weight evenly placed on all five toes and the heel.

3. You will begin to move the front (right) leg into a cou-de-pied position while maintaining external rotation of both the gesturing leg and the supporting leg. Feel opening in the front of the hips and a deep, strong contraction of the hip external rotators of both legs. The pelvis should remain stable. The gluteus medius and quadratus femoris of the supporting hip should feel a deeper contraction to maintain your balance and pelvic stability. Continue to feel the anterior hip bones and the pubic bone aligned directly along the frontal plane.

4. Point your right foot so that your fifth toe can be placed lightly along the inside of your supporting leg and continue to slide the right leg up the inside of the supporting leg. While maintaining turnout, think about the femur moving directly along the frontal plane. Think about the outside of the left leg contracting to keep the thigh turned out and the knee directly out to the side.

5. Once your leg is in the passé position, begin to lift your right knee into attitude à la seconde. Slightly shift your weight laterally over the supporting leg while maintaining a strong contraction of the gluteus minimus, gluteus medius, and deep outward rotators.

6. The psoas major is contracting to elevate your right thigh, and the quadratus femoris is contracting to help maintain turnout of your left leg. Your right femur must rotate externally so that the greater trochanter does not come into contact with the ilium. Maintain a strong, deep abdominal contraction along with the multifidus to stabilize your spine.

7. Once your right thigh has reached an elevated attitude à la seconde, contract your rectus femoris to begin to extend your knee while floating the lower leg up without allowing the femur to drop. Continue visualizing the iliopsoas contracting to keep your femur elevated. Your sartorius will also contract to help hold the turnout. Feel lengthening in your right hamstring and maintain a deep abdominal and multifidus contraction to stabilize your spine.

Muscles Involved

Trunk placement: Deep transversus abdominis, internal oblique, external oblique, multifidus, pelvic floor muscles (coccygeus, levator ani), adductor longus, adductor magnus, gracilis, pectineus, gluteus medius

Cou-de-pied to passé: Obturator internus, obturator externus, piriformis, quadratus femoris, gemellus inferior, gemellus superior, posterior gluteus medius, sartorius

Supporting leg: Obturator internus, obturator externus, piriformis, quadratus femoris, gemellus inferior, gemellus superior, posterior gluteus medius, lower fibers of the gluteus maximus, adductor longus, adductor brevis, adductor magnus, gracilis

Attitude to extension: Iliopsoas, tensor fasciae latae, transversus abdominis, sartorius

Knee extension: Quadriceps (rectus femoris, vastus lateralis, vastus medialis, vastus intermedius), sartorius

CHAPTER 10

Ankles and Feet

Strong and balanced feet provide the foundation for the whole body. You can give your ankles and feet the power you need to be quick and fearless by gaining knowledge of lower-leg alignment together with core and pelvic strength. As a dancer, you need basic understanding of accurate alignment and muscle action to improve your functional technique. The human foot contains 26 bones and 33 joints, thus creating multiple movement possibilities. When bearing weight, any joint movement has a direct relation to the other joints in your feet; therefore, to dance effectively, you must move with all joints working in harmony as a unit.

The foot and ankle movements used by dancers are similar in modern, jazz, ballroom, Irish, and most folk-dance styles. You must be able to travel quickly on your feet and rise on the balls of your feet and the tips of your toes. In addition, basic skills such as turning, jumping, pointing, relevé, and plié are needed for all dance techniques. Each style requires certain foot positions, not to mention various certain types of footwear that may be used more for aesthetic appearance than for true support. For instance, you may need to run and jump in heels or pivot and push with bare feet. Tappers, cloggers, and flamenco dancers perform challenging, percussive footwork requiring intense power. Classical ballet dancers need extreme foot and ankle range of motion for pointe work.

This chapter applies to all dance styles and illustrates the importance of understanding fundamental anatomy. You need to know the supporting structures that keep your arches alive and strong. To reduce the risk of ankle sprains, you need to know where ankle stability comes from. It's also useful to understand basic muscle movement so that you can benefit from strengthening exercises. Functional, quick, and fearless feet don't just happen—they need training, care, and maintenance.

Bony Anatomy

The malleoli (ankle bones) are the projections at the base of the tibia and fibula bones; the ankle bones are the sites of some of the ankle's strong supporting ligaments. The base of the tibia and fibula meets the talus bone, which snugly fits between the malleoli. The talus bone is somewhat responsible for transmitting your weight to the rest of your foot. It meets the calcaneus bone (heel bone) posteriorly and the navicular bone anteriorly (figure 10.1). The calcaneus provides the base for the attachment of the Achilles tendon, whereas the navicular bone provides the base for the tibialis posterior tendon. Both tendons activate to point the foot and ankle.

`Located along the midfoot region are three cuneiform bones and the cuboid, which meet the five metatarsal bones. This midfoot gives you mobility for a beautiful point and firmness for support. The metatarsals meet the phalanges, or toe bones, and flexibility in these joints is needed for the highest possible half pointe. All the bones in your feet are connected by ligaments and muscle tendons, which provide support.

The rest of this chapter considers the foot in terms of segments. The forefoot contains the phalanges and metatarsals; the midfoot contains the navicular, the three cuneiforms, and the cuboid; and the rearfoot contains the calcaneus and the talus.

The bones in your feet are not organized in a flat formation. The inside, or medial, border forms a long arch, which is referred to as the medial longitudinal arch (figure 10.2). When instructors say "don't roll in," they are usually advising you not to flatten out this arch. Typically, weight should be placed evenly over the medial border, lateral border, and calcaneus. Even though the outside border of your foot is on the floor, it forms a lateral longitudinal arch. If some of your weight is placed along the outside arch, the inside arch can activate and lift. The transverse arch runs across from the inside to the outside. The arches of your feet are supported by the bones in your feet; they need to be strong and active to support your weight, jumping activities, balance poses, and twisting movements. The arches are also supported by fascia and ligaments.

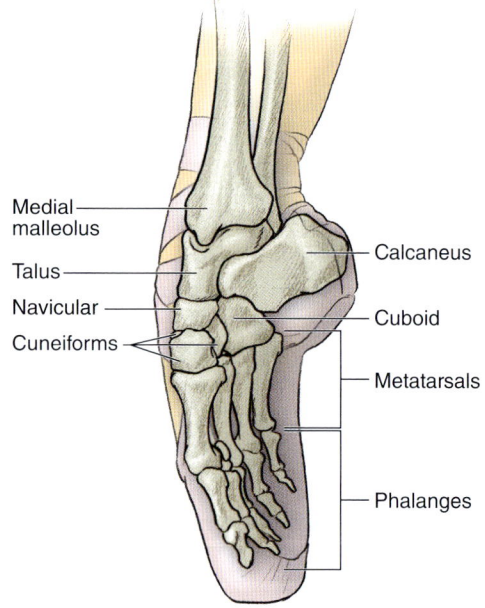

Figure 10.1 Bones of the foot.

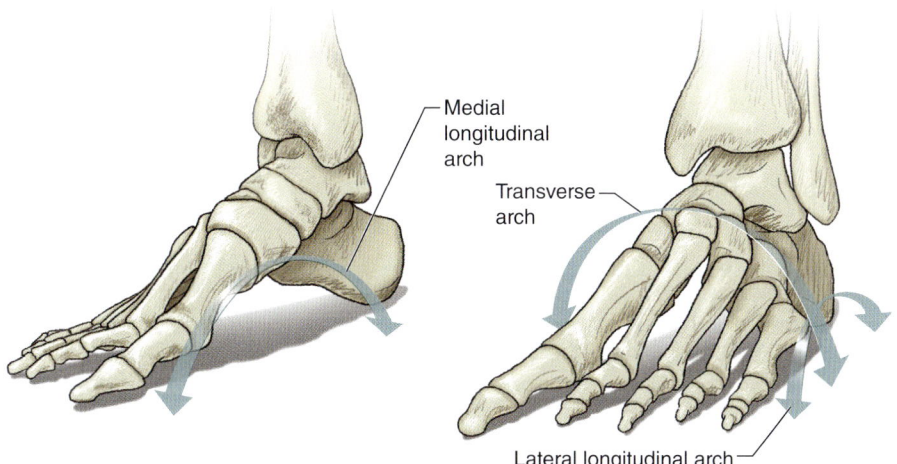

Figure 10.2   Three arches of the foot: medial longitudinal, lateral longitudinal, and transverse.

The plantar fascia is a very tough, thick band of connective tissue on the sole of your foot; it runs between your forefoot and your calcaneus. It provides support for your arch. Overuse, strenuous exercise, and lack of supportive shoes can result in plantar fasciitis, or inflammation of the fascia. Normal tension is placed on the plantar fascia when you bear weight or push off when walking, running, jumping, and dancing. Overscheduling and overworking your feet can cause small tears along the medial aspect of the calcaneus into the arch. To minimize the risk of this overuse syndrome, maintain strength of the small intrinsic muscles and lower leg muscles as well as flexibility of the gastrocnemius, soleus, and arches of the feet. The doming exercise and fascia stretch later in this chapter can be helpful.

Foot and Ankle Motion

The ankle joint, or talocrural joint, can point and flex, or, in medical terms, perform plantar flexion and dorsiflexion that occur along the sagittal plane. The ankle joint can also move inward and outward, or invert and evert, which occurs along the frontal plane.

The talus sits snugly in a boxlike space or mortise. In plié, the talus moves posteriorly, providing stability. In some cases, when the demi-plié is too deep, the talus can meet the base of the tibia. This contact can cause pain and swelling and can eventually lead to bone spurs. To help prevent your plié from causing this impingement, maintain strength and eccentric muscle control in your lower legs. The elevé-with-ball and seated soleus pump exercises are good choices for maintaining lower leg strength.

During demi-pointe, the talus moves slightly forward, out of the security of its space, thus causing instability. Ankle support is the focus of the inversion press, winging, and elevé-with-ball exercises presented in this chapter. Some dancers struggle with a fully pointed position because the back of the talus bone has an abnormal bony projection that meets the heel bone. This unfortunate posterior impingement limits full-height relevé, creates an unstable ankle, and leads to a weight-back situation—that is, your body weight remains too far back when you are unable to transfer your center of gravity completely over your half-pointe or full-pointe position. This incorrect weight placement can create overuse and stress injuries; specifically, it compromises balance and overworks the posterior lower-leg musculature due to compensation.

The subtalar joint in the rearfoot is located where the talus meets the calcaneus. This joint allows for adequate pronation in plié and supination in relevé, whether you are working in parallel or turned out. Pronation involves a combination of dorsiflexion, eversion, and abduction; supination is just the opposite. In adequate pronation, the medial arch drops slightly when bearing weight to take on stress and then releases once the weight is removed. This movement is needed for propulsion in relevé and jumps and for shock absorption on landing. Excessive pronation, however, leads to rolling in and undue stress on the arch. Rolling in sometimes occurs from forcing the turnout at the feet rather than using the deep hip external rotators and adductors.

Good movement through the rearfoot dictates needed movement for the midfoot. With plié, for example, the medial portion of the heel bone moves slightly inward so that the talus can do the same. This small movement must happen to open the joints of the midfoot. When the midfoot joints loosen, the resulting flexibility enables shock absorption and a soft plié. The exact opposite occurs for relevé; that is, the heel and talus lift slightly so that the midfoot joints can tighten, thus providing the necessary firm arch. Strengthening the muscles of the midfoot region aids relevé by enabling excellent weight transfer onto the first, second, and third metatarsals, thus enabling the arches to become rigid and help stabilize the movement.

The joints where the metatarsals meet the phalanges must be strong and mobile for toeing-off during jumping movements. Eccentric lengthening under the toes is also required to provide an adequate base for relevé; the eccentric lengthening allows the small muscles under the forefoot and toes to be long but strong and active. Even in a standing position, your toes should be lengthened and your arch musculature activated to provide a firm anchor. The first exercise presented in this chapter, doming, activates your foot's intrinsic muscles to improve arch support.

Support Ligaments

You probably know a dancer who has had an ankle sprain, which is a very common ligament injury. Numerous ligaments are contained in the foot and ankle; here, we look at five that provide support (figure 10.3). The medial, or deltoid, ligament complex originates on the medial malleolus and fans out to attach on the navicular, talus, and calcaneus bones. This complex is an extremely strong combination of four ligaments that provide vital stability. The spring ligament is also located on the medial side of the foot and connects the calcaneus with the navicular bone; its principal job is to provide a sling for the talus, which helps support the weight of the body. Weakness or lengthening along this ligament can cause flattening of the foot.

Located on the outside of your ankle are three ligaments that together provide stability: the anterior talofibular (ATF) ligament, the calcaneofibular ligament, and the posterior talofibular (PTF) ligament. The ATF and calcaneofibular ligaments are not as strong as the deltoid ligament and are usually the first ligaments to be injured in a lateral ankle sprain. In this injury, the sole of the foot turns inward and damages the supporting ligaments; dancers sometimes refer to this injury as "rolling the ankle."

The ATF ligament runs between the talus and the fibula; when in relevé, it moves into a vertical stable position and tightens (Russell et al. 2008). It is the weakest of the three ligaments and usually the first to be injured in the lateral ankle sprain. The calcaneofibular and PTF ligaments, as you might guess by their names, run between the calcaneus, talus, and fibula bones and help maintain critical alignment and ankle stability. The PTF is the strongest of the three lateral ankle ligaments. While your ankle is moving through its full range of motion, these lateral ligaments create different levels of tension forces to stabilize your ankle.

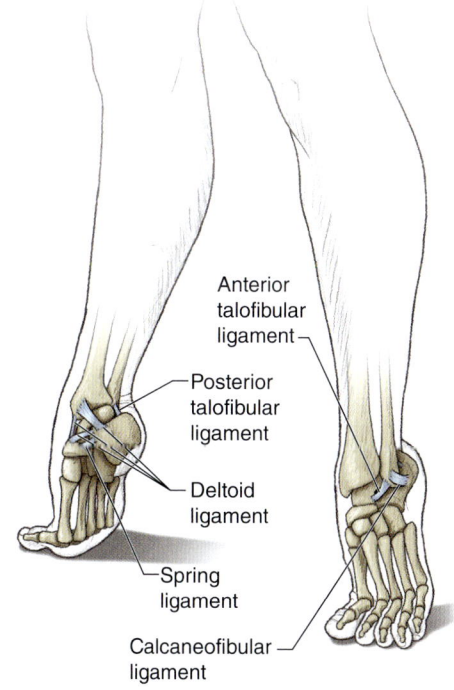

Figure 10.3 Five ligaments that support the foot and ankle.

Muscle Mechanics

Foot and ankle action is allowed by about 12 intrinsic muscles located within the foot itself and about 12 extrinsic muscles that originate outside of the foot and have multiple actions. The gastrocnemius is the large muscle that originates behind the knee and inserts in the calcaneus bone by way of the Achilles tendon (figure 10.4). It is a two-joint muscle—it can flex the knee and point or plantarflex the foot. Located underneath the gastrocnemius is the soleus, which also connects into the Achilles tendon. The soleus can also point the foot and helps maintain balance. Together, these two muscles serve as the primary movers for relevé and pointing. The soleus is valuable in rising from half pointe to full pointe and in securing control on landings from jumps. The seated soleus pump exercise presented in this chapter offers two variations for soleus-specific strengthening.

Other muscles that originate behind the tibia or fibula include the tibialis posterior, the flexor digitorum longus, and the flexor hallucis longus, which help with plantar flexion and inversion. The tibialis posterior inserts mainly into the navicular bone and provides added support for the inner arch. The flexor digitorum longus inserts into digits (toes) 2 through 5.

The flexor hallucis longus deserves more attention. This muscle originates along the back of the fibula and runs along the back of the lower leg through

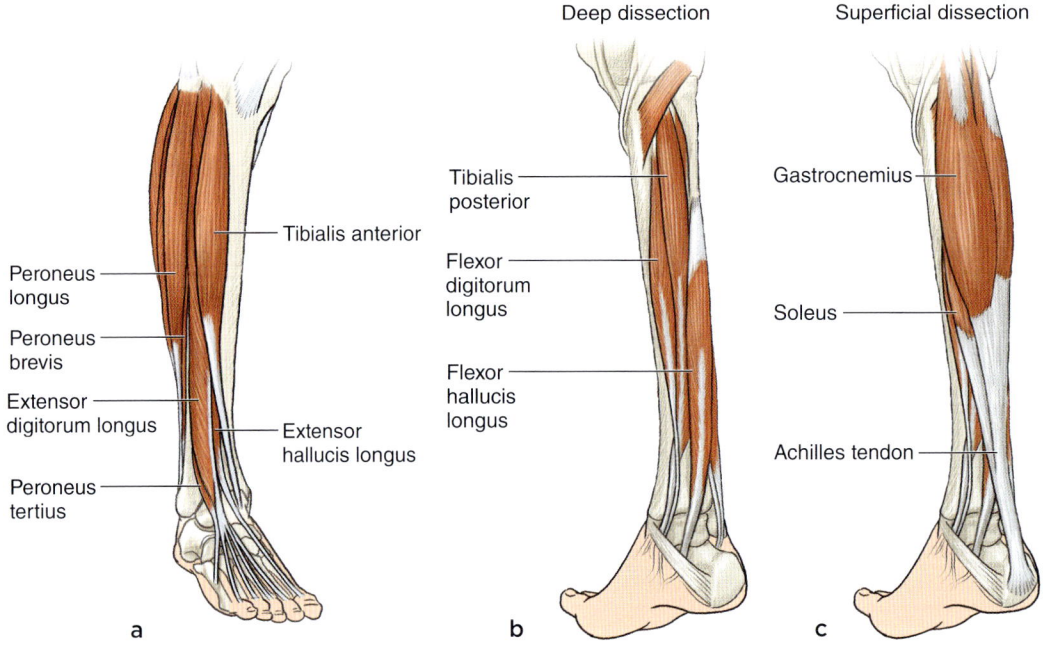

Figure 10.4 Muscles of the lower leg and foot: *(a)* front; *(b)* and *(c)* back.

a small tunnel beneath the inside ankle bone; its tendon then inserts into the base of the big toe. The flexor hallucis longus muscle serves multiple purposes: flexion of the big toe, push-off power for jumps, and support for the inner arch. Repetitive overuse of the flexor hallucis longus tendon with pointing and relevé can lead to discomfort and inflammation, a condition that has been referred to as dancer's tendinitis. This tendon can also become trapped in the tunnel and cause triggering, which can lead to fraying or tearing. To avoid overuse of the tendon, it is imperative to strengthen all the muscles responsible for pointing. Exercises for doing so are included in this chapter.

The muscles along the lateral lower leg are the peroneal muscles, which originate at the upper fibula. One inserts into the fifth metatarsal, and one continues under the foot to insert into the first metatarsal. Their job is to provide strength for the outside of your lower legs and stability for your ankle joints, which can help reduce the risk of lateral ankle sprains. Located along the front of the tibia are the tibialis anterior, extensor hallucis longus, and extensor digitorum longus. These muscles pull the toes upward and invert the ankle. All extrinsic muscles work to hug your lower leg and provide support and stability for your foot and ankle.

The soles of your feet are also layered with supportive muscles (figure 10.5). These intrinsic muscles connect the heel with the tarsal and metatarsal bones and are solely responsible for lengthening the toes. The abductor hallucis is

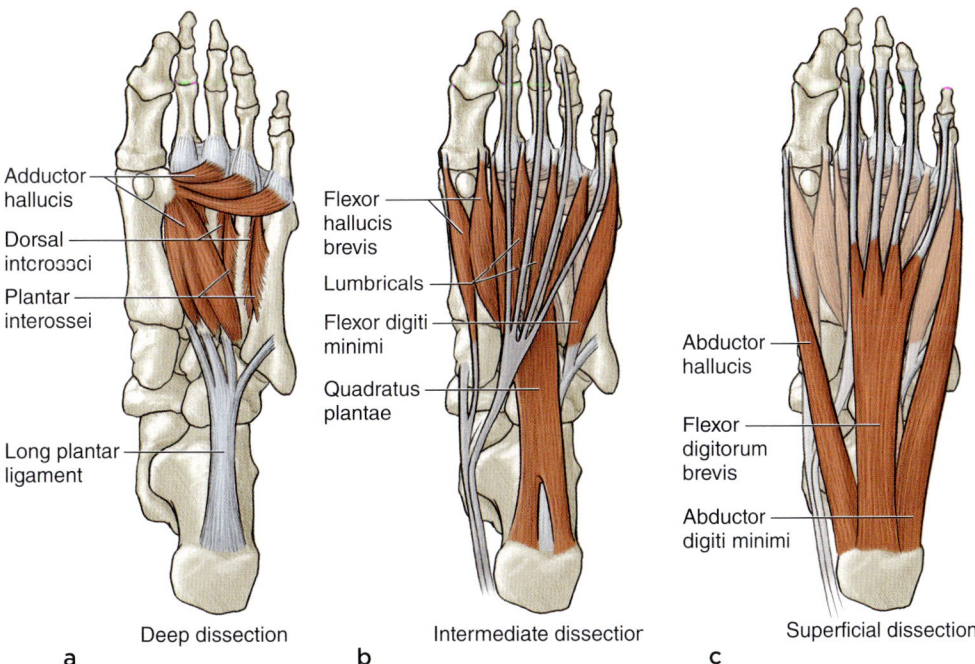

Adductor hallucis

Dorsal interossei

Plantar interossei

Long plantar ligament

Flexor hallucis brevis

Lumbricals

Flexor digiti minimi

Quadratus plantae

Abductor hallucis

Flexor digitorum brevis

Abductor digiti minimi

Deep dissection

Intermediate dissection

Superficial dissection

a

b

c

Figure 10.5 Intrinsic muscles of the foot: (a) deep dissection; (b) intermediate dissection; (c) superficial dissection.

a small muscle that supports the inner arch and runs from the big toe to the inside of the heel. You can train this muscle to activate to provide strength in the inner-arch area; it can be strengthened through the big-toe abduction exercise presented in this chapter. Deep muscles are also located between the metatarsals and phalanges. Weakness in these intrinsic muscles can cause clawing of the toes; the toes must stay lengthened to enable push-off skills for jumping.

Dance-Focused Exercise

While executing the following series of exercises, visualize the muscles hugging your ankle for support. (Use table 10.1 as a reference.) Each time you flex or demi-plié, visualize the talus bone situated securely in its space for support. Think about energy all along your arches. Each time you point your foot, align the second and third metatarsals with the tibia bone for a perfect line. Remember to lengthen under the toes to avoid clawing; this lengthening gives you a wider base during half pointe, thus providing a better platform for balancing. Try to perform all the exercises at various speeds and work with control throughout the range of motion. The chapter concludes with a detailed description of a forced arch relevé.

Table 10.1   **Intrinsic Muscles of the Foot**

Muscle, deep dissection	Origination	Insertion	Action
Adductor hallucis	Base of metatarsals 2, 3, 4	Base of proximal big toe	Flexes and adducts big toe
Dorsal interossei (DI)	Sides of metatarsals	First DI inserts into medial second toe; other three DI insert into lateral toes 2, 3, 4	Abducts, flexes, and extends middle toes 2, 3, 4
Plantar interossei	Medial 3, 4, 5 metatarsals	Base of proximal toes 3, 4, 5	Adduct toes 3, 4, 5
Long plantar ligament	Bottom surface of calcaneus	Base of metatarsals 2, 3, 4, 5 and cuboid bone	Maintains and supports longitudinal arch
Muscle, intermediate dissection	**Origination**	**Insertion**	**Action**
Flexor hallucis brevis	Lateral cuneiform, cuboid bones	Base of proximal big toe	Flexes big toe
Lumbricals	Tendons of flexor digitorum longus	Medial proximal toes 2, 3, 4, 5	Flexes and extends toes 2, 3, 4, 5
Flexor digiti minimi	Base of 5th metatarsal	Base of proximal little toe	Flexes little toe
Quadratus plantae	Medial and lateral calcaneus	Tendons of flexor digitorum longus	Assists flexor digitorum longus to flex toes 2, 3, 4, 5 when ankle is plantar flexed
Muscle, superficial dissection	**Origination**	**Insertion**	**Action**
Abductor hallucis	Calcaneal tuberosity	Proximal big toe	Abducts and flexes big toe
Flexor digitorum brevis	Calcaneal tuberosity	Middle of toes 2, 3, 4	Flexes toes 2, 3, 4
Abductor digiti minimi	Calcaneal tuberosity	Proximal little toe	Abducts and flexes little toe

DOMING

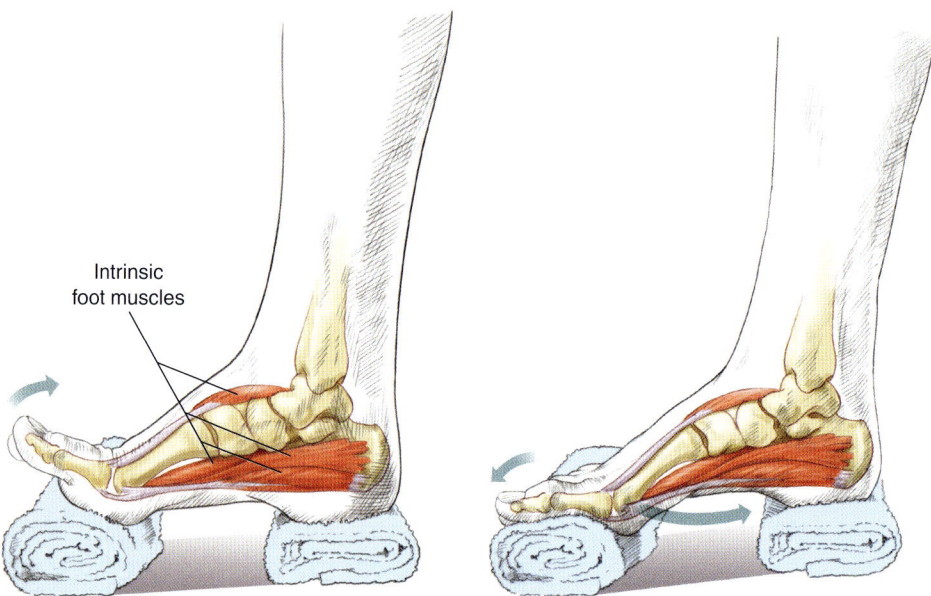

Toes up.   Toes down.

EXECUTION

1. Begin in a seated position. Place your forefoot on one small rolled towel and your heel on another. Use the rolls to balance the placement of your foot evenly across the metatarsal heads and the heel.

2. Lift all toes upward without lifting the forefoot off the towel roll. Reestablish equal weight placement. Lengthen under the toes as you begin to press them toward the floor.

3. Engage the deep intrinsic muscles throughout your arch and draw the metatarsal heads toward the heel. The movement is initiated from the metatarsal phalangeal joint. Do not curl the toes; allow the intrinsic musculature to draw the metatarsal heads toward the heel. Perform 15 times; work up to 30 repetitions.

MUSCLES INVOLVED

Intrinsic foot muscles: Lumbricals, interossei

DANCE FOCUS

Some choreographic requirements can really take a toll on your feet. Numerous small muscles are located along the soles of your feet; they play a role in pointing your feet, in moving from half pointe to full pointe, and in pushing off for jumps. The intrinsic muscles and the anatomy of the bones support the various arches in your feet and help resist curling of the toes. The intrinsic muscle group must feel active to create support. Close your eyes and focus on this specific area of your feet. Visualize the strong fascia along with the numerous muscle fibers contracting to provide control. Whether you are dancing barefoot, in pointe shoes, or in character shoes, the intrinsic musculature must be strong to give you the power and spring needed for jumps and pointe work. Limited education about this specific area of your foot may be provided in some warm-up and basic technique classes. It is up to you, however, to maintain the quality of your arches by taking extra time to strengthen your feet.

VARIATION

Doming With Resistance

Place your foot on the floor with a resistance band underneath your entire foot. Hold on to the opposite end of the band as it runs from underneath your toes. Lift your toes upward without lifting the forefoot off the resistance band. Lengthen under your toes as you begin to press them into the band. Engage the deep intrinsic muscles and draw the metatarsal heads toward your heel against the resistance of the band. Do not curl the toes. As you press the toes down, engage the deep intrinsic muscles to create a strong dome of your arches. Perform 15 times; work up to performing 30 repetitions.

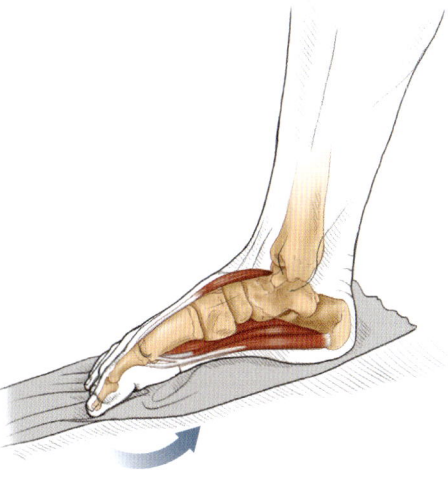

BIG-TOE ABDUCTION

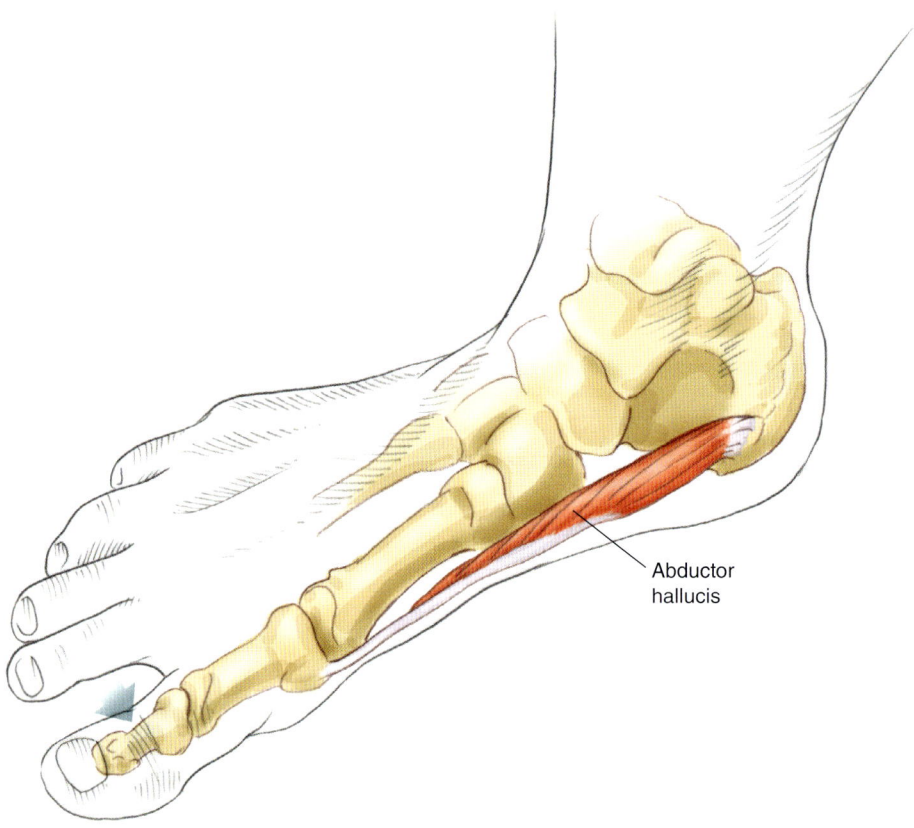

Abductor
hallucis

EXECUTION

1. Begin in a seated position. Place your feet on the floor, emphasizing equal weight placement between the metatarsal heads and the heel.

2. Try to open the big toe away from the other toes. Hold for 2 to 4 counts, then slowly return. Feel the boost of the medial arch as the big toe moves.

3. Perform 10 to 12 times to feel the muscle contraction; work up to 3 sets of 12 repetitions.

MUSCLES INVOLVED

Abductor hallucis

DANCE FOCUS

The medial arch should have a beautiful dome shape; the lack of an arch is typically what instructors focus on when cueing you to not roll in. Over time, a flattened medial arch will result from weakness in the abductor hallucis and laxity in the ligaments. Collapse of the arch can lead to numerous injuries and can be caused by exaggerating turnout from the feet instead of executing turnout from the hips. To help provide the correct spring needed along the medial arch, place equal weight along the lateral arch to help organize the muscles.

Once you let go of the barre to begin center work, your medial arch musculature fires to help you maintain balance. All dance styles require constant shifting of body weight, thus causing the arches to change form; to tolerate these changes, your arches must be strong. You can use the abductor hallucis muscle to provide support for your medial longitudinal arch, whether you are dancing barefoot, en pointe, or in character shoes. The medial arch needs to become rigid and secure in relevé, lengthened but active in plié, and alive and toned for balancing.

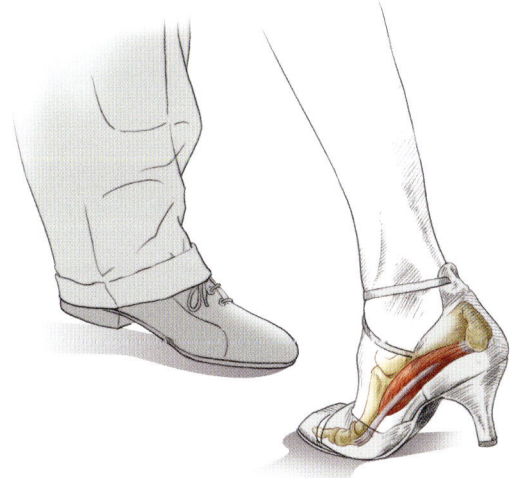

FASCIA STRETCH

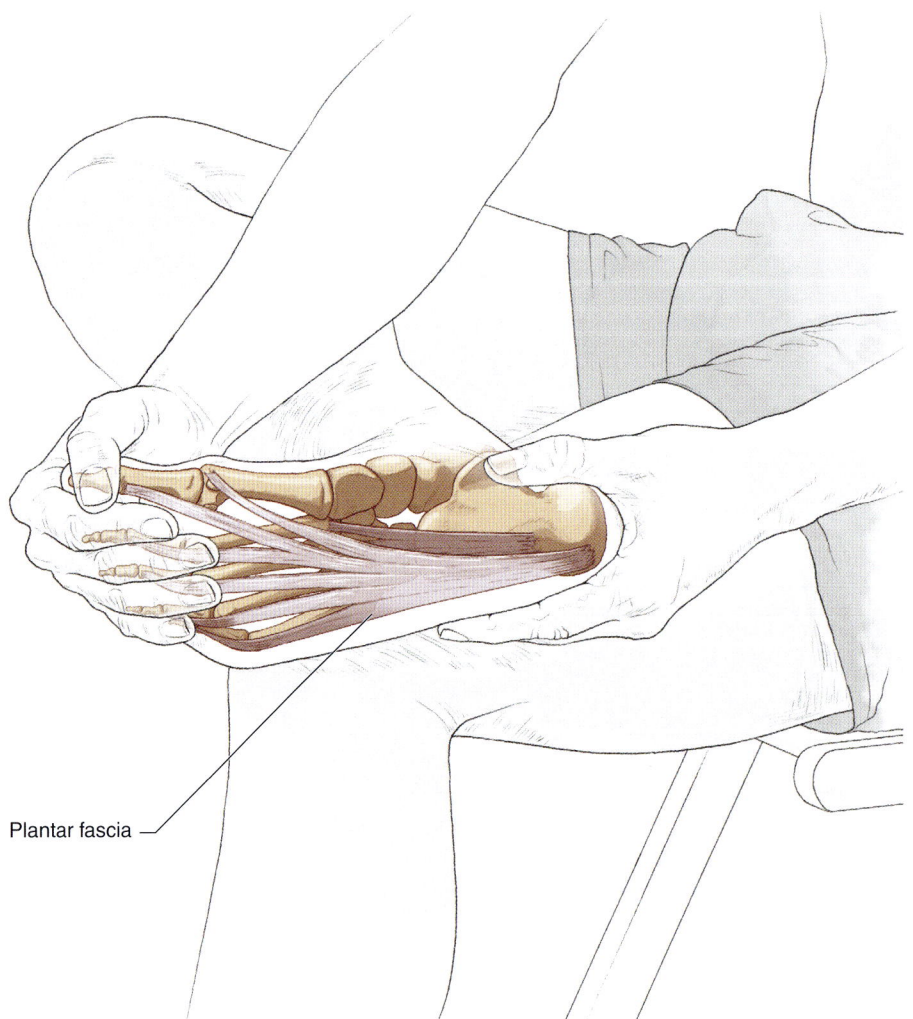

Plantar fascia

EXECUTION

1. While seated, cross your right ankle over your left thigh.
2. With your right hand, gently pull your right great toe and ankle into dorsiflexion.
3. Hold this position for 10 to 15 seconds, then release.
4. Repeat 3 to 5 times to feel a gentle stretch along the sole of the foot.
5. Repeat the manual stretch on the other foot.

MUSCLES INVOLVED

Plantar fascia

DANCE FOCUS

Plantar fasciitis can be concerning for dancers, and maintaining healthy, strong, and flexible feet will be key to reducing the risk of fasciitis. While the fascia stretch will also stretch other intrinsic muscles of the foot, the focus here is on the plantar fascia. Since the plantar fascia originates at the medial calcaneus and inserts along the metatarsals and toes, it provides arch support and shock absorption.

While there are other causes of plantar fasciitis, such as poorly fitted shoes, overrehearsing can cause strain, tears, or inflammation. The first steps when you get out of bed are usually the most painful. It is important to maintain strength in the intrinsic muscles of the feet and flexibility in the fascia. Excessive foot pronation and poor conditioning can cause fasciitis to become chronic. Many modern and contemporary dancers enjoy the freedom of dancing barefoot. They feel a good connection with the floor, can balance more efficiently, and move through space with more fluidity. However, the lack of arch support can make you more prone to plantar fasciitis. Consult with a health care provider if you have symptoms related to plantar fasciitis.

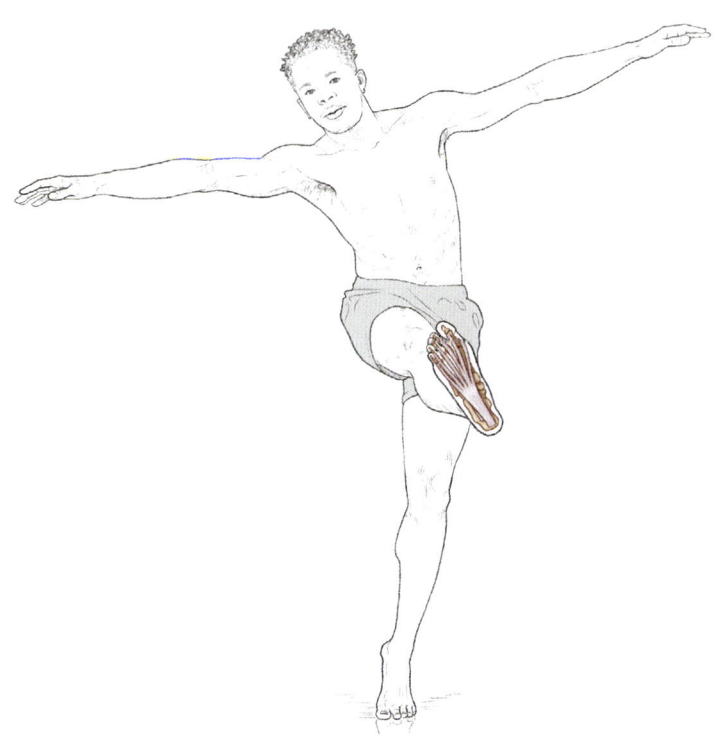

INVERSION PRESS

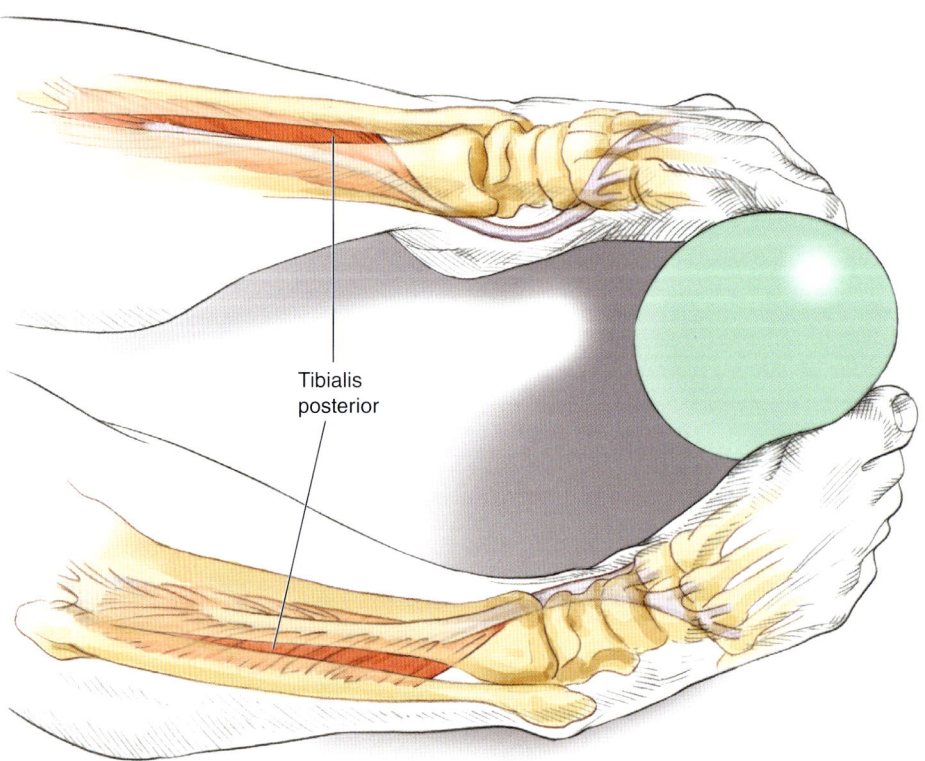

Tibialis posterior

EXECUTION

1. Sit with both knees bent and the soles of the feet parallel on the floor. Place a medium-sized ball between your feet in the area of the forefoot.

2. With the heels of both feet remaining on the floor, begin to press the forefeet into the ball, lifting the inner arches of both feet.

3. As you move your forefeet inward, press into the ball and maintain an isometric contraction for 2 to 4 counts. Perform 10 to 12 times; work up to 3 sets.

SAFETY TIP: Avoid overstretching the outside of the ankle. Use this exercise to focus on the arch-lifting aspect and to strengthen the inside of the ankle.

MUSCLES INVOLVED

Tibialis posterior with assistance of the flexor hallucis longus and flexor digitorum longus

DANCE FOCUS

The tibialis posterior supports the medial arch and helps resist pronation. Although the tibialis anterior also contracts, focus on the tibialis posterior pulling the foot inward and lifting the arch. The talus bone needs to stay in a relatively neutral position to provide the most stability for the foot and ankle. Some natural pronation occurs with each plié (and some natural supination with relevé), but excessive pronation leads to numerous overuse injuries. In relevé, feel deep support of the tibialis posterior tendon by visualizing its many insertions into the navicular bone and tarsal bones.

Maintaining strength of the tibialis posterior tendon also helps provide stability for the foot and ankle when landing from jumps. The foot begins to articulate as it meets the floor from a jump; the tibialis posterior can help your arch feel lifted, giving you smoother, more cushioned landings.

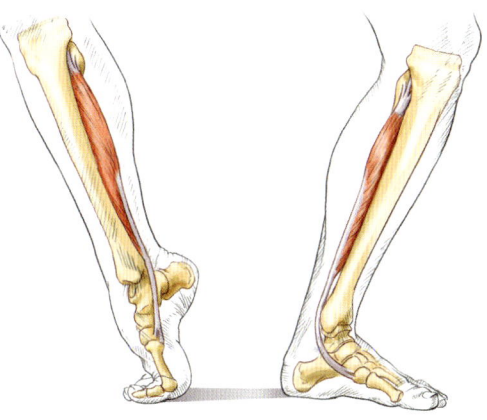

Vary the tempo with this exercise: Move into inversion quickly and return slowly, then reverse the tempo change. Doing so provides changes in velocity of the muscle contraction, which simulates challenges and changes in choreography.

VARIATION

Resisted Inversion

Wrap a resistance band around the sole of your foot. Stabilize the band or hold it to the outside of your foot. Pull the forefoot inward against the resistance of the band. Continue to move through the full range of motion. Do this in a pointed position and in a flexed position. Perform at least 10 times with control; work up to 3 sets of 10 in both a pointed and flexed position and repeat on the other foot.

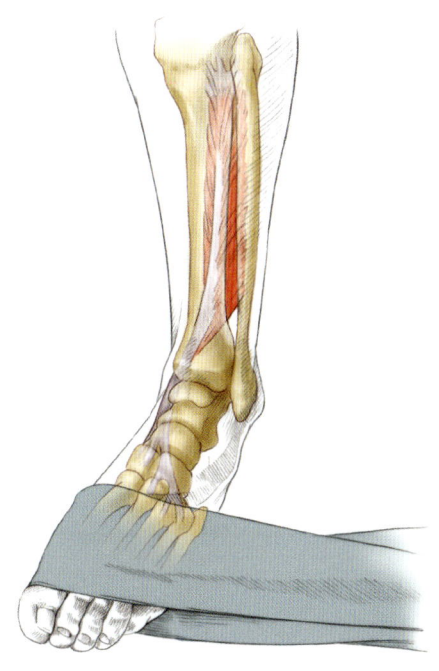

WINGING

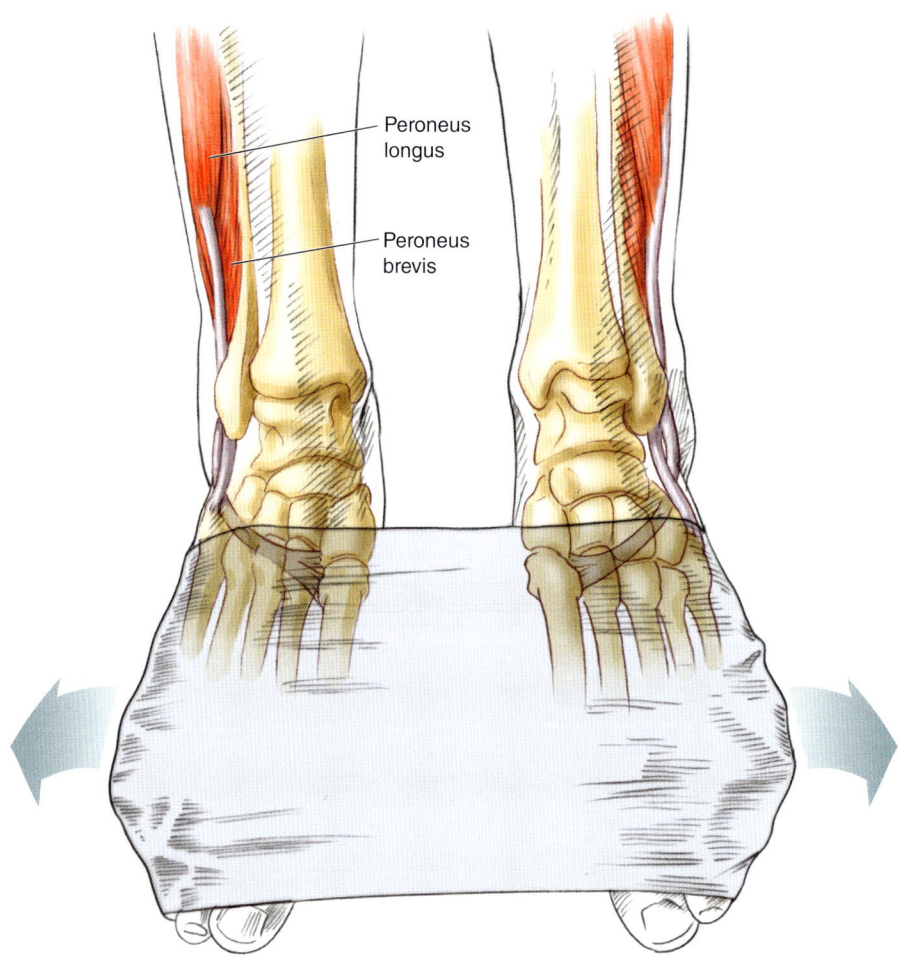

Peroneus
longus

Peroneus
brevis

EXECUTION

1. Sit with an elastic band tied together and wrapped around the fore-feet. Breathe comfortably and push the forefeet outward against the resistance of the band.

2. Perform 10 repetitions in a pointed position and in a flexed position; work up to 3 sets of 10 in each position. Reemphasize control through the full range of motion.

SAFETY TIP: Avoid creating torque through the knees; isolate the movement in the feet and ankles.

MUSCLES INVOLVED

Peroneus longus, peroneus brevis, peroneus tertius, with help from the extensor digitorum longus

DANCE FOCUS

The combination of the muscles along the lateral lower leg and the tibialis posterior gives you support through the effect of a stirrup. With excessive range of motion in relevé, you need security to avoid twisting your ankle and damaging the ligaments. Without adequate strength of the peroneus muscles, the ankle continues to twist, thus leaving the joint unstable. This caveat applies to every style of dance movement and every pointing position, as well as relevé, push-off, and jump landings. Visualize a stirrup holding your ankle secure so that you are free to point your foot through extreme ranges. Many injuries sustained by dancers occur in the lower leg and foot; therefore, it is imperative that you strengthen the ankles to reduce your risk of traumatic injury.

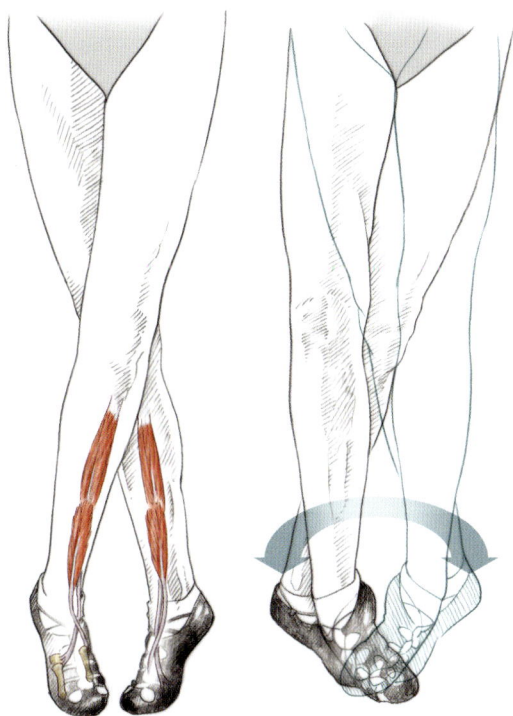

ELEVÉ WITH RESISTANCE

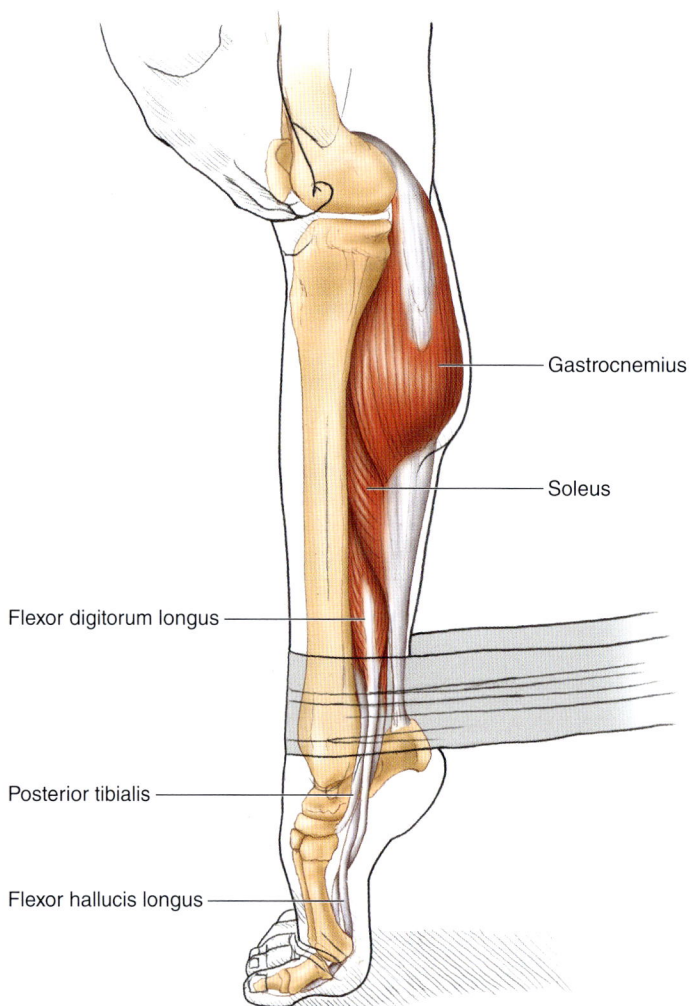

Gastrocnemius

Soleus

Flexor digitorum longus

Posterior tibialis

Flexor hallucis longus

EXECUTION

1. Place a strong resistance band around a stable support, such as the leg of a strong table. Step into the band with your right foot so that the band is snug across the anterior aspect of the foot and ankle.

2. While focusing on alignment, rise directly over your second and third toes against the resistance of the band.

3. Hold the isometric contraction for 10 to 12 counts before slowly returning to the start position. Perform 10 to 12 times, then switch sides. Focus on alignment and lifting the ankle into the highest possible elevé without compromising alignment.

MUSCLES INVOLVED

Elevé: Concentric contraction of gastrocnemius, soleus, peroneus longus, peroneus brevis, posterior tibialis, flexor hallucis longus, flexor digitorum longus

Descent: Eccentric contraction of gastrocnemius, soleus, peroneus longus, peroneus brevis, posterior tibialis, flexor hallucis longus, flexor digitorum longus

DANCE FOCUS

This exercise serves as great preparation for any dancers wanting to increase the height of their elevé en pointe. Because you are working against the resistance of the band in a full weight-bearing position, the exercise challenges your strength, range of motion, and balance skills.

Remember to think about alignment, focusing your elevé over your second and third toes but lifting as high as you can in your half pointe. The way you present your feet to your audience is extremely important. Sloppy footwork, low heel height, and limited range of motion are not attractive for dancers, especially ballet dancers. Even your preparation and transition steps are very important during performances. Paying attention to detail in these areas sets you apart from other dancers. As you rise, think about releasing the flexors under your toes eccentrically but widening the metatarsals to give you a solid base of support. Strongly engage your gastrocnemius into the Achilles tendon to help you elevate your heel for the highest half-pointe position you can achieve. If it becomes easy to do 10 to 20 repetitions daily, move up to 25 to 30 repetitions to build strength.

ELEVÉ WITH BALL OVER THE EDGE

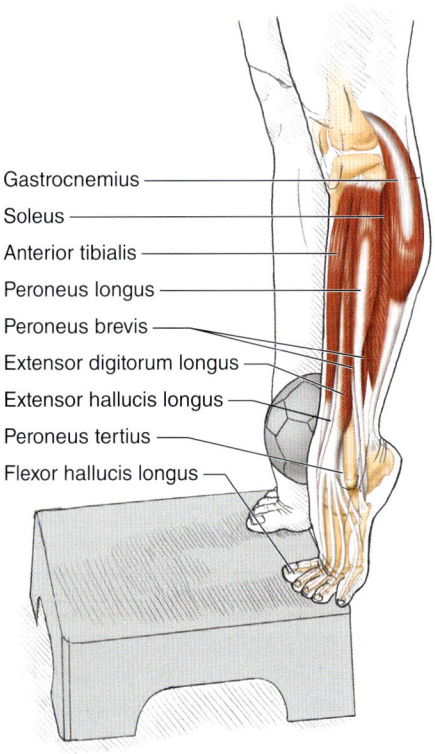

Gastrocnemius
Soleus
Anterior tibialis
Peroneus longus
Peroneus brevis
Extensor digitorum longus
Extensor hallucis longus
Peroneus tertius
Flexor hallucis longus

SAFETY TIP: To improve ankle support and control, maintain the ball squeeze through the entire range of motion to help you avoid sickling the ankle. Focus on rising directly over the second and third metatarsals.

EXECUTION

1. While facing the barre with your legs parallel, stand on the edge of a stable step. Place a small ball between the heels. Reorganize your trunk to maintain neutral postural alignment. Align the tibia over the second toe.

2. Begin to rise with gentle pressure against the ball and align the middle of the talus over the second toe.

3. Hold for 2 to 4 counts before lowering your heels with control. Allow your heels to drop gently as low as you can while maintaining good alignment and holding the ball between your heels. Perform 10 times.

4. Now, hold the heels down for a static stretch of the gastrocnemius for 30 to 45 seconds, then begin the elevé series again. Perform the entire sequence 3 to 5 times. You can use this exercise as a strengthening and stretching tool.

MUSCLES INVOLVED

Rise: Concentric contraction of gastrocnemius, soleus, peroneus longus, peroneus brevis, posterior tibialis, flexor hallucis longus, flexor digitorum longus

Descent: Eccentric contraction of gastrocnemius, soleus, peroneus longus, peroneus brevis, posterior tibialis, flexor hallucis longus, flexor digitorum longus

Dorsiflexion: Anterior tibialis, extensor digitorum longus, extensor hallucis longus, peroneus tertius

DANCE FOCUS

Exercising against your own body weight heightens your body awareness and increases the dynamic challenge. Use this exercise to reinforce the relationship of the talus and the heel during elevé. Feel the lateral and posterior lower leg giving you incredible support. Try it one time while allowing the heels to supinate slightly; notice that you are unable to hold the ball and that the ankles feel unstable.

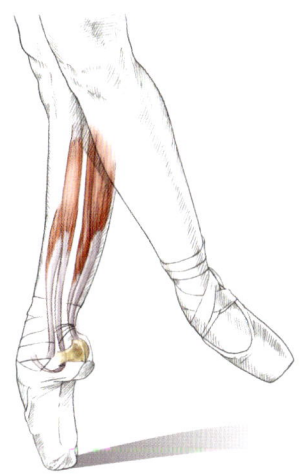

Traveling movements that involve pivots require power to push off in a horizontal direction; to execute such movements, you need strength in the lateral lower leg combined with strength in the gastrocnemius and soleus. The muscles along the outside of your lower leg also provide strength as well as the ability to wing your feet in a coupé-type position.

Remember to control your movements on landing. We tend to use all our efforts and momentum on the upward phase and then let gravity bring us down. This loss of control on the down phase puts us at risk for injury. To avoid injury, you must have adequate strength to recover from an extreme off-balance accident.

ADVANCED VARIATION

Eccentrics With Ball Over the Edge

To advance this exercise, continue to rise while holding the ball between the heels. Organize your placement to maintain your balance, but release and create dorsiflexion with one foot while holding on to the ball. Slowly lower the other heel with control while working on your control and eccentrics. Once you have lowered the heel as far possible with control, place your other foot back on the edge of the step and elevé with both feet. Repeat on the other side. In other words, rise with both legs, release one, and come down with only one leg. Perform on each side 10 to 15 times; work on descending with control and alignment.

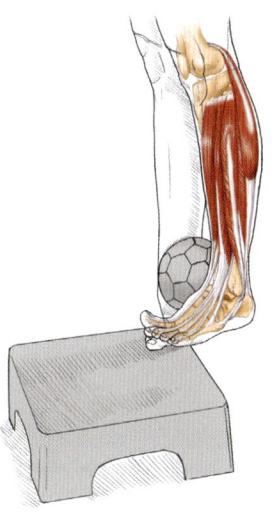

SEATED SOLEUS PUMP

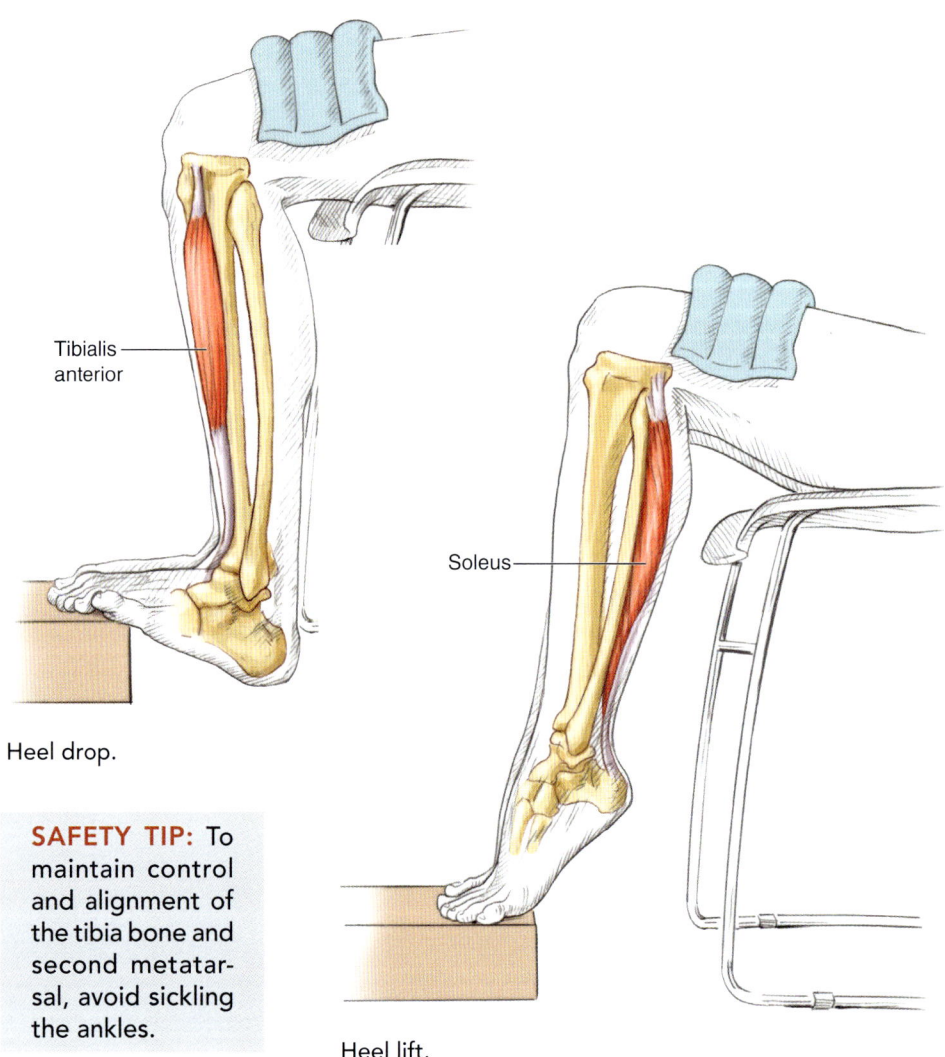

Tibialis anterior

Soleus

Heel drop.

Heel lift.

SAFETY TIP: To maintain control and alignment of the tibia bone and second metatarsal, avoid sickling the ankles.

EXECUTION

1. While seated in a chair with your legs parallel, place your forefeet on a ledge while your heels remain on the floor. Check that your knees are at a 90-degree angle and place a small weight (5 to 10 pounds, or about 2.5 to 4.5 kg) on top of each thigh to give you more resistance.

2. Begin to elevate the heels (plantar flexion), rising onto the balls of the feet, aligning the second toe with the center of the talus. Lengthen under the toes and widen the metatarsals.

3. Return to the start position with control. Perform 15 to 30 times; work up to 3 sets. Engage the deep soleus. While other muscles participate in lifting the heels, the focus is placed on the deep soleus.

MUSCLES INVOLVED

Dorsiflexion: Tibialis anterior, extensor digitorum longus, extensor hallucis longus

Plantar flexion: Soleus

DANCE FOCUS

It is crucial to land from jumps with control. Strengthening the lower-leg muscles enables you to control your body when coming down from relevé, small jumps, and grand allegro movements, which enables you to avoid injury while appearing to defy gravity. To that end, this exercise requires you to maintain muscular strength as the muscle lengthens. When your toes first contact the floor, articulation is needed to cushion the landing, and muscular endurance is needed to support your body weight against gravity. The gastrocnemius typically fires more on the landing phase of jumping, and strengthening the soleus provides it with better assistance.

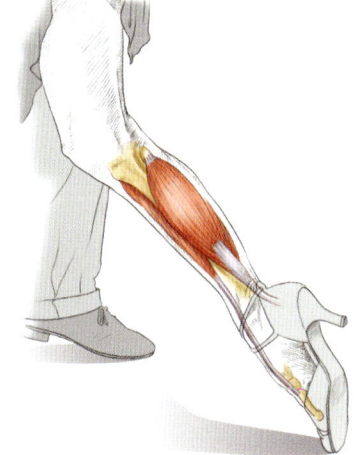

In addition, the soleus muscle contains more type I (slow-twitch) muscle fibers, which helps provide awareness of the balance and security of the lower leg on the ankle. Most of the muscles in your body contain both type I and type II fibers, but the soleus contains mostly type I, providing postural stability of your lower legs. For instance, the soleus helps keep you from falling forward when standing and helps maintain balance whether you are dancing in character shoes or en pointe. Because of the higher content of type I fibers, the soleus is more resistant than the gastrocnemius to fatigue; to improve strength, you will need to increase your number of repetitions.

VARIATION

Resistance Band Pump

Sit on the edge of a table or hang your leg over the barre so that the barre hits just above the back of your knee. Wrap a resistance band around the metatarsal heads. Keeping the toes covered, hold the band from above. Without activating your quadriceps, push your foot into plantar flexion against the resistance of the band. You don't even have to point your toes; just point the ankle. Alternate between plantar and dorsiflexion at the ankle while emphasizing deep soleus contraction. Perform 30 times or more; work up to 3 sets.

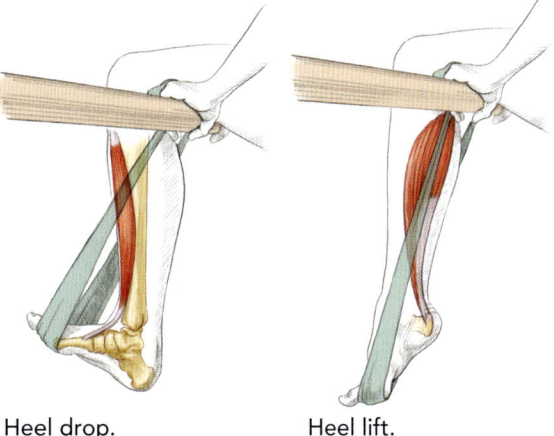

Heel drop.   Heel lift.

TOE PRESS

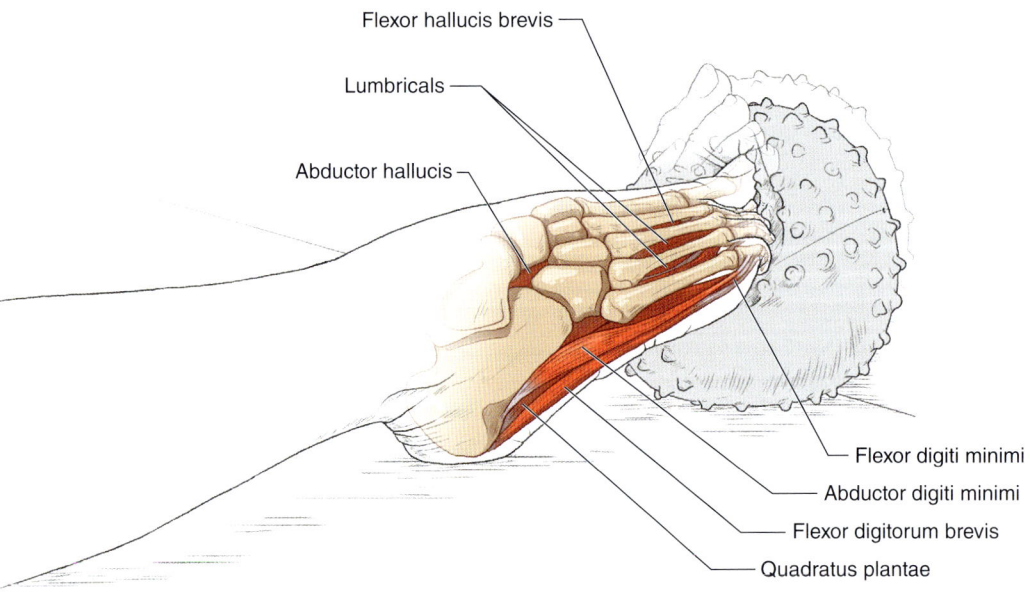

Flexor hallucis brevis
Lumbricals
Abductor hallucis
Flexor digiti minimi
Abductor digiti minimi
Flexor digitorum brevis
Quadratus plantae

EXECUTION

1. Stand facing a wall with the ankle slightly pointed and the heel resting on the floor.
2. Place a small ball under your toes against the wall.
3. Lengthen the toes and press the toes into the ball. Hold for 3 to 5 seconds, relax, and repeat 10 to 15 times.
4. Isolate the deep muscles of the plantar surface of the foot while holding the ankle stable.
5. Repeat on the other foot.

MUSCLES INVOLVED

Flexor hallucis brevis, lumbricals, flexor digiti minimi, quadratus plantae, abductor hallucis, flexor digitorum brevis, abductor digiti minimi

DANCE FOCUS

Moving from a half-pointe position to a full-pointe position requires strength and articulation through the forefoot and midfoot. It is important not to curl or claw the toes. The toe press exercise is a great pre-pointe exercise to help transition the foot from half-pointe to full-pointe position. While executing this exercise, press the toes into the ball without curling the joints of the toes, which can also improve foot dexterity or coordination of the various foot muscles. Quality range of motion, strength, and control are all needed for pointe work. The toe press exercise can be used as a warm-up for muscles on the plantar surface of the foot or for a strength training exercise.

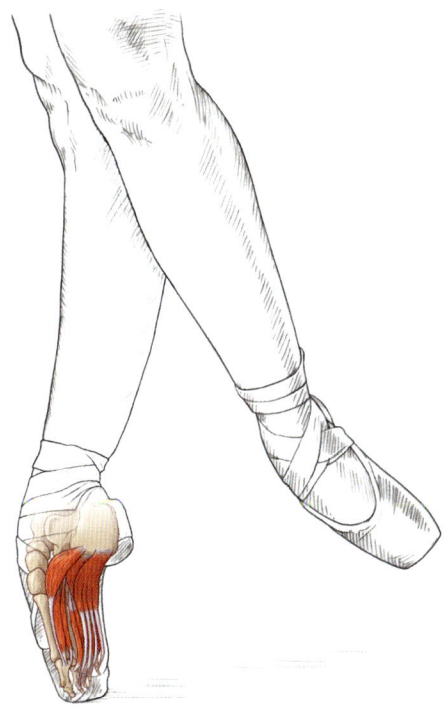

ANKLE DORSIFLEXION

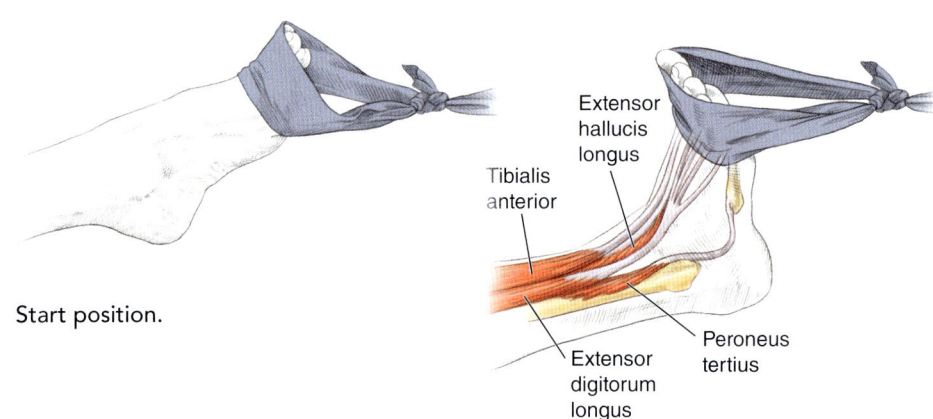

Start position.

Extensor hallucis longus

Tibialis anterior

Peroneus tertius

Extensor digitorum longus

EXECUTION

1. Sit with a resistance band wrapped around your forefoot. Secure the other end to a stable base in front of you. Begin with the ankle in a softly pointed position; the band must be taut at the start of the exercise.

2. Lift the toes against the resistance of the band and continue to increase the resistance by flexing the ankle. Focus on the muscles of the anterior tibia contracting and the posterior tibia lengthening.

3. Hold the contraction for 2 to 4 counts, then slowly return to the start position. Maintain tautness of the band throughout the entire range of motion. Perform 15 to 30 times; work up to 2 or 3 sets.

SAFETY TIP: To avoid sickling or winging of the foot, focus on a neutral position of the ankle, aligning the second toe with the tibia.

MUSCLES INVOLVED

Tibialis anterior, extensor digitorum longus, extensor hallucis longus, peroneus tertius

DANCE FOCUS

Keeping the front of the tibia strong gives you more security when dancing or turning on your heels. Your warm-up includes a significant amount of relevé and pointing of the toes, but it probably doesn't include rocking back on your heels, which may be called for by some choreographers. As a result, the muscles in the back of your lower legs get more work than the muscles along the front. This imbalance can create overuse injuries and inhibit your technique. Strengthening the muscles along the front of the tibia bone may reduce the risk of shin splints.

Every grande plié you execute requires contraction of the tibialis anterior muscle to support your tibia bone. This muscle also works to transfer your weight forward to prepare for relevé and helps to maintain a nice lift in your arch. Don't forget about it in your conditioning schedule.

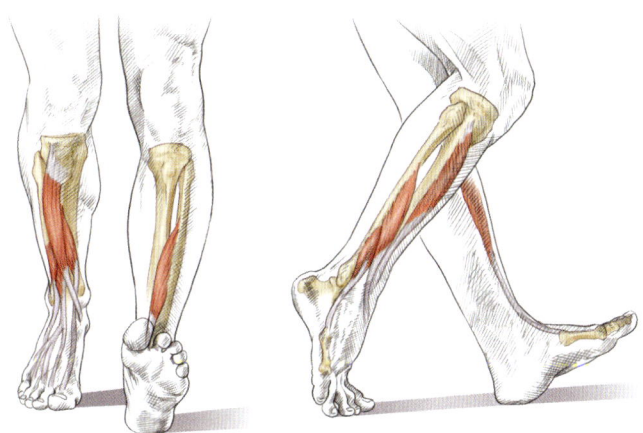

VARIATION

Heel Walks

Stand with your legs parallel. Elevate your forefeet and feel the anterior tibialis muscles activating. Hold for an isometric contraction of 10 seconds, then maintain control while bringing the forefeet back down. Now repeat, and this time take small steps on your heels. Again, feel the anterior tibialis activating. Maintain strong lift of the forefoot while aligning the second and third toes with the tibia. While executing heel walks, place your weight directly in the center of your heel. Walk at least 10 steps; perform at least 3 times.

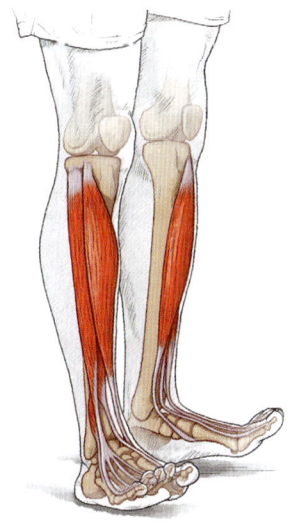

FORCED ARCH RELEVÉ

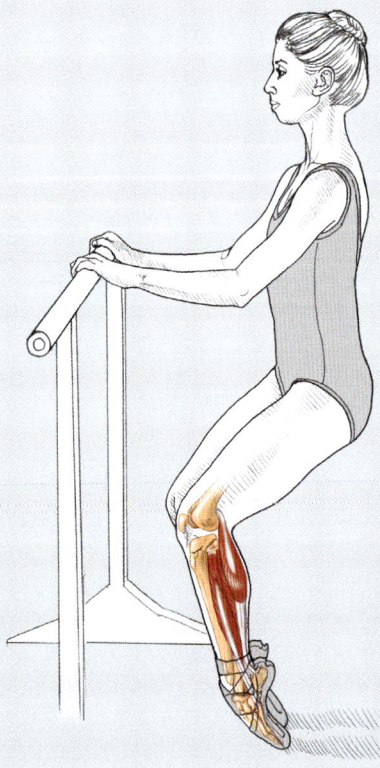

This exercise can help you warm up the feet and stretch the ankle. Let's break it down.

1. Begin with the feet in either a parallel or a turned-out position. Hold on to the barre for balance. Find your neutral spine placement and support it with your deep abdominals.

2. Roll up into a half-pointe position, focusing on rising over the second and third toes. Feel the posterior lower leg muscles engaging to lift your heel as high as you can. Feel the muscles of the lateral lower leg engaging for stability. Visualize your medial and lateral ankle ligaments working to stabilize your ankle joint. Allow your trunk to transfer weight directly over your toes.

3. While maintaining a high relevé, carefully bend your knees, aligning them directly over your toes, and gently stretch the anterior portion of your ankle. Feel lengthening under your toes for an eccentric stretch. Visualize the soleus helping to control your lower leg, and keep your heel lifted as high as possible.

4. With the plié, maintain your neutral pelvis position; don't tuck your pelvis under. Movement in your pelvis should be very limited. Separate your femur, moving in the hip joint, from your pelvis. Continue feeling a stretch along the anterior portion of your ankle.

5. Slowly begin to straighten your knees, engaging your quadriceps and trying to maintain heel height. Again, visualize the medial and lateral ankle ligaments supporting your ankle joint. Keep focusing on the posterior lower leg muscles contracting to maintain heel height. Relevé only as high as you can while maintaining alignment over your second and third toes. Do not allow your ankle to sickle or supinate.

6. Remember your plumb line alignment and make sure that your trunk is directly over your relevé. Continue stretching under your toes and widening through the metatarsal heads to create a wide base for balance. Once you have acquired your high relevé, slowly return to your starting position with control.

Muscles Involved

Relevé: Concentric contraction of gastrocnemius, soleus, posterior tibialis, flexor hallucis, flexor digitorum, peroneus longus, peroneus brevis

Forced arch relevé: Soleus (to be noted while other muscles contract)

CHAPTER 11

Whole-Body Training for Dancers

The field of dance and performing arts medicine continues to grow. This development is inspiring for those who work in the field, and excellent dance medicine specialists are now located throughout the world. But the greatest benefits of this growth accrue to you, the dancer, as well as your instructors. Basic dance medicine science can help you become a better dancer and help your instructors become better teachers. Athletes all over the world are embracing what science has to offer to enhance their athletic skills; dancers can do the same.

Research published in medical journals gives dance medicine specialists crucial information you can use in helping your own dancing. For example, Grossman and Wilmerding (2000) showed that if you incorporate simple hip flexor conditioning exercises into your daily routine, you can improve the height of your développé. Bowerman et al. (2015) concluded that knee pain and injury can result from faulty turnout, excessive lumbar lordosis, screwing at the knee, and twisting through the tibia.

You can use such information to integrate dance-specific exercises into your training that enhance your performance and decrease your risk of injury. For instance, improvement in arabesque could be as simple as strengthening the abdominals and hip extensor muscles while improving movement of the thoracic spine. Improving turnout could be as simple as understanding good neutral pelvic alignment while activating the true hip rotators. By incorporating the principles of body placement, you can improve coordination. When your muscles and bones are more aligned, you require less overall muscle action! Therefore, you can perform dance movements without straining and overusing muscles.

Periodization

Dance is an artistic discipline, with preprofessional dancers spending 20 to 25 hours a week working on skill. Physical performance demands include excellent technique with hip and trunk strength to move through space in various planes of motion. You also need to have flexibility skills that take the hip and lumbar spine into extreme ranges. Competitive dance includes jumping exercises at different speeds (both explosive and slow) and heights, some bilateral and some unilateral, along with multiple variations of those just listed. Strength is needed in the trunk, hips, and legs for the repeated slow and controlled movements associated with ballet training; power is needed for jumping efforts in all dance movement; and anaerobic fitness is needed for dynamic and intense stop-and-go movement patterns. Since daily dance classes may not provide enough specific strength or aerobic training, supplemental training is important. The goals are to increase muscle strength, improve core strength, and improve balance skills for motor control and proprioception.

How can periodization help? Periodization refers to an organized, planned gradual increase in training while alternating exercise and rest periods. The goal is to increase strength, flexibility, and balance leading up to the dance performance. The yearly training regimen is broken into blocks: microcycles, mesocycles, and macrocycles. A microcycle represents approximately a one-week block, while the mesocycle could be composed of 4-week blocks, and the macrocycle represents your full season. Periodization includes two different models: linear and nonlinear. Linear refers to slowly increasing training, volume, and intensity during each block. Nonlinear refers to changing your training regimen for each microcycle.

As an example, collegiate dancers aim to improve technique by the end of each semester for their live performances. It would be advantageous to design a yearly program (macrocycle) with training blocks that coincide with each semester in preparation for the live performance, just as other athletes prepare for a competition. Each mesocycle would be broken down into 4-week blocks with dance-specific target goals for each block. A linear periodization model for a freshman dancer starting the first semester might be a good choice. The focus would be on strengthening, but the program would take into consideration that the dancer is also taking 6 dance classes a week, which would be considered in-season. The dancer could participate in an in-season strength training program twice a week, a plyometric training program twice a week, and a balance training program twice a week. Each training program would be no longer than 30 minutes and include specific rest and recovery times between exercises. With strategic planning, the training would taper one to two weeks prior to the performance. The blocks would be designed specifically to the needs of the dancer and can help prevent overtraining. Organizing supplemental training around the dance schedule can be beneficial and help reduce the risk of injury.

Static and Dynamic Stretches

Flexibility is the intrinsic property of body tissues that determines the range of motion achievable without injury at a joint or series of joints. It can be developed through static stretching or dynamic stretching, and a combination of the two will be the most effective for you as a dancer. Let's briefly review both types.

Static stretching involves isolating one muscle group and holding a stretch; no joint movement occurs. This type of stretch involves elongating the muscle fibers and taking the slack out of the muscle–tendon complex; it should not create intense pain. Breathe comfortably and relax into the stretch; again, do not allow any joint movement. This is a safe and gentle type of stretch that will not injure your muscle fibers. Static stretching is most effective when performed after a workout when you are warm. Hold the stretch for 30 seconds, which gives your muscle plenty of time to relax and stretch.

Dynamic stretching, on the other hand, involves performing controlled, active, warm-up stretches without creating aggressive force on the tissue. In this type of stretching, your body moves, and you can focus on multiple body parts. These stretches may include dance movements that you will be rehearsing or performing. Dynamic stretching also increases blood flow to keep your muscles warm, which is very important between rehearsals.

This chapter includes illustrations of both static and dynamic stretches to help you improve your flexibility and reduce your risk of injury.

Small Props

The exercises presented in this chapter review musculature that has been discussed earlier in the book and use props for added resistance or neuromuscular feedback. Of course, dance class involves working against the resistance of your own body, but this may not be enough to increase your strength. By adding small apparatus and resistance tools, you can build strength beyond the limits of gravity, vary your conditioning plan, and challenge your balance skills.

Resistance bands and free weights have already been introduced, and you can also use other props to improve your technique and keep your training fresh. For example, you can increase your body awareness (proprioception) by performing exercises on a stability ball, minitrampoline, foam roller, or rotating disc. You can also add hand weights or ankle weights to increase resistance. These small props make the exercises more challenging, thus enabling you to develop improved overall balance, which you can transfer to your dance experiences.

Your ability to maintain balance derives from three sources of input: your eyes, sensory receptors in the inner ear, and receptors in the muscles and joints that help with postural control. Whenever you try to maintain your balance on an uneven or unstable surface, you challenge your sensory receptors to work harder. To advance any of the exercises throughout this book, close your eyes at various times to focus on integration of mind and body. Have you ever lost your balance when the stage lights suddenly change or go to a blackout? Have you noticed that your balance is weak after an injury? Balance can also be compromised by adolescent growth spurts, which can cause fatigue as well. In other words, your proprioception is weakened by any abrupt change in your sensory system. The good news is that training your balance skills will improve your acuity and your precision of movement.

Training Specifics

If you're concerned about fitting all these exercises into your busy schedule, focus on a few exercises at a time and slowly incorporate some into your warm-up and others into your cool-down. Try taking one concept at a time, working on it for a week, and then gradually adding others. Execute several exercises from chapters 4, 5, and 6 every other day and perform the extremity exercises on the days in between. Use the exercises to make positive changes in the way you work.

To perform each exercise with efficiency, organize your thoughts. For instance, alignment is essential for precision of movement; it is a whole-body sensation. Continue to visualize each movement of your dancing along the various planes of your body. Notice that you can gradually change poor habits and improve your lines.

Maintain spinal stability while releasing unnecessary tension. Improve your lung capacity by incorporating good breathing patterns while dancing. Deeper breathing enhances core control and supports moving from your center. Imagine your breath reaching every muscle in your body to enhance every movement.

Improvement in proprioception includes mind–body integration while advancing the functional work. Maintain balance awareness while the base of support changes during various floor exercises. Continue to focus on postural awareness while moving from the floor to the barre and into center. Imagine your new balance skills working while you're turning, jumping, and balancing in relevé.

While many of the exercises in this book focus on individual muscle groups to enhance certain dance movements, functional training takes it a step further. Functional training involves continuing the movement to work in all planes of motion by using multiple joints and challenging your nervous system. After all, this is what you do every day in your classes, rehearsals, and performances! You train for whole movement using all the muscles. Working individual muscle groups can still improve your strength and restore any muscle imbalance that might be present.

When using the exercises presented in this chapter, think functionally; think about your whole body, about how your joints are working through the entire range of motion, and about efficient movement. Remember that whichever style of dance you embrace will involve variations, such as choreography changes and floor and costume changes. You will adapt to such variations more easily if your fundamental skills are strong.

To gain muscular strength, warm up your body and repeat the exercises to fatigue without compromising alignment. To progress, you can increase either the number of repetitions or the amount of resistance; either way, vary your speed to correlate with changing dance tempos. Practice your favorite dance steps with the same attack and vigor. For instance, repeat basic jump variations to focus on controlling the landing. To improve your cardiorespiratory endurance, increase the number of repetitions. To avoid faulty compensations, focus on the muscle group creating the movement, but continue to think about whole-body function as you work.

Plyometric Training

In a nutshell, plyometrics involves jump training designed to develop more power and height in your jumping combinations. You know that you must be strong both physically and neurologically, but let's take a moment to focus on the nature of jumping power. Power is the combination of strength and speed. In dance, for example, you need both speed and power for petit allegro and grand allegro movements. This chapter includes several plyometric exercises to help you both with eccentric contraction (i.e., control) when landing and with amortization, which is the ability to get yourself together, prepare, and go right into the next jump. We will also look at concentric contractions, in which your muscles shorten strongly to elevate you into the air.

Dance-Focused Exercise

The following exercises are full of challenges. You will add props and execute full-body, functional movements. Envision applying the principles of each exercise to your specific dance style. For best results, memorize the correct desired movement. In these exercises, you are taking your work to the next level by increasing the challenges for your core and your balance skills. The chapter concludes by examining arabesque.

Your mind is a powerful tool. Be selective in what you focus on. Quiet your mind so that you can concentrate on the specific area of the body that you are working on. Before each exercise, zero in on the starting position and movement execution while maintaining a feeling of ease. Speak to yourself using only positive reinforcement! Keep the flow of mental talk inspiring and optimistic.

WALL PLIÉ

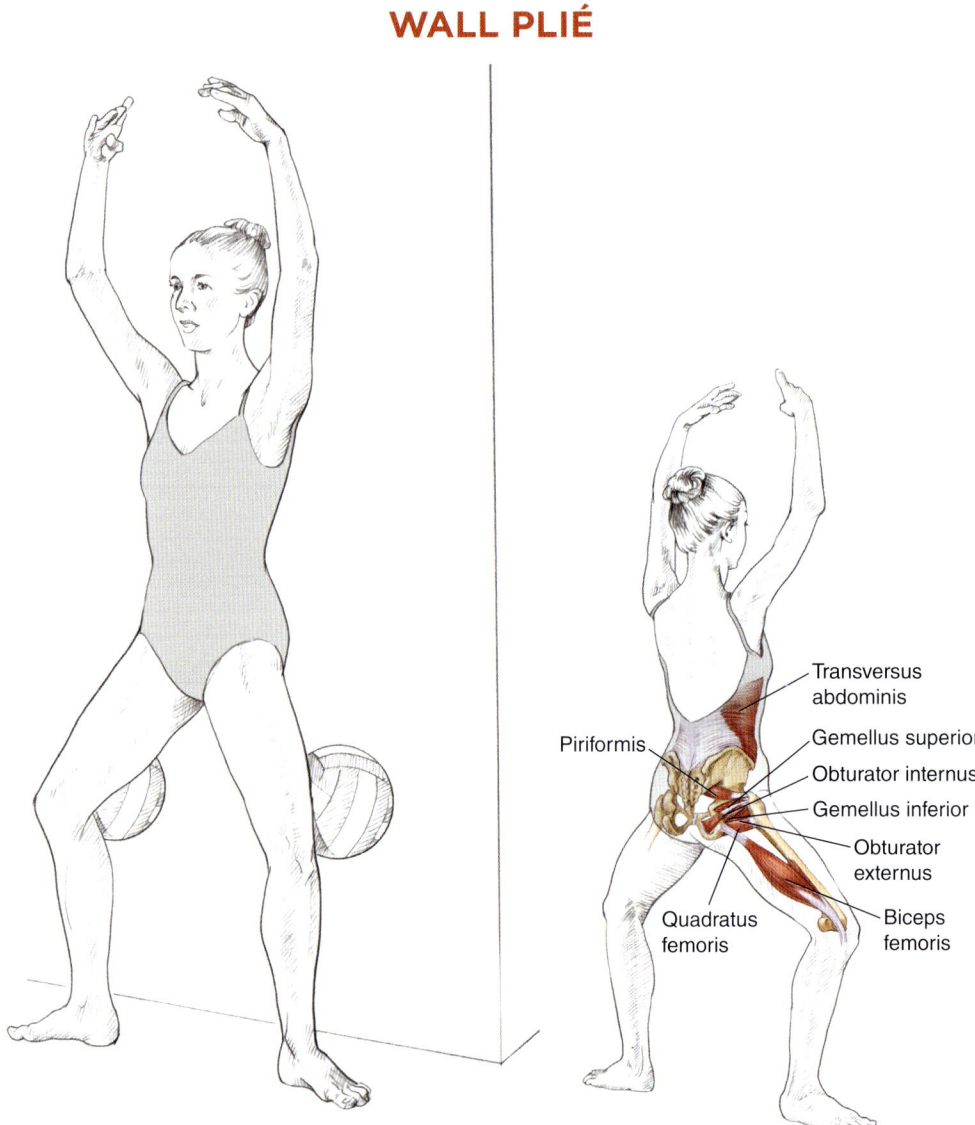

Transversus abdominis

Gemellus superior

Obturator internus

Gemellus inferior

Obturator externus

Biceps femoris

Quadratus femoris

Piriformis

EXECUTION

1. Place your back against the wall. Turn out your legs and place your feet more than hip-width apart (aligned according to what your personal turnout allows). Place a ball between each thigh and the wall.

2. Inhale to prepare; maintain a neutral spine and pelvis.

3. On forced exhalation, press your thighs into the balls by contracting the deep rotators. Focus on maintaining a neutral pelvis. In each leg, align the femur over the mid talus and second metatarsal and hold for 2 to 4 counts. Perform 8 times.

MUSCLES INVOLVED

Transversus abdominis, gluteus medius, quadriceps (rectus femoris, vastus lateralis, vastus medialis, vastus intermedius), sartorius, biceps femoris, piriformis, gemellus superior, gemellus inferior, obturator internus, obturator externus, quadratus femoris, anterior tibialis, gastrocnemius, soleus, peroneals

DANCE FOCUS

Maintaining ease in the hips without straining through the trunk allows for a better turnout. Use this exercise to focus on the deep hip rotators while maintaining a neutral and secure position of the pelvis. Memorize the feeling of true external rotation in the hip without overuse of the lateral thigh or tilting of the pelvis. Focus on alignment of the femur over the second toe; avoid any torque throughout the knee. The long line of the tibia should be placed directly over the center of your foot. Close your eyes for a moment and visualize the deep external obturator as it contracts and pulls the femur outward to increase the turnout. Now relax the rotators. Repeat the detail of the contraction again until you feel how firm and supportive this muscle is in rotating your thigh outward. To reemphasize, focus on hip disassociation: Let the movement occur in the hip joint so that the thighs open into external rotation while the pelvis and spine are stable.

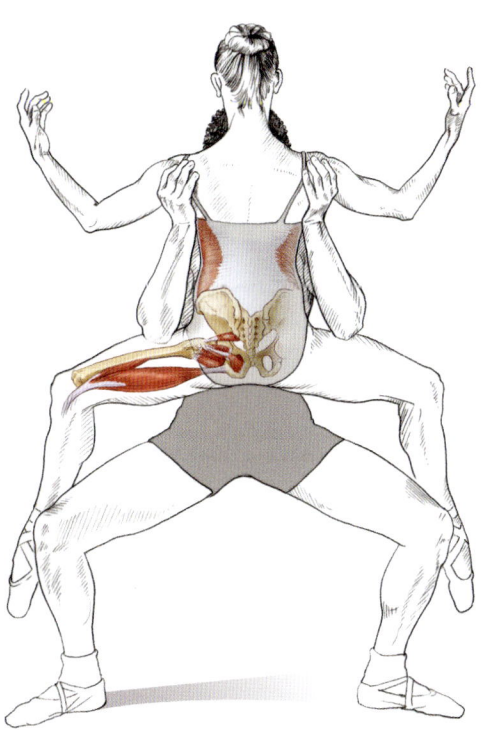

BRIDGE WITH FEET ON ROLLER

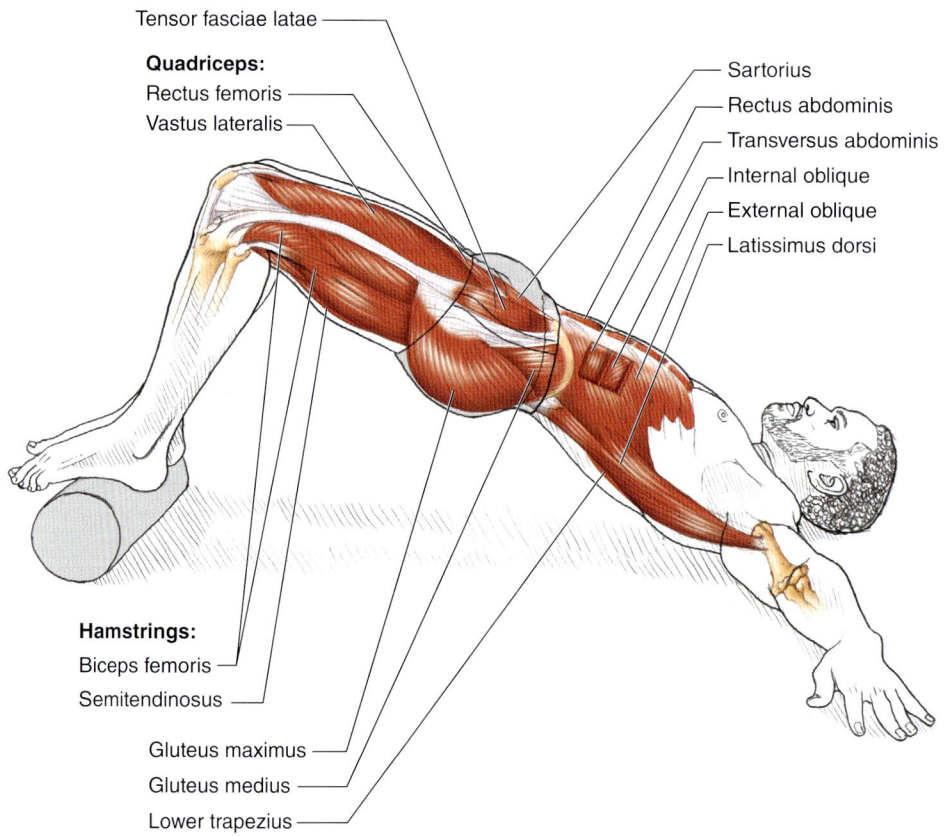

Tensor fasciae latae

Quadriceps:
Rectus femoris
Vastus lateralis

Sartorius
Rectus abdominis
Transversus abdominis
Internal oblique
External oblique
Latissimus dorsi

Hamstrings:
Biceps femoris
Semitendinosus

Gluteus maximus
Gluteus medius
Lower trapezius

EXECUTION

1. Lie supine with your arms extended to the sides at shoulder level. Locate your neutral spine position. Bend your knees and place your feet on a roller.

2. Inhale to prepare widening through the rib cage. On exhalation, engage your deep abdominals and rectus abdominis as you move into a posterior pelvic tilt.

3. Continue rolling segmentally through your spine into hip extension, elevating your hips off the floor. Move through your sagittal plane until your shoulders, hips, and knees align. You should be resting on the upper segments of your thoracic spine.

4. Hold for 3 to 5 counts. Press your heels into the roller, keeping the roller as stable as possible. Lengthen through the hip flexors. Inhale to prepare without elevating the shoulders. On forced exhalation, slowly begin to roll down with control, beginning with the sternum and then the thoracic spine, the lumbar spine, and finally the hips, until you return to your neutral spine position.

5. Perform 10 to 30 times.

SAFETY TIP: Try not to overextend the lumbar spine. Keep working the abdominals, gluteus maximus, and hamstrings to maintain a neutral pelvis position once you have reached the top of the bridge.

MUSCLES INVOLVED

Transversus abdominis, rectus abdominis, external oblique, internal oblique, hamstrings (semitendinosus, semimembranosus, biceps femoris), gluteus maximus, latissimus dorsi, lower trapezius, quadriceps (rectus femoris, vastus lateralis, vastus medialis, vastus intermedius), tensor fascia latae, sartorius, gluteus medius, iliopsoas, pelvic floor muscles (coccygeus, levator ani)

DANCE FOCUS

If you have weakness in the gluteus maximus, hamstrings, or spinal extensors, this is a great exercise for learning to activate or contract your hip extensors to improve trunk stabilization, pelvic stabilization, and hip extension—all of which are needed for arabesque, attitude derrière, and tour jeté. This exercise also helps you work eccentrically through the hip flexors, quadriceps, and sartorius. This exercise provides an excellent way to warm up before performances because it doesn't require much space yet is useful for your whole body. You can advance it further by lifting one leg to challenge sensory receptors and trunk stabilizers.

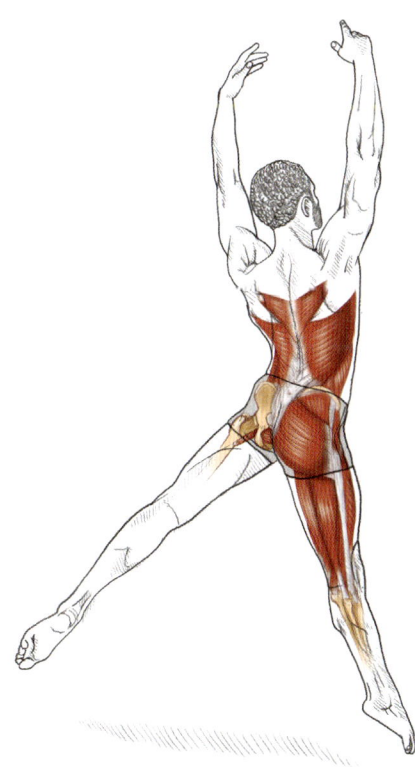

WEIGHTED SIDE-BEND

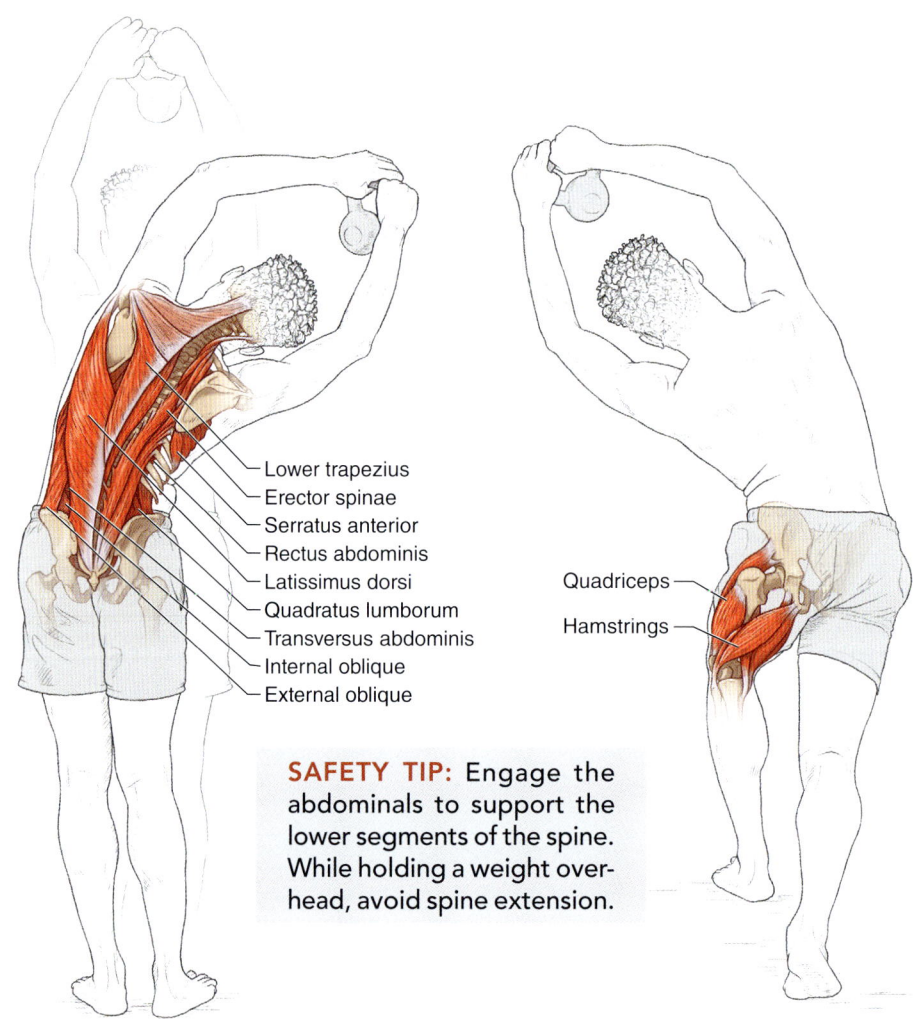

Lower trapezius
Erector spinae
Serratus anterior
Rectus abdominis
Latissimus dorsi
Quadratus lumborum
Transversus abdominis
Internal oblique
External oblique

Quadriceps
Hamstrings

SAFETY TIP: Engage the abdominals to support the lower segments of the spine. While holding a weight overhead, avoid spine extension.

Right side-bend.

Left side-bend with lunge.

EXECUTION

1. Stand with the legs hip-width apart. Hold a small hand weight in each hand or one weight with both hands and lift your arms overhead. Elongate through the spine and inhale to prepare.

2. On exhalation, tighten the abdominals while leaning into a right side-bend. Securely hold the weight and the arms overhead.

3. Hold as you inhale. Feel the external obliques working to stabilize your spine.

4. On exhalation, bring your upper body back up before moving upright into a left side-bend. While moving into the left side-bend position, slide the right foot back into a lunge.

5. Hold as you inhale and continue to engage the external obliques to stabilize the spine with the lunge.

6. On exhalation, return to the upright beginning stance before repeating the entire set again. Perform 3 to 5 repetitions before resting the arms and repeating on the other side.

MUSCLES INVOLVED

Side-bend: Internal oblique, external oblique, transversus abdominis, rectus abdominis, quadratus lumborum, erector spinae (iliocostalis, longissimus, spinalis), latissimus dorsi, serratus anterior, lower trapezius

Lunge: Front-leg quadriceps (rectus femoris, vastus lateralis, vastus medialis, vastus intermedius), front-leg hamstrings (semitendinosus, semimembranosus, biceps femoris)

DANCE FOCUS

The principle of axial elongation applies throughout the entire length of the movement. This elongation creates height along the spinal column to increase motion while your head balances with ease on top. Feel as though you are moving each vertebra separately to achieve a flexible but secure spine. Lengthening along the spine provides more space between the vertebrae and less compression on the discs. The added hand weight provides a more dynamic challenge while the trunk moves along the frontal plane and the legs move along the sagittal plane. Your external oblique muscles are important for cambré side movement and spine rotational movement. The external and internal oblique muscles also help create balance and stability while turning. Remember the discussion from chapter 6 on the core: the external oblique primary action is spinal flexion and side-bend, but it also contracts in rotation from the opposite side. The external obliques help you feel the connection between the ribs and the pelvis. They originate off ribs 5 to 12 and insert into the iliac crest. As you are moving into side-bend, feel lengthening through your quadratus lumborum. Your pelvis should feel anchored and stable to resist the trunk's upward pull. Visualize a half-moon and imagine soaring sideways!

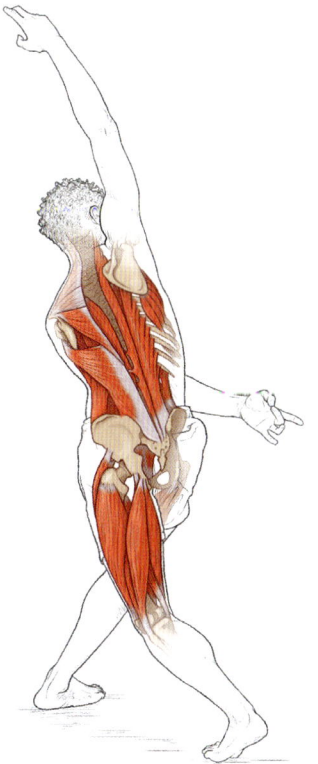

DIAGONAL TWIST

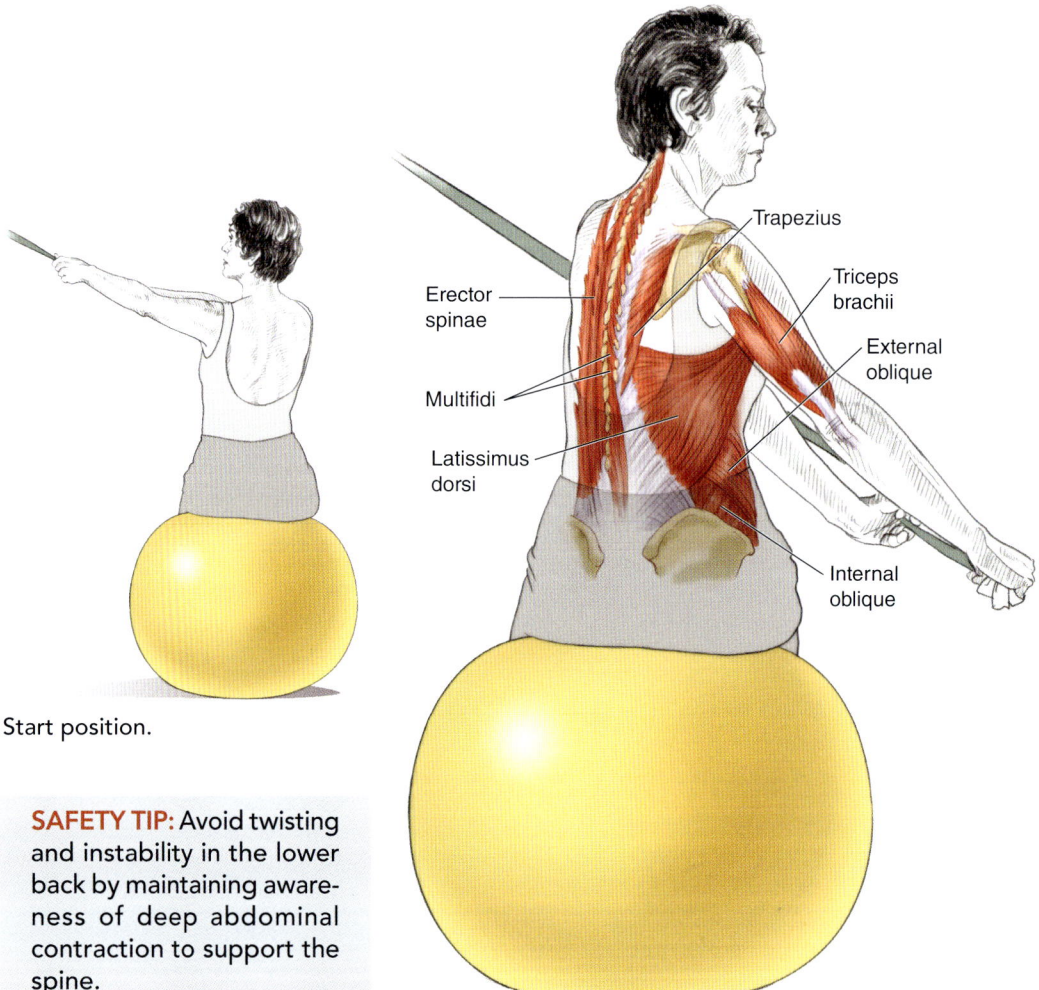

Start position.

Erector spinae

Multifidi

Latissimus dorsi

Trapezius

Triceps brachii

External oblique

Internal oblique

SAFETY TIP: Avoid twisting and instability in the lower back by maintaining awareness of deep abdominal contraction to support the spine.

EXECUTION

1. Sit on a stability ball with your hips and knees flexed at 90 degrees and your feet on the floor. Wrap a resistance band high over your left shoulder; both hands hold the end. Your pelvis remains neutral on the stability ball while your trunk rotates to the left. Your hands, holding the resistance band, remain aligned with your sternum. Inhale to prepare.

2. On exhalation, engage the deep abdominals, obliques, and scapular stabilizers to rotate your trunk to the right. Your arms pull against the resistance of the band in a downward right diagonal pattern.

3. Hold this position for 2 to 4 counts. Feel the oblique musculature working to support your center. Maintain alignment of the hands and extended elbows with the sternum. Return slowly with inhalation. Perform 10 to 12 times on each side.

MUSCLES INVOLVED

Latissimus dorsi, lower trapezius, triceps brachii, transversus abdominis, internal oblique, external oblique, erector spinae (iliocostalis, longissimus, spinalis), multifidus

DANCE FOCUS

Coordinating strength in rotational and spiraling movements requires strength in the core and deep multifidus muscles of the spine. To allow for more rotation, release tension in the neck and shoulders before the spiral occurs. To secure the lower spine, engage the lower abdominals, which also allows for more rotation. The diagonal twist is wonderful for the ballroom dancer who has been practicing for hours in upper-back extension and left trunk rotation. Remember that the obliques are working for you on both sides; the internal oblique is contracting on the same side as the rotation, and the external oblique is contracting on the opposite side. The same muscle assistance applies with the deep erector spinae muscles: While muscles contract to produce movement on one side, you also have muscles contracting on the opposite side. This reinforces the need to move from your center; you must initiate spiral movement from deep in the core and close to the spine.

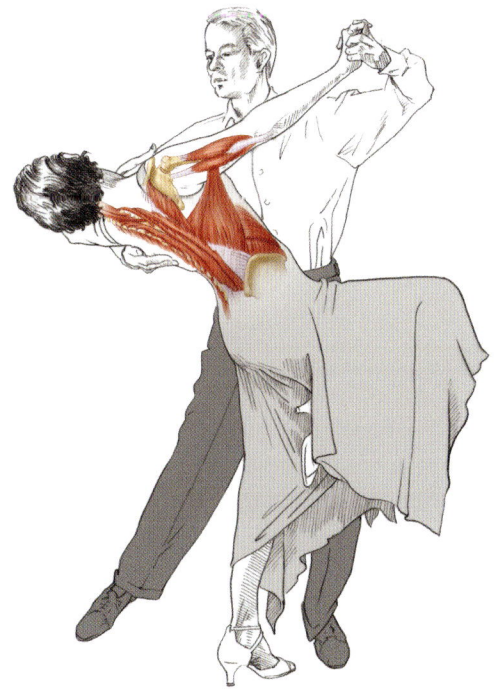

HIGH KICK WITH RESISTANCE

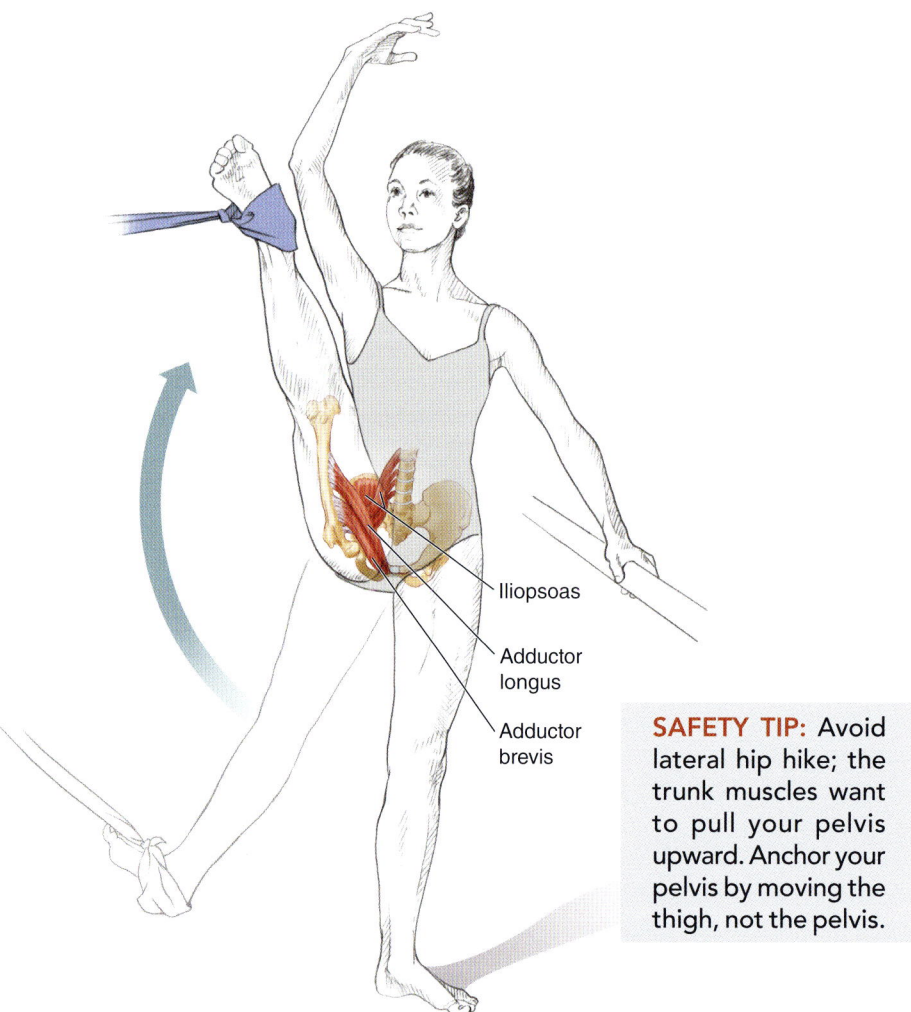

Iliopsoas

Adductor longus

Adductor brevis

SAFETY TIP: Avoid lateral hip hike; the trunk muscles want to pull your pelvis upward. Anchor your pelvis by moving the thigh, not the pelvis.

EXECUTION

1. Begin with the left hand on the barre and the right leg in a turned-out tendu position to the side. Tie one end of a resistance band around the ankle of the right leg and the other end to an immovable object to the side. Reorganize your neutral placement. Secure the turned-out supporting leg by engaging the gluteus medius.

2. Bring the leg quickly through first position and cross through fifth into a battement devant against the resistance of the band, holding firmly along the gluteus medius of the supporting leg. Coordinate your breathing so that you inhale as the leg goes up.

3. Initiate the movement from the core and inner thighs in the low range. Use the brush through first to fifth to emphasize hip adduction, then engage the iliopsoas as soon as possible to elevate the leg. Return slowly with control.

4. Lengthen through the spine and quadratus lumborum. Maintain turn-out throughout the exercise. Perform 6 to 8 times, then repeat 6 to 8 more times without the resistance.

MUSCLES INVOLVED

Gesture leg: Adductor longus, adductor brevis (low level), iliopsoas (higher level)

Standing leg: Hip external rotators, gluteus medius, gluteus maximus, hamstrings (semitendinosus, semimembranosus, biceps femoris), quadriceps (rectus femoris, vastus lateralis, vastus medialis, vastus intermedius), sartorius

Trunk: Transversus abdominis, internal oblique, external oblique, rectus abdominis

DANCE FOCUS

Lifting the legs with ease and grace requires avoiding extra adjustments, unnecessary weight shifts, and overuse of the quadriceps. Working effectively the first time reduces the risk of injury and improves your technique. The higher your leg goes, the harder the deep iliopsoas must contract. Keep working to maintain turnout as much as you can. When the working leg begins to turn in, the anterior fibers of the gluteus minimus and medius begin to take over and elevate your hip. Visualize the attachment of the iliopsoas on the inside of the femur. Initiate the movement from that area of the thigh and let your leg float up to your chest. With each leg lift, lengthen the hamstrings, buttocks, and lower-spine musculature. Train your inhalation to help elevate your leg and your exhalation to secure your spine as the leg lowers. Your legs can fly!

ATTITUDE ON DISC

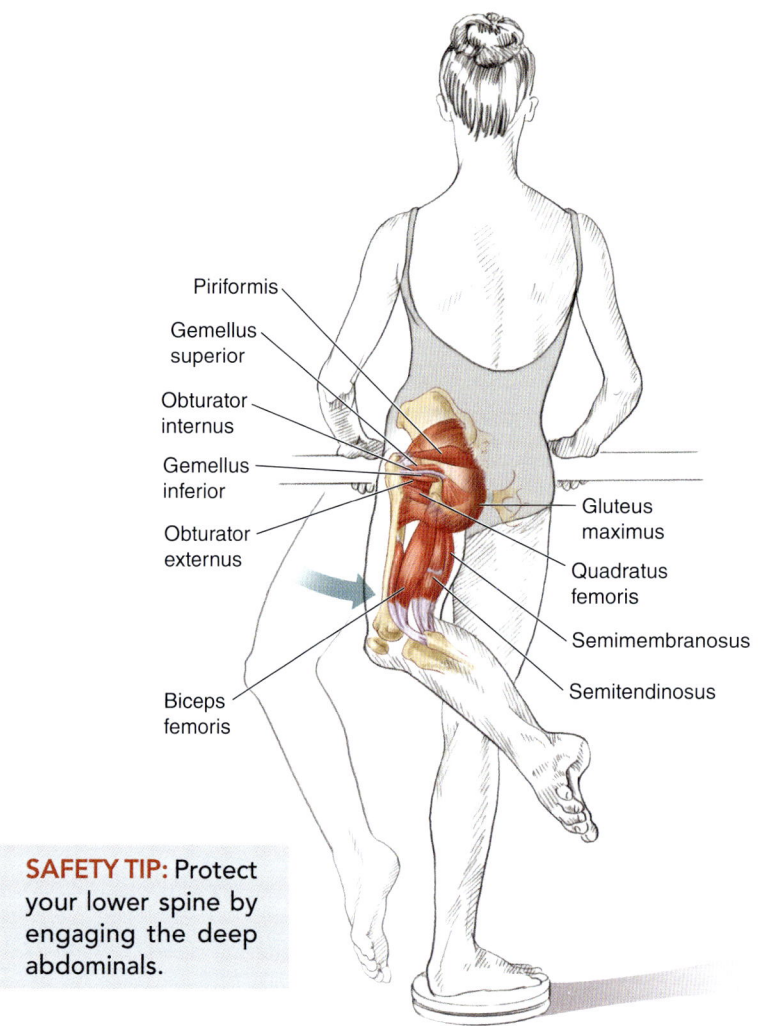

Piriformis

Gemellus
superior

Obturator
internus

Gemellus
inferior

Obturator
externus

Gluteus
maximus

Quadratus
femoris

Semimembranosus

Semitendinosus

Biceps
femoris

SAFETY TIP: Protect
your lower spine by
engaging the deep
abdominals.

EXECUTION

1. Face the barre with the right leg turned out on a disc. Your left leg is in coupé position. Organize your placement and balance.

2. Coordinate inhalation with hip extension, moving from coupé to attitude derrière. As the leg elevates, there must be a slight accommodating forward shift of your body and pelvic rotation. Reemphasize the deep rotators turning out the attitude derrière leg. Engage your deep abdominals to support the lower spine. Lengthen the thoracic spine into a long arch.

3. Hold for 2 to 4 counts, focusing on the gluteus maximus and hamstrings. With exhalation and control, reverse the movement to return to coupé. Perform 10 to 12 times on each side.

MUSCLES INVOLVED

Gesture leg: Piriformis, gemellus superior, gemellus inferior, obturator internus, obturator externus, quadratus femoris, gluteus maximus, hamstrings (semitendinosus, semimembranosus, biceps femoris)

Standing leg: Hip external rotators, hamstrings (semitendinosus, semimembranosus, biceps femoris), gluteus medius, gastrocnemius, peroneals

DANCE FOCUS

The quality of your technique improves if you initiate extension to the back with the muscles that are primarily responsible for that movement. Your arabesques improve when you protect your lower spine and develop more strength in the hamstrings and gluteus maximus. Practice moving your leg to the back and see how far it will go before your lower spine moves. You may be capable of only 15 degrees of movement; in that case, shift forward slightly to accommodate, but continue to lift the leg from the hamstrings and gluteus maximus contraction.

Whether you are moving into a low attitude derrière or a full arabesque, engage the abdomen to support your spine. Incorporate movement along your thoracic spine. As you maintain a strong lift in the abdomen, visualize the vertebrae in your midback moving into extension. You have more movement capabilities in the upper back and chest area than you think; it's not about arching your lower back. Use the deep turnout muscles to avoid twisting in the pelvis. Remember that your spine is elongated and moving in the longest possible arch. Coordination and beautiful alignment reduce tension in the neck and shoulders.

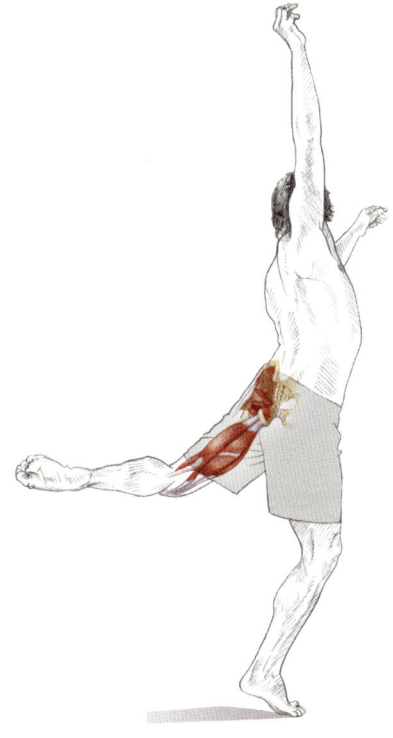

PLANK AND PIKE

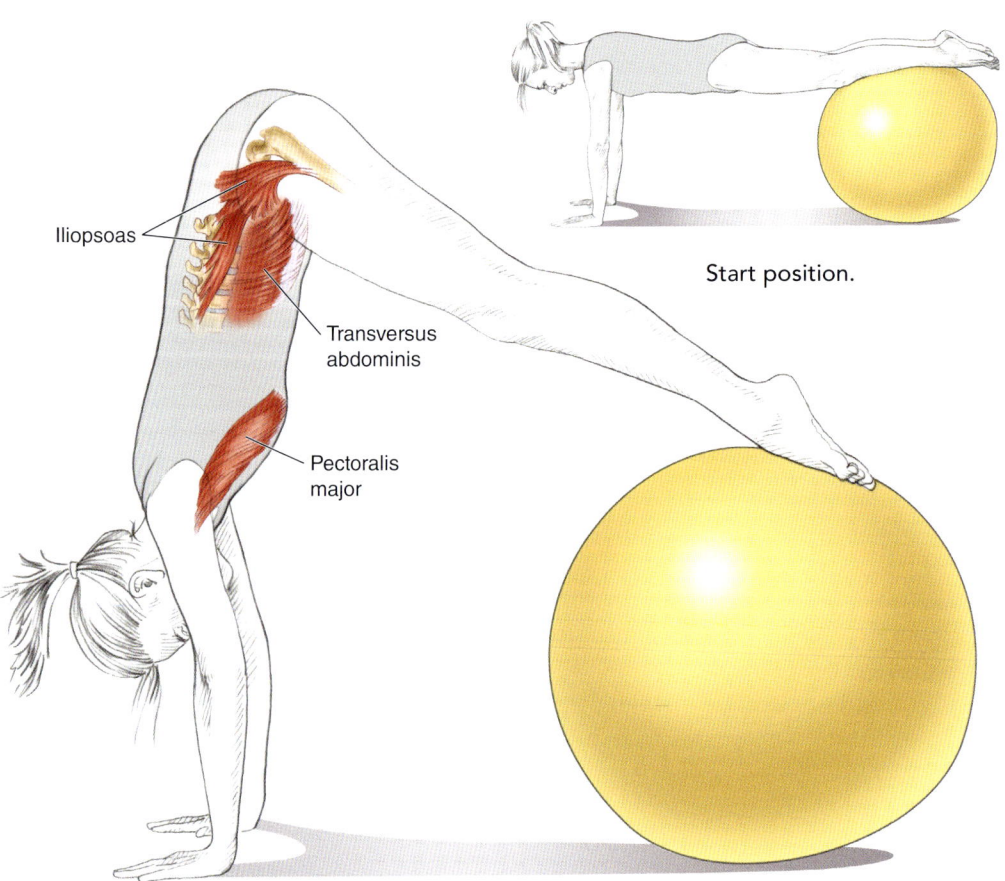

Start position.

Iliopsoas

Transversus abdominis

Pectoralis major

EXECUTION

1. Lie prone over a stability ball. Walk your hands out until you reach a straight-arm plank position with the tibias resting on top of the ball. Your knees are straight; your elbows are straight but not locked. Engage the scapular stabilizers and all trunk stabilizers.

2. On inhalation, initiate the movement with a slight posterior tilt as well as deep contraction of the abdominals and hip flexors to elevate your hips into a pike. Lengthen through your spine as you pull the ball toward your chest, pointing your feet.

3. Hold this position for 2 to 4 counts with inhalation. Reemphasize scapular depression and adduction. Slowly return to the beginning plank position with exhalation. Hold the trunk firm to protect your spine. Perform 6 to 8 times.

SAFETY TIP: Maintain scapular stability; avoid winging of the scapula. Engage the deep abdominals to resist gravity tending to pull your spine into extension.

MUSCLES INVOLVED

Transversus abdominis, internal oblique, external oblique, iliopsoas, tensor fascia latae, quadriceps (rectus femoris, vastus lateralis, vastus medialis, vastus intermedius), hamstrings (semitendinosus, semimembranosus, biceps femoris), gluteus maximus, gluteus medius, erector spinae (iliocostalis, longissimus, spinalis), multifidus, pectoralis major, anterior deltoid, lower trapezius, latissimus dorsi

DANCE FOCUS

Some of the most captivating and challenging choreography calls for dancing on the hands, which might involve, for example, cartwheels, back handsprings, push-ups, or falling on one hand. Regardless of the movement, you must be prepared and strong, but most dance technique classes won't work your upper body and core sufficiently. Therefore, it's up to you to put it all together.

The plank-and-pike exercise is a fully integrated mind-to-body skill. It involves contraction of the small postural muscles close to your spine as well as your larger muscles. To assist with any movement of this kind, awaken your breathing skills. Practice deep lateral inhalation to prepare yourself and forced exhalation on the movement to support yourself. If you find yourself losing stability in the lower back, increase your lower-abdominal training. If you find that you are unable to maintain stability with the scapulae, increase your shoulder exercises. If you are weak, you will find choreography using plank-type poses to be both challenging and risky. However, conditioning will give you a powerful, accomplished look.

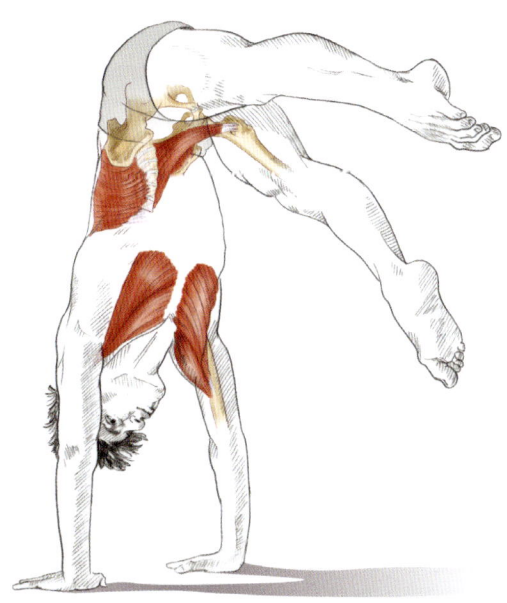

PLYOMETRICS

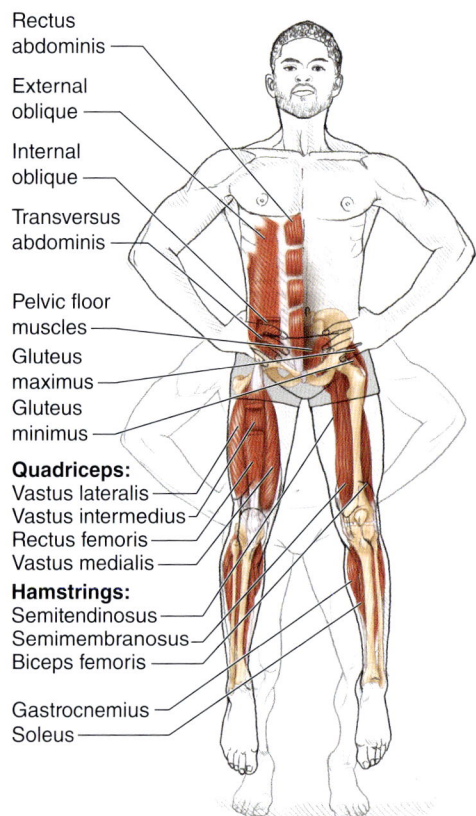

Rectus abdominis

External oblique

Internal oblique

Transversus abdominis

Pelvic floor muscles

Gluteus maximus

Gluteus minimus

Quadriceps:
Vastus lateralis
Vastus intermedius
Rectus femoris
Vastus medialis

Hamstrings:
Semitendinosus
Semimembranosus
Biceps femoris

Gastrocnemius
Soleus

Parallel jump squat.

PARALLEL JUMP SQUAT

Stand facing a mirror with your legs parallel and your feet hip-width apart. For now, just place your hands on your hips. Breathing comfortably, perform at least 10 basic squats while focusing on hip stability, keeping your heels on the ground, and making sure to keep your knees directly over your toes. If you are able to maintain hip stability and good leg alignment, follow the 10 squats by continuing into another squat and then jumping as high as you can. Land in your parallel squat position. If you can maintain pelvic stability and leg alignment, perform 10 more jumps. If you can continue maintaining stability and alignment, then perform 2 more sets of 10 jumps, for a total of 3 sets of 10 jumps. If you are maintaining good alignment and control, you don't need to repeat the squats.

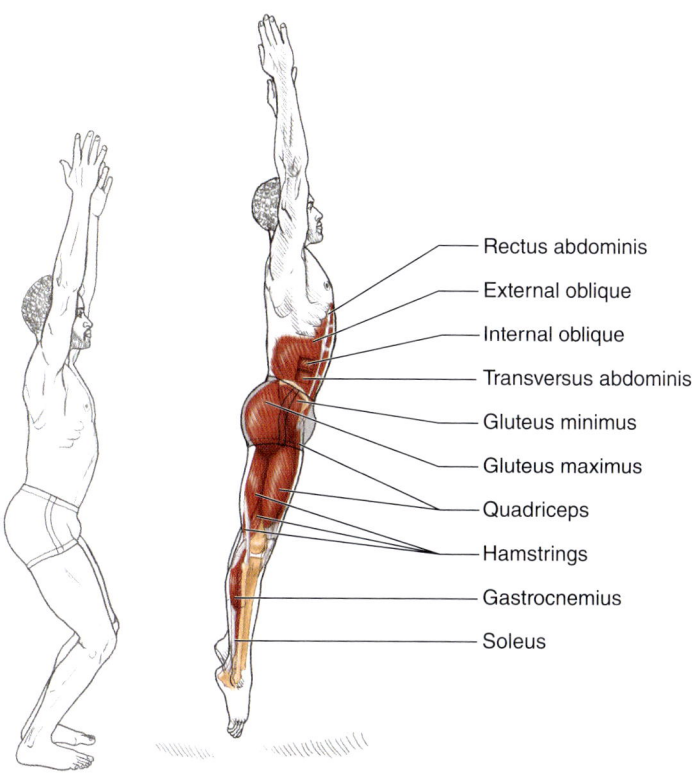

Rectus abdominis

External oblique

Internal oblique

Transversus abdominis

Gluteus minimus

Gluteus maximus

Quadriceps

Hamstrings

Gastrocnemius

Soleus

Traveling jump squat.

TRAVELING JUMP SQUAT

Begin in the same parallel squat position with your arms overhead. Again, focus on maintaining pelvic stability and keeping your knees aligned over your toes. Breathing comfortably, go into your squat position. Now jump as high as you can while traveling forward 10 to 12 inches (25 to 30 centimeters). Land with control and maintain alignment. If you are able to maintain alignment, perform 10 more traveling jumps. Next, repeat the same traveling-and-jumping exercise, but make the first jump slower and the second jump quicker; alternate slow and quick jumps for a total of 10 traveling jumps, 5 quick and 5 slow.

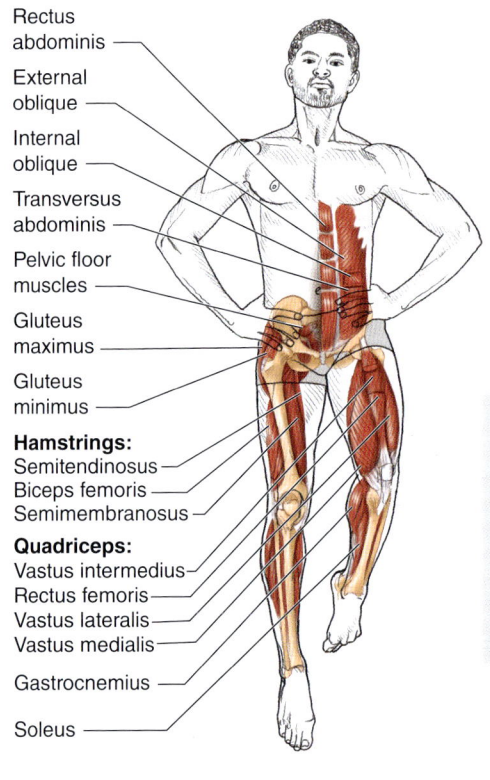

Rectus abdominis

External oblique

Internal oblique

Transversus abdominis

Pelvic floor muscles

Gluteus maximus

Gluteus minimus

Hamstrings:
Semitendinosus
Biceps femoris
Semimembranosus

Quadriceps:
Vastus intermedius
Rectus femoris
Vastus lateralis
Vastus medialis

Gastrocnemius

Soleus

SAFETY TIP: It is imperative to maintain hip and leg alignment to avoid injury. Plyometric training requires practice, proper alignment, and strength. Before participating in plyometric exercises, be sure to warm up.

Alternating jump squat.

ALTERNATING JUMP SQUAT

To advance your power jumping to the next level, begin in a parallel squat position on one foot with your hands on your hips. Breathing comfortably, jump as high as you can and land on the opposite foot. If you are able to maintain pelvic stability and leg alignment, alternate legs, work up to a total of 10 jumps per set, and repeat 3 times.

MUSCLES INVOLVED

Elevation: Concentric contraction of gastrocnemius, soleus, quadriceps (rectus femoris, vastus lateralis, vastus medialis, vastus intermedius), hamstrings (semitendinosus, semimembranosus, biceps femoris), gluteus maximus, gluteus minimus

Trunk stabilization: Transversus abdominis, internal oblique, external oblique, rectus abdominis, pelvic floor muscles (levator ani, coccygeus), multifidus

Descent: Eccentric contraction of gastrocnemius, soleus, quadriceps (rectus femoris, vastus lateralis, vastus medialis, vastus intermedius), hamstrings (semitendinosus, semimembranosus, biceps femoris), gluteus maximus, gluteus minimus

DANCE FOCUS

This type of training develops powerful jumping skills and improved landing skills. As you begin the muscular contractions necessary for jumping into the air, strong concentric contraction is created in the quadriceps, hamstrings, gluteus maximus, anterior tibialis, and gastrocnemius. When landing with control, these same muscles create an eccentric contraction, which is extremely important for safe landings. While practicing any of these exercises, work through your entire range of motion to get the most benefit, especially on the landing. You are working through a range of quick, strong, concentric contractions to a quick, strong, eccentric stretch before another explosive contraction to repeat the jump. To challenge the amortization (transition) phase, perform the exercises quickly, as in the traveling jump squat variation. If you can maintain the stabilization and alignment, then repeat that exercise as quickly as possible instead of alternating the speed. Focus on landing control.

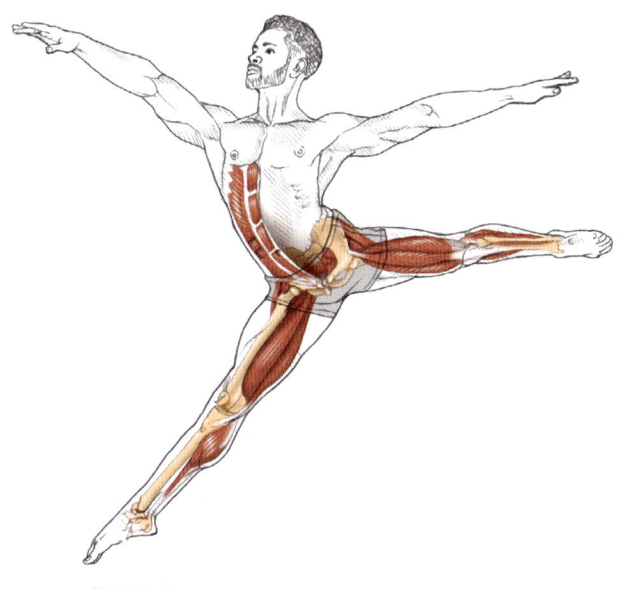

VARIATIONS

Advanced Plyometric Training

To advance your plyometric training, practice this type of jumping with just one foot; you can also jump onto a low, secure box. When these variations have become more routine, add jumping onto a minitrampoline or other unstable surface—but only when you have mastered the other variations.

BOUNDING

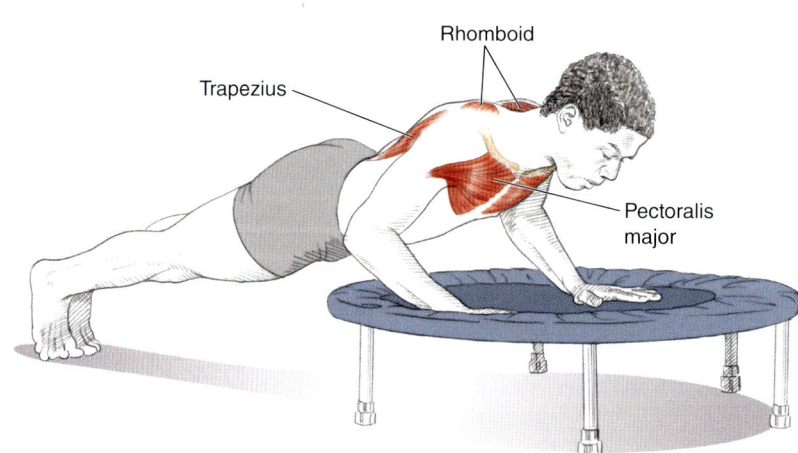

Rhomboid

Trapezius

Pectoralis major

Start position.

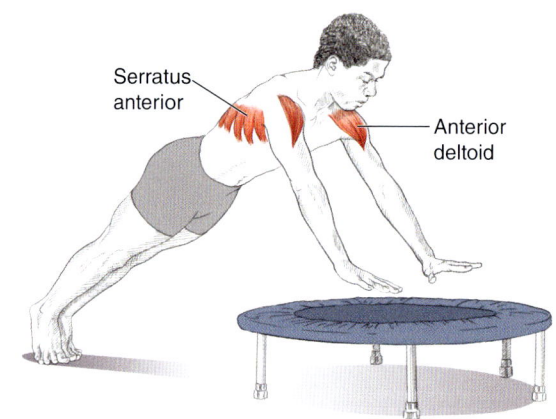

Serratus anterior

Anterior deltoid

Finish position.

EXECUTION

1. Begin in a classic push-up position with your hands farther than shoulder-width apart on a minitrampoline. Your legs are extended, and your feet are on the floor (you can also begin with your knees on the floor). Reorganize your trunk for core control.

2. While breathing comfortably, bend your elbows with control to initiate a push-up. Maintain scapular stability.

3. Press into the trampoline and push into the air, returning with control. Perform 10 to 12 times.

SAFETY TIP: Maintain lower-back stability with trunk control. Maintain scapular control and engage the wrist flexors to avoid hyperextension of the wrists.

MUSCLES INVOLVED

Pectoralis major, anterior deltoid, serratus anterior, lower trapezius, rhomboid, transversus abdominis, internal oblique, external oblique, rectus abdominis, multifidus

DANCE FOCUS

This rebounding exercise gives you an excellent way to challenge your core and shoulders to build strength for executing almost any tricky choreography. It is also an excellent exercise for building dynamic stability. Though it may not be obvious at first, bounding on the trampoline is another form of resistance training. Your muscles lengthen under the load in the down phase (the eccentric contraction), then follow with a quick, strong concentric contraction to push you into the air. This combination helps you develop greater muscular power, which can make the controlled falls associated with the Graham technique seem effortless. Greater muscular power also reduces the tension in all fall-and-recover movements in jazz styles. More generally, training safely with organized rebounding-type exercises prepares you for the complexity of atypical choreographic falls.

AIRPLANE BALANCE

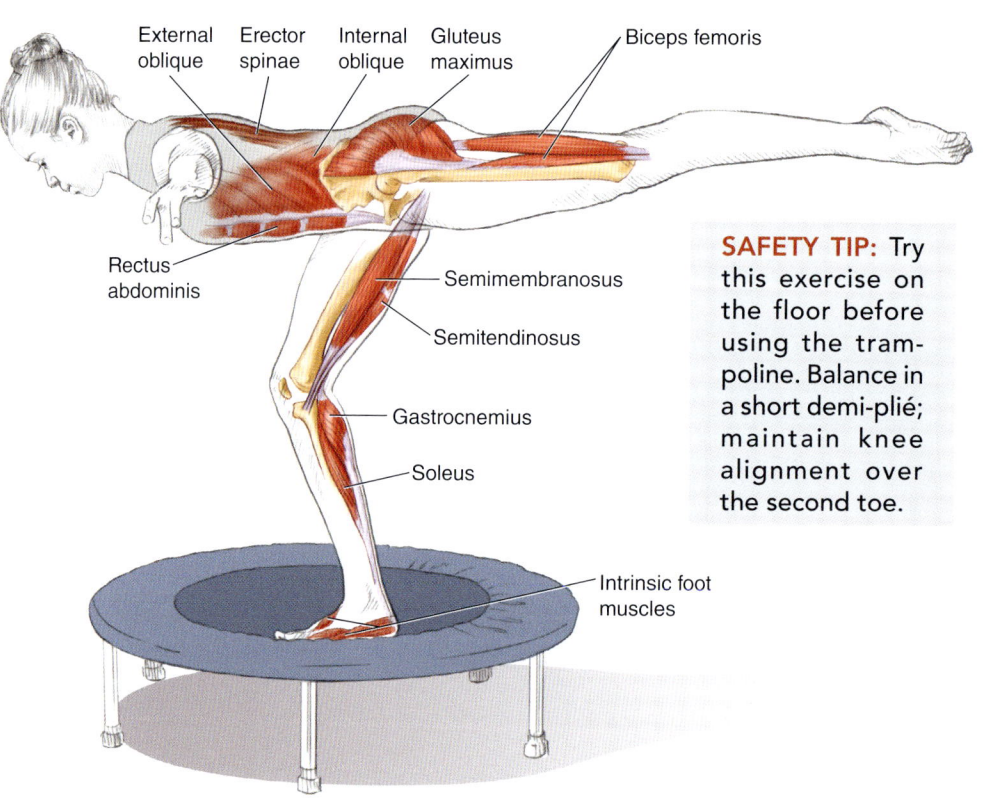

External oblique
Erector spinae
Internal oblique
Gluteus maximus
Biceps femoris
Rectus abdominis
Semimembranosus
Semitendinosus
Gastrocnemius
Soleus
Intrinsic foot muscles

SAFETY TIP: Try this exercise on the floor before using the trampoline. Balance in a short demi-plié; maintain knee alignment over the second toe.

EXECUTION

1. Stand in the middle of a minitrampoline on one leg in parallel position. The other leg is in parallel arabesque. Lengthen along your spine and move into a flat-back position. Bring your arms out to the sides.

2. Organize your balance skills and place your weight between the ball of the foot and the heel. Using the intrinsic muscles of your foot, add a small demi-plié.

3. Maintain your balance for 10 to 30 seconds. Rest and then repeat. Perform 3 times on one side before repeating on the other side. Breathe comfortably. Release tension in the neck and shoulders. Use abdominal control and the principle of axial elongation.

MUSCLES INVOLVED

Trunk: Rectus abdominis, internal oblique, external oblique, erector spinae (iliocostalis, longissimus, spinalis)

Standing leg: Intrinsic foot muscles, gastrocnemius, soleus, hamstrings (semitendinosus, semimembranosus, biceps femoris), gluteus maximus, gluteus minimus

Arabesque leg: Hamstrings (semitendinosus, semimembranosus, biceps femoris), gluteus maximus

DANCE FOCUS

Performing exercises to improve your balance skills engages your nervous system and can reduce your risk of injury. Challenging your proprioceptive system relieves unnecessary muscle tension and improves your jumps and turns. Take a little time each day to practice balancing. If you don't have a minitrampoline, then balance in the sand or on a pillow. Find your center and placement, beginning along the arches of your foot. Align your weight over the first and fifth metatarsals and the heel. Feel the deep intrinsic muscles supporting you. Focus on your deep postural muscles along your spine and down the leg. When you are truly balanced, you will need less muscular effort, which means you can work more efficiently. Throughout the balancing process, breathe comfortably. Let your breathing quiet your center and release tension. Gather your thoughts and organize your body to maintain a healthy balance between body, mind, and spirit.

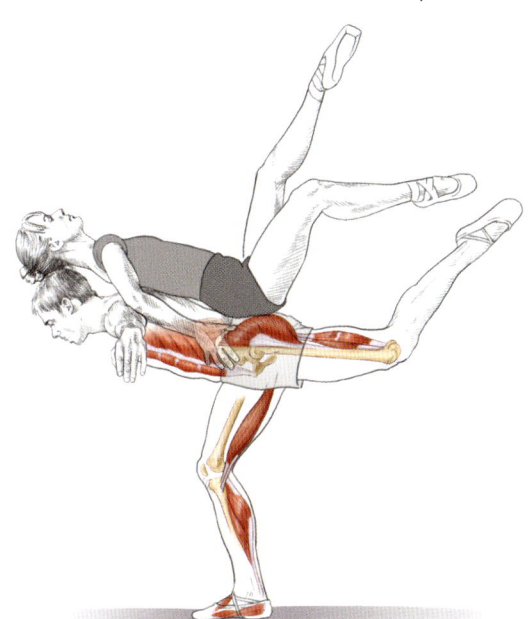

VARIATION

Développé Balance

Step onto a minitrampoline or another unstable surface. Slightly turn out your standing (right) leg. Locate your secure trunk placement and hold your arms in first position. Begin with your left leg in a coupé position. While maintaining your balance, slowly move your left leg into passé as your arms move to high fifth. Hold your spine and pelvis stable. Hold for 6 to 8 seconds before slowly returning your arms and legs to start position. Perform 5 times before switching to the other side. Once you feel comfortable bringing the leg into passé, try to move through passé to attitude and développé second. Work hard to maintain trunk and pelvic stability. Try the développé 4 times before switching sides.

LATERAL LEG LIFT

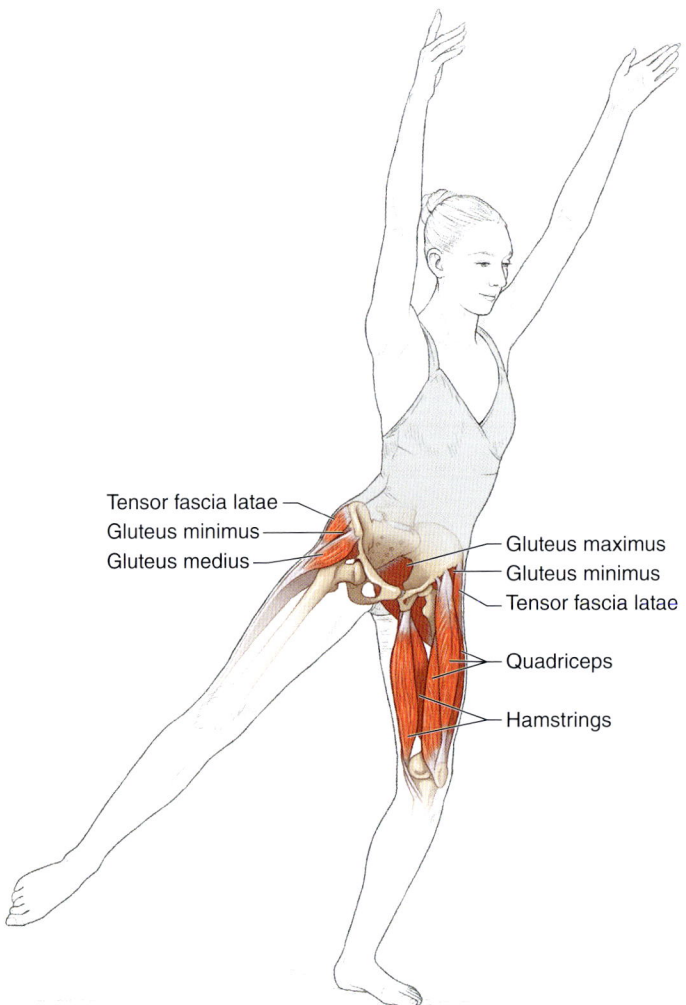

Tensor fascia latae
Gluteus minimus
Gluteus medius

Gluteus maximus
Gluteus minimus
Tensor fascia latae

Quadriceps

Hamstrings

EXECUTION

1. Stand with the legs hip-width apart. Shift your weight to the left as you drop the hips back into a lunge. The right leg moves directly to the side. The arms can be overhead. Inhale to prepare.

2. On exhalation, engage the deep abdominals to stabilize the pelvis and spine. Perform right leg lifts to the side, being mindful of lifting your leg only as high as you can maintain a stable pelvis. Think about aligning the left knee directly over the toes.

3. Perform 15 side leg lifts while also focusing on pelvis stability, balance, and alignment of the left (standing) leg.

4. Repeat on the other side. Work up to 3 sets of 15 leg lifts each leg. Once this exercise begins to feel less challenging, consider adding a small ankle weight to the lifting leg.

MUSCLES INVOLVED

Gesture leg: Gluteus medius, gluteus minimus, tensor fascia latae

Standing leg: Gluteus medius, gluteus minimus, tensor fascia latae, quadriceps (rectus femoris, vastus medialis, vastus intermedius, vastus lateralis), gluteus maximus, hamstrings (semitendinosus, semimembranosus, biceps femoris)

DANCE FOCUS

Remember, to gain strength, you must add more repetitions or resistance training to your conditioning program. More specifically, pelvic stability is one of the keys to improving your posture and technique. This exercise helps you get in tune with the outside of your pelvis to assist you in gaining strength. The gluteus medius will help you on the supporting leg with développé and grand battement work. It will also help your gesture leg during all side-layout positions, traveling steps to the side, and jumping combinations.

Lateral movement matters! The gluteus medius has the important job of keeping your hips level while you perform all dance choreography. Weakness in the gluteus medius will cause your hips to move side to side along the frontal plane with a lack of control. Landing from a jump without security in the pelvis will allow valgus stress on the knee or cause the knee to buckle inward. While performing the lateral leg lift exercise, stay aware of the standing leg and focus on aligning the knee over the toes while holding the pelvis steady. Although the gluteus medius is the main muscle worked, this exercise engages the whole body due to its focus on stability.

HAMSTRING STATIC STRETCH

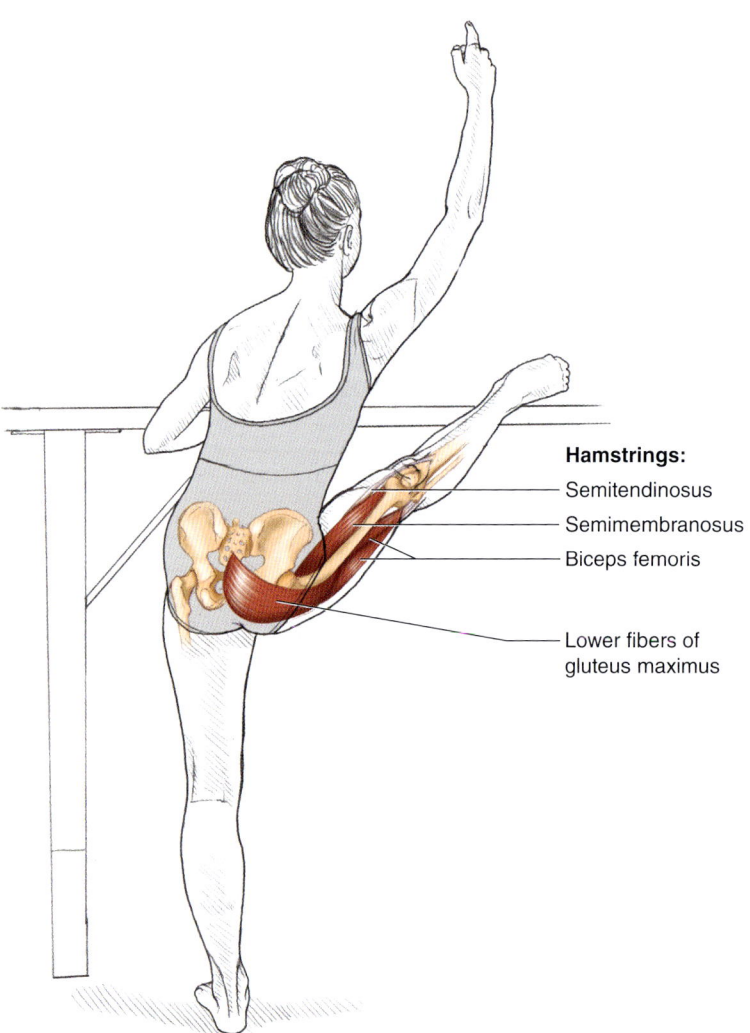

Hamstrings:

Semitendinosus

Semimembranosus

Biceps femoris

Lower fibers of
gluteus maximus

EXECUTION

1. Place your right leg on the barre in a devant position. Angle yourself so that you can hold on to the barre with your left hand while your leg is slightly rotated externally in the sagittal plane. Your right arm is overhead.

2. On inhalation, lengthen through your spine by incorporating axial elongation. Begin to lean forward, without allowing your chest to drop, until you feel a strong but comfortable stretch.

3. Release and lengthen in the hamstrings and gluteus maximus. Feel as though your pelvis is tilting anteriorly along your sagittal plane. Breathe comfortably as you maintain a lift in the chest. Relax into the hamstring stretch and hold for 30 seconds. Perform 3 times before moving to the other side.

MUSCLES INVOLVED

Lower fibers of gluteus maximus, hamstrings (semitendinosus, semimem-branosus, biceps femoris)

DANCE FOCUS

Static stretching benefits your muscles, tendons, and joints by lengthening the muscle fibers. Try to relax into the stretch; it should not be painful. As a dancer, you need a good balance between flexibility and strength. Prolonged stretching, however, is not recommended; in fact, it creates weakness. The most beneficial approach to stretching for you as a dancer is to combine static and dynamic stretching when you are warmed up. Dynamic stretching helps you get ready for rehearsals and performances by simulating your dance movements slowly; it also helps strengthen the muscles that are contracting. If you are looking to increase your range of motion without losing strength, implement six to eight weeks of dynamic and static stretching variations after you have done a good warm-up, and you will be pleased with the results.

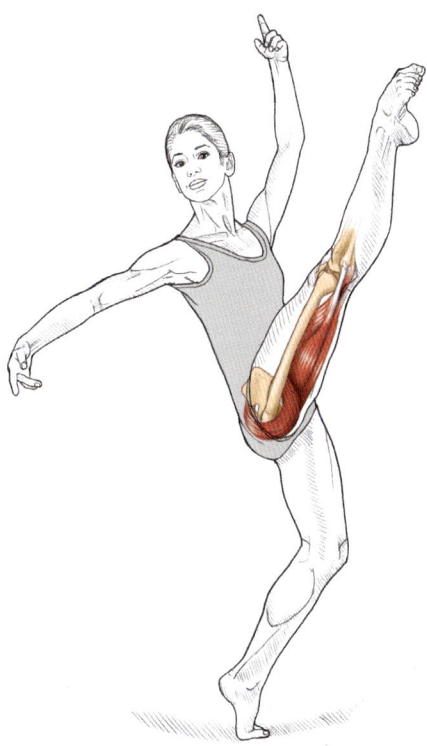

VARIATIONS

Dynamic Thigh-to-Chest Stretch

Stand, breathing comfortably. Focus on your trunk placement as you lift your left leg. Hold a steady balance on your right leg. Use both arms under your left thigh to gently pull the thigh to your chest. Feel lengthening in your lumbar spine and posterior hip. Slowly release your left leg while maintaining your balance. Step forward to lift your right leg. Using both arms under your right thigh, pull it toward your chest, lengthening your lumbar spine and posterior hip. Perform 10 to 12 times on each side.

Dynamic Devant Stretch

Stand in parallel position. Find your trunk placement. Hold your arms out to the sides in second position. Slowly brush your right leg devant to feel a stretch in the right hamstrings, maintaining pelvic placement. This is a controlled lift and stretch. Slowly lower the leg. Maintaining balance, step forward to repeat on the other side. Allow your leg to slowly lift and stretch while holding a stable spine and pelvis. Perform 10 to 12 times on each side.

ARABESQUE

Arabesque is a posture in which the body bends forward from the hip of one leg with one arm extended forward while the other arm and leg are back. Pretty generic, don't you think? Yet everyone knows that an arabesque is more than just a posture. Little girls who watch a performance for the first time try to execute arabesque. Young students take photos of themselves in arabesque for auditions. Professional dancers of all genres work extensively to perfect their arabesque. A recently retired principal ballet dancer told me that she felt like she had executed about 98,000 arabesques in classes alone during her 18-year professional career. Imagine the amount of physical work she performed at the spine and hip!

Arabesque is, in fact, one of the most widely used movements in dance, and it can be one of the most beautiful. Here is an anatomical breakdown of a fundamental first arabesque.

1. Begin with the right arm forward in first arabesque and the left leg in tendu derrière. Your right leg, the supporting leg, is turned out, with external rotation coming from the deep lateral rotators and a strong contraction of the hamstrings and quadriceps. Weight on the supporting foot and ankle is evenly distributed over all five metatarsals and the heel; pay close attention to tone along the medial arch and intrinsic muscles of the foot.

2. The left leg, the gesture leg, is also externally rotated while extended into tendu derrière. The pelvis and spine have begun to move into slight left rotation.

3. All four layers of the abdominals and the deeper spinal extensors are toned and supporting the spine. The hip extensors are taut and the posterior lower leg muscles are contracting to maintain a strong pointed foot.

4. Your right arm is executing shoulder flexion to the front slightly higher than your shoulder while maintaining strong scapula depression to activate the stabilizers. Your left arm is in shoulder abduction with slight shoulder exten-

sion but also maintaining strong scapula depression to engage the stabilizers. Both palms face down.

5. As the left leg begins to lift, a strong contraction occurs in the deep abdominals in coordination with a contraction of the gluteus maximus and hamstrings to support the lumbar spine, the spine and pelvis rotation, and hip extension. Begin to flex the right hip to allow for more forward movement of the trunk along the sagittal plane.

6. Maintain lumbar and pelvic stability while lifting the leg toward 90 degrees. Incorporate a lengthening and lifting feeling through your thoracic spine. Lift the sternum and extend through your entire spine. You should have a long, beautiful arc of your entire spine as your leg reaches 90 degrees.

Muscles Involved

Hamstrings (semitendinosus, semimembranosus, biceps femoris), quadriceps (rectus femoris, vastus lateralis, vastus medialis, vastus intermedius), transversus abdominis, rectus abdominis, external oblique, internal oblique, gluteus maximus, erector spinae (iliocostalis, longissimus, spinalis), piriformis, gemellus superior, gemellus inferior, obturator internus, obturator externus

EXERCISE FINDER

REFERENCES

Abichandani, D., and Hule, V. 2016. "Common Musculoskeletal Injuries Faced by B-Boydancers." *International Journal of Science and Research* 5 (7): 929-34.

Bergland, C. 2013. "Why Is Dancing So Good for Your Brain?" *Psychology Today*, October 1, 2013. www.psychologytoday.com/blog/the-athletes-way/201310/why-is-dancing-so-good-your-brain.

Bowerman, E.A., Whatman, C., Harris, N. & Bradshaw, E 2015. "A Review of the Risk Factors for Lower Extremity Overuse Injuries in Young Elite Female Ballet Dancers." *Journal of Dance Medicine & Science* 19 (2): 51-56.

Challes, J., and A. Stevens. 2019. "Nutrition Resource Paper." https://iadms.org/media/3589/iadms-resource-paper-nutrition-resource-paper.pdf.

Franklin, E. 2019. *Conditioning for Dance: Training for Whole-Body Coordination and Efficiency.* 2nd ed. Champaign, IL: Human Kinetics.

Friel, K., N. McLean, C. Myers, and M. Caceres. 2006. "Ipsilateral Hip Abductor Weakness After Inversion Ankle Sprain." *Journal of Athletic Training* 41 (1): 74-78.

Gildea, J.E., J.A. Hides, and P.W. Hodges. 2013. "Size and Symmetry of Trunk Muscles in Ballet Dancers With and Without Low Back Pain." *Journal of Orthopaedic and Sports Physical Therapy* 43 (8): 525-33.

Grossman, G., and M.V. Wilmerding. 2000. "The Effects of Conditioning on the Height of Dancer's Extension in à la Seconde." *Journal of Dance Medicine & Science* 4 (4): 117-21.

Haputhanthirige, N.K.H., K. Sullivan, G. Moyle, S. Brauer, E.R. Jeffrey, and G. Kerr. 2023. "Effects of Dance on Gait and Dual-Task Gait in Parkinson's Disease." *PLoS One* 18 (1): e0280635.

Hodges, P. 2003. "Core Stability Exercise in Chronic Low Back Pain." *Orthopedic Clinics of North America* 34:245-54.

Hodges, P., and S. Gandevia. 2000. "Changes in Intra-Abdominal Pressure During Postural and Respiratory Activation of the Human Diaphragm." *Journal of Applied Physiology* 89:967-76.

Irvine, S., E. Redding, and S. Rafferty. 2011. "Resource Paper—Dance Fitness." International Association for Dance Medicine & Science. iadms.org.

Kline, J.B., J.R. Krauss, S.F. Maher, and W. Qu 2013. "Core Strength Training of Home Exercises and Dynamic Sling System for the Management of Low Back Pain in Pre-Professional Ballet Dancers." *Journal of Dance Medicine and Science* 17 (1): 24-25.

Koutedakis, Y., and A. Jamurtas. 2004. "The Dancer as Performing Athlete." *Sports Medicine* 34 (10): 651-61.

Koutedakis, Y., A. Stavropoulos-Kalinoglou, and G.O. Metsios. 2005. "The Significance of Muscular Strength in Dance." *Journal of Dance Medicine & Science* 9 (1). https://doi.org/10.1177/1089313X0500900106.

Krasnow, D., and V. Wilmerding. 2011. "Resource Paper—Turnout for Dancers: Supplemental Training." International Association for Dance Medicine & Science. iadms.org.

Mainwaring, L.M., and C. Finney. 2017. "Psychological Risk Factors and Outcomes of Dance Injury: A Systematic Review." *Journal of Dance Medicine and Science* 21 (3): 87-96.

Mirkin, G. 2015. "Why Ice Delays Recovery." DrMirkin.com. www.drmirkin.com/fitness/why-ice-delays-recovery.html.

Office of Dietary Supplements. 2022. "Calcium." National Institutes of Health. November 17, 2022. https://ods.od.nih.gov/factsheets/Calcium-HealthProfessional.

Nicholls, C. 2022. "How to Use Nutrition to Enhance Your Sleep—and Your Dancing. *Dance Magazine*, August 12, 2022. www.dancemagazine.com/nutrition-to-enhance-sleep.

Ramkumar, P.N., J. Farber, J. Arnouk, K.E. Varner, and P.C. McCulloch. 2016. "Injuries in a Professional Ballet Dance Company: A 10-Year Retrospective Study." *Journal of Dance Medicine & Science* 20 (1): 30-37.

Richardson, C., P. Hodges, and J. Hides. 2004. *Therapeutic Exercise for Lumbopelvic Stabilization.* New York: Churchill Livingstone.

Rodrigues-Krause, J., M. Krause, and A. Reischalk-Oliveira. 2015. "Cardiorespiratory Considerations in Dance: From Classes to Performances." *Journal of Dance Medicine & Science* 19 (3): 91-102.

Russell, J., I. McEwan, Y. Koutedakis, and M. Wyon. 2008. "Clinical Anatomy and Biomechanics of the Ankle in Dance." *Journal of Dance Medicine and Science* 12 (3): 76-77.

Sandberg, E., M. Möller, S. Särnblad, P. Appelros, and A. Duberg. 2021. "Dance Intervention for Adolescent Girls: Effects on Daytime Tiredness, Alertness and School Satisfaction. A Randomized Controlled Trial." Journal of Bodywork and Movement Therapies 26:505-514. https://doi.org/10.1016/j.jbmt.2020.09.001.

Teixeira-Machado, L., R. Mario Arida, and J. Mari. 2019. "Dance for Neuroplasticity: A Descriptive Systematic Review." *Neuroscience and Biobehavorial Reviews* 96:232-40.

Tao, D., Y. Gao, A. Cole, J.S. Baker, Y. Gu, R. Supriya, T.K. Tong, Q. Hy, and R. Awan-Scully. 2022. "The Physiological and Psychological Benefits of Dance and Its Effects on Children and Adolescents: A Systematic Review." *Frontiers in Physiology* 13:925958. https://doi.org/10.3389/fphys.2022.925958.

Tao, D., R. Supriya, Y. Gao, F. Li, W. Liang, J. Jiao, W.Y. Huang, F. Duheil, and J.S. Baker. 2021. "Dementia and Dance: Medication or Movement?" *Physical Activity and Health* 5 (1): 250-54.

Watson, T., J. Graning, S. McPherson, E. Carter, J. Edwards, I. Melcher, and T. Burgess. 2017. "Dance, Balance and Core Muscle Performance Measures Are Improved Following a 9-Week Core Stabilization Training Program Among Competitive Collegiate Dancers." *International Journal of Sports Physical Therapy* 12 (1): 25-41.

Willard, F.H., A. Vleeming, M.D. Schuenke, L. Danneels, and R. Schleip. 2012. "The Thoracolumbar Fascia: Anatomy, Function and Clinical Considerations." *Journal of Anatomy* 221 (6): 507-37.

ABOUT THE AUTHOR

Jacqui Greene Haas recently retired as performing arts medicine program manager for Mercy Health and Cincinnati Ballet, serving as the company's athletic trainer for almost 30 years. She is a former professional ballet dancer who danced with Cincinnati Ballet, New Orleans City Ballet, Cleveland Ballet, and Southern Ballet Theatre and was an apprentice with Boston Ballet. Haas has a BA in dance from the University of South Florida, an athletic training certificate through the University of Cincinnati, and an MA in integrative studies from Northern Kentucky University.

Haas has certifications in Pilates instruction from St. Francis Memorial Hospital in San Francisco and in Pilates for rehabilitation from Polestar Education in Miami. She is also the director of Dance Medicine Academic Seminars, speaking to dancers throughout the United States on performing arts medicine, injury risks, and screenings for dancers. She enjoys sharing her workshops at Planet Dance Cincinnati, McGing Irish Dancers, University of Cincinnati College–Conservatory of Music, Bluegrass Youth Ballet, and Dancer's Pointe, to name a few. She has been a performing arts athletic training resource for the greater Cincinnati area since 1989 and is a member of the International Association for Dance Medicine and Science, the National Athletic Trainers' Association, and The Bridge Dance Project. Haas also teaches dance physiology, Pilates, and ballet at Northern Kentucky University's School of the Arts.

ANATOMY SERIES

Each book in the *Anatomy Series* provides detailed, full-color anatomical illustrations of the muscles in action and step-by-step instructions that detail perfect technique and form for each pose, exercise, movement, stretch, and stroke.

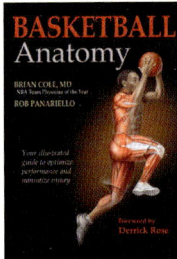
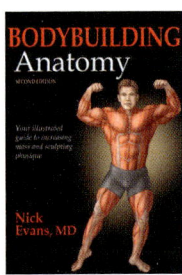

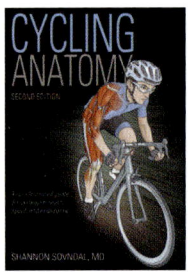
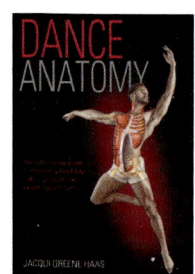

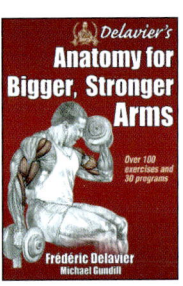

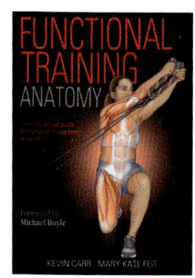

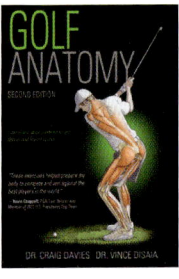
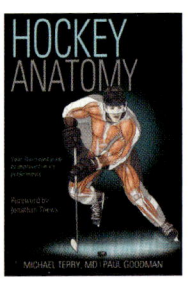
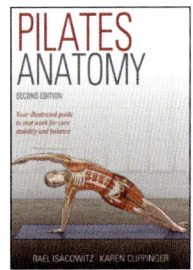
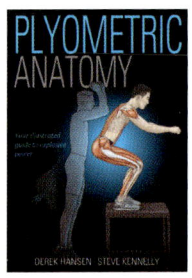
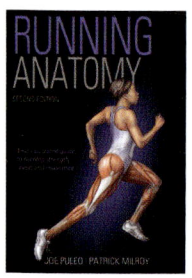

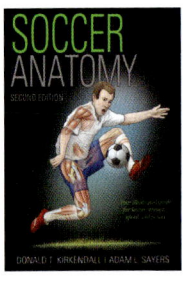
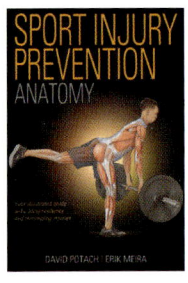

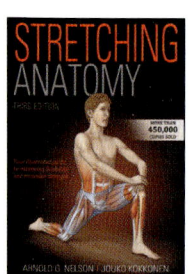

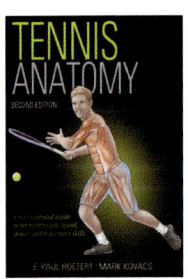

HUMAN KINETICS

U.S. 1-800-747-4457 • US.HumanKinetics.com/collections/anatomy
Canada 1-800-465-7301 • Canada.HumanKinetics.com/collections/anatomy
International 1-217-351-5076